T0256429

EARTH, LIFE, AND SYSTEM

EARTH, LIFE, AND SYSTEM

Evolution and Ecology on a Gaian Planet

EDITED BY BRUCE CLARKE

Fordham University Press : New York 2015

Copyright © 2015 Fordham University Press

All rights reserved. No part of this publication may be reproduced, stored in a retrieval system, or transmitted in any form or by any means—electronic, mechanical, photocopy, recording, or any other—except for brief quotations in printed reviews, without the prior permission of the publisher.

Fordham University Press has no responsibility for the persistence or accuracy of URLs for external or third-party Internet websites referred to in this publication and does not guarantee that any content on such websites is, or will remain, accurate or appropriate.

Fordham University Press also publishes its books in a variety of electronic formats. Some content that appears in print may not be available in electronic books.

Visit us online at www.fordhampress.com.

Library of Congress Cataloging-in-Publication Data

Earth, life, and system : evolution and ecology on a Gaian planet / edited by Bruce Clarke.
     pages cm. — (Meaning systems)
  Includes bibliographical references and index.
  ISBN 978-0-8232-6524-4 (cloth : alk. paper) —
ISBN 978-0-8232-6525-1 (pbk.)
    1. Convergence (Biology)   2. Gaia hypothesis.
3. Biodiversity.   4. Evolution (Biology)—Philosophy.
I. Clarke, Bruce, 1950– editor.
  QH373.E27 2015
  570.1—dc23

                                        2014030497

Printed in the United States of America

17 16 15   5 4 3 2 1

First edition

MEANING SYSTEMS

SERIES EDITORS
Bruce Clarke and Henry Sussman

EDITORIAL BOARD
Victoria N. Alexander, Dactyl Foundation
    for the Arts and Humanities
Erich Hörl, Leuphana University at
Lüneberg
John H. Johnston, Emory University
Hans-Georg Moeller, Philosophy and
    Religious Studies Program, University
    of Macau, China
John Protevi, Louisiana State University
Samuel Weber, Northwestern University

For Lynn

CONTENTS

Color plates follow page 224

PLATES AND FIGURES

ACKNOWLEDGMENTS

This volume emerged from the scholarly meeting Earth, Life & System: An Interdisciplinary Symposium on Environment and Evolution in honor of Lynn Margulis, held at Texas Tech University on September 13–14, 2012. The symposium was sponsored at TTU by the College of Arts and Sciences, the Office of the Vice President for Research, the Transdisciplinary Research Academy, and the TTU chapter of Sigma Xi. It was cosponsored by the TTU/Howard Hughes Medical Institute Science Education Program at the Center for the Integration of Science Education and Research (CISER) and the President Donald R. Haragan Lecture Series Endowment. I would like to thank everyone at TTU who helped to make the symposium a success: Lou Densmore, Donald Haragan, Julie Isom, Maria O'Connell, Curtis Peoples, Michael San Francisco, Lawrence Schovanec, Lesley Shelton, Bob Smith, and Josie Solis. Thanks also to Manuela Rossini for moderating the symposium, to Henry Sussman and Victoria Alexander for moral support, and to Donna Clarke for everything she does.

At Fordham University Press, thanks to Tom Lay for all his efforts on behalf of this volume. It would not have come about without the editorial guidance and goodwill of the late Helen Tartar, who not only shepherded *Earth, Life, and System* from conception to submission and then to contract but also participated in the creation of the "Meaning Systems" series of which it is a part. It is a sad thought that neither Helen Tartar nor Lynn Margulis will see it, lightened some by the hopeful notion that they would have been happy with the result.

EARTH, LIFE, AND SYSTEM

INTRODUCTION: EARTH, LIFE, AND SYSTEM

BRUCE CLARKE

The late evolutionary thinker Lynn Margulis produced four decades of controversial, paradigm-shifting science. She also authored an unusual amount of popular materials for students and nonscientists. In this regard, Margulis and her son Dorion Sagan collaborated on a series of works for general readers, including *Microcosmos: Four Billion Years of Microbial Evolution, Acquiring Genomes: A Theory of the Origins of Species, Dazzle Gradually: Reflections on the Nature of Nature,* and *What Is Life?* These remarkable volumes narrate her cutting-edge evolutionary ideas through strikingly graceful prose of literary caliber. In addition, along the way Margulis and Sagan articulated a refreshing posthumanism, that is, a viewpoint placing humanity not as the measure of all things but as an obligatory participant within a wider planetary ecology. In its broadest strokes, Margulis's science offers a compelling cosmic-evolutionary view of the processes that couple solar energy and the material Earth together with the integrated operations of all living systems (Plate I).

This volume is the culmination of a conversation with Lynn Margulis that began in the fall of 2005 when paleontologist Sankar Chatterjee invited her to Texas Tech University as part of the Paul Whitfield Horn Professor Lecture Series. Chatterjee had an ulterior motive: he wanted her views on his theoretical work on the origin of life, work that has its general debut in this collection. Lynn and Sankar pursued that discussion, and in the meantime, in addition to the Horn Lecture, while on campus she gave several other talks. After one of these, presented to five hundred students in the biology lecture hall, with *What Is Life?* in hand and poised for an autograph, I introduced myself to its lead author.

I had first heard of Lynn Margulis in the early 1980s, in the references Lewis Thomas listed at the back of his popular volume first published in 1974, *The Lives of a Cell.* I discovered her own writings around 2000, now as a professor of literature and science seeking materials to introduce serious but reasonably accessible biological and evolutionary content into literature courses. By 2005, *What Is Life?* had become the keystone text in a series of my literature and science classes. Additionally, I found myself increasingly committed to its philosophical message, one facet of which can be summed up in this statement: "All extant species are equally evolved."[1] Lynn received me warmly in that lecture hall, and a year later, I spent several weeks of my faculty development leave visiting her Environmental Evolution lab at the University of Massachusetts and, on her insistence, joining the throng of humanity that passed through her three-story home and virtual boardinghouse at 20 Triangle Street. Flash forward five more years. I learned belatedly that my TTU colleague microbiologist Michael San Francisco had been a student of Lynn's at Boston University. Before long we were conspiring to bring Lynn back to Texas Tech for a second visit, not as a solo speaker this time, but as the headliner of a symposium devoted to her kind of science. Lynn agreed to come.

An academic symposium on Lynn's kind of science would have to be an interdisciplinary meeting gathering together readers of both her technical works and her popular expositions—Earth and life scientists along with social scientists and humanities scholars. And holding the conversation together would be the Gaian parameter of systems thinking, since that broad set of discourses can move down from planetary geobiological systems to local ecosystems and biological systems. From there, various dialects of systems thinking are available for any natural or cultural realm in which, to be comprehensively grasped, the phenomena to be understood must have their systemic natures placed into their environmental contexts. In consultation with Lynn, I sent out invitations to colleagues in paleontology, molecular biology, evolutionary theory, and geobiology, in psychology and developmental systems theory, history of science, cultural science studies, environmental ethics, and literature and science. The symposium's invitees were all tentatively signed on to the event, pending the funding that eventually materialized in full, when Lynn Margulis, aged seventy-three, died of a massive stoke in November 2011.

We carried on with our plans for "Earth, Life & System: An Interdisciplinary Symposium on Environment and Evolution in honor of Lynn Margulis." From its inception the symposium was designed to yield a wide-ranging essay collection inspired by Lynn Margulis's scientific practice. This is that book. It is neither a eulogy nor a posthumous festschrift.[2] While it is anchored by and dedicated to Lynn Margulis's scientific career, it offers open-ended treatments that span the breadth of the themes that animated her science: astrobiology and the origin of life, ecology and symbiosis from the microbial to the planetary scale, molecular genetics and developmental epigenetics, cybernetics and systems theories, thermodynamics and technology, the coupled interactions of Earthly environments and evolving life in Gaia theory and its companion discourse—Earth system science, and the repercussions of these newer scientific ideas on cultural and creative productions. Contributors to this volume will occasionally assess the current discussion and standing of her evolutionary ideas or interrogate the coherence and resilience of the worldview that she conceived and championed throughout her career. And while they are obviously sympathetic to this work and its wider outlook, at times they express their solidarity with Lynn through critiques of her positions. Lynn was as notorious for her combative spirit as she was beloved for her generosity, and so, for this volume not to pay her the respect of challenging her ideas at certain points would not be Margulisian!

Dorion Sagan's "Life on a Margulisian Planet: A Son's Philosophical Reflections" begins with a meditation on his unique inheritance as the son of two world-class scientists and on his role as the literary and philosophical side of a significant collaboration in scientific popularization. Although he does not make the claim himself, I would say that the exposition of Margulis's work enjoyed such cogent and pointed statements of its cultural applications and implications because of Sagan's good offices as a literary mediator. Here Sagan reflects on both the scientific and the cultural contexts of his mother's work as a field and laboratory scientist doing revisionary evolutionary theory and as a public intellectual making out the meanings of her findings. In the process, Sagan acquaints the reader with salient issues in Margulis's work, the controversies they raised, and the vocabulary necessary to follow the arguments. These issues include the planetary ubiquity and ecosystemic necessity of the microcosm—the realm of the microbes, the evolutionary origin of the eukaryotic cell as an assemblage of preevolved precursors (serial endosymbiosis theory), and the subsequent debates regarding the proper enumeration of that cast of archaean and eubacterial (or prokaryotic) endosymbionts.

Adding in Lynn's signature collaboration with atmospheric chemist James Lovelock, another set of issues for her kind of science concerns the promotion of Gaia as a theoretical description for a planetary system carrying out operations that integrate geological and biological elements and that interlock abiotic and biotic subsystems. In the Gaian view, all living systems are necessarily in the midst of and communicating with social consortia of other organisms, and are, at the same time, coupled into the feedback dynamics of ecosystemic niches. In the evolution of their survival strategies, then, living systems bring their own autopoietic cognitions to bear on the ongoing modulation and recomposition of the environment. Natural wisdom starts with an attunement to the microbes. Even though we supposedly superior metazoans receive, as Sagan writes, "biotic innovation, metabolic diversity, evolvability, time-tested hardiness, and sustainable biogeochemical recycling" from "our tiny derided fellows and planet-mates," the microbes, we tend to do so without thanks or even acknowledgment. Margulis insisted that we outgrow such parochial complacency and get right with the Gaian program.

In "The RNA/Protein World and the Endoprebiotic Origin of Life," Sankar Chatterjee addresses one of the oldest and most elusive problems in science. From evidence that meteorites during the Archean period (before life began) deposited both organic chemicals and water on the early Earth, he argues for two distinct processes behind the origin of life: the exogenous delivery of carbon-based molecules from space and the endogenous synthesis of life on Earth. Comets deposited water and the key ingredients of life, which were then concentrated, catalyzed, and polymerized in the pores and pockets of mineral substrates of the crater basins created by asteroid impacts. Chatterjee complicates a currently popular origin-of-life scenario, the "RNA-world hypothesis," by pointing to evidence for the coevolution of protein and RNA components. In a process he calls, by analogy with Margulis's "serial endosymbiosis theory," *serial endoprebiosis*, protein molecules would catalyze the chemical reactions that eventuated in living metabolism, while RNA molecules would store genetic information to allow the replication of prebiotic relationships, and both would percolate together within prebiotic lipid membranes. From there he details a series of hypothetical stages in the transition from such prebiotic assemblages to the first living cells. In a final move, Chatterjee suggests that the same primordial RNA and protein molecules that interacted to bring about the first cellular forms of life were also the precursors to viruses and prions, which entities of our world may thus represent evolutionary legacies of the origins of the microbial world.

In "Exobiology at NASA: Incubator for the Gaia and Serial Endosymbiosis Theories," James Strick tells the intertwined origin stories of James Lovelock's Gaia hypothesis, which in its original form suggests that life controls its own environment, and Margulis's serial endosymbiosis theory (SET), which accounts for the origin of the eukaryotic or nucleated cell not by classical Darwinian accumulation of gradual alterations, but by the wholesale integration of microbial genomes. Both theories were developed in and funded by the NASA Exobiology Program, first begun in 1960. Excerpts from the extensive oral histories Strick has taken from Lovelock, Margulis, and other key players in the NASA exobiology milieu provide his reader with firsthand accounts both of the experiences of the main actors and of the social and institutional dynamics of this particular coupling of science in action. Gathering this oral documentation together with the published record, Strick details how NASA exobiology gave crucial support, both intellectual and financial, for the development of both theories. At the same time, given various prejudices about NASA science in the wider research community, this context also conditioned some of the skepticism with which both theories were initially received. Examining the implications of these theories and of their receptions, Strick also explores the cultural valences resulting from the systemic or integrative nature of the theories themselves and from the tenor of the cultural period in which they first appeared, and to which they have arrived at the present moment. In this regard, Strick surveys the closely coupled impress of Gaia and/or SET in Carolyn Merchant's ecofeminist, anticapitalist book *The Death of Nature*, the British moral philosopher Mary Midgley's defense of Gaian thinking in *Science and Poetry*, and the American philosopher of religion Jacob Needleman's *An Unknown World: Notes on the Meaning of the Earth*.[3]

In "On Symbiosis, Microbes, Kingdoms, and Domains," Jan Sapp notes that the neo-Darwinian evolutionary biology of the last century, which is to say, mainstream evolutionary thought, said nothing about the evolution of microorganisms. And yet, this is where the great diversity of the great kingdoms of life lay hidden. Throughout most of the twentieth century, phylogenetic kingdoms were limited to "animals and plants." The place of microorganisms within those kingdoms was banished to the same peripheries of biological thought where the heavily microbial notions of symbiosis and speciation through symbiogenesis also languished. As a result, microbiology in particular developed without an evolutionary context and mainly as an applied science associated with the medical study of disease. While long sought by some microbiologists, a phylogenetic understanding of bacterial taxa based on classical comparative morphology was deemed to be impossible. Later in the century, biological doctrine established separate microbial kingdoms along a prokaryote-eukaryote dichotomy. And although the details were in flux, by the 1980s, a five-kingdom system first proposed by Robert Whittaker and then developed and championed by Lynn Margulis—monera (prokaryotic microbes, bacteria), protoctists (eukaryotic microbes such as algae), animals, fungi, and plants—met with considerable acceptance.

However, the controversies over the number and distinctions among biological kingdoms also came to confront an entirely different scheme of life. In the last two decades of the twentieth century, based on a comparative sequencing of ribosomal RNA, biologists led by Carl Woese developed new molecular methods and concepts to investigate early microbial evolution with the aim of creating a universal phylogenetic tree. These developments led to a taxonomic proposal of three domains representing three fundamental lineages of life: archaea, eubacteria, and eukarya. Sapp's essay tracks the arrival of these new phylogenetic concepts, explains their logic, and surveys the subsequent controversies to which they give rise—issues concerning the relative conceptual integrity of, in particular, Woese's domains and Margulis's kingdoms. Margulis vigorously resisted Woese's new phylogenetic scheme, with cogent arguments drawn from her orientation to phenotypes in the field. Nonetheless, with the new molecular-phylogenetic evidence his research team established, Woese helped to confirm that the microbial events of horizontal gene transfer and symbiosis studied by Margulis are indeed major modes of evolutionary innovation. While illuminating the real stakes of the dispute between Margulis and Woese, Sapp's essay also shows us how, in the end, their key ideas do not clash but effectively coexist in a world of both kingdoms and domains.

Susan Squier's "The World Egg and the Ouroboros: Two Models for Theoretical Biology" examines the work of embryologist and developmental biologist C. H. Waddington—now primarily remembered for his concept of the epigenetic landscape—at a crucial moment in its later development. In 1967, Waddington convened the second of four Serbelloni Symposia. Charged with formulating the structure of a discipline of "general theoretical biology," the symposium participants ultimately affirmed the model of language as the most appropriate for explaining biological self-assembly and self-organization. However, an alternative, spatial and visual model can also be recovered from Waddington's Serbelloni talk. There, he explained that his work as a scientist was influenced by his life-long interest in two ancient metaphysical notions: the World Egg and the Ouroboros. Waddington did not develop the emergent and recursive properties of these images fully at Serbelloni, but he would give the productive spatial logic of such visual images a fuller elaboration in the other project preoccupying him at that time, *Behind Appearance: A Study of the Relations between Painting and the Natural Sciences in This Century*.[4] Although both the Serbelloni Symposia and this massive coffee-table study were generally viewed as quirky, ambitious, but unproductive attempts at interdisciplinary synthesis, Squier suggests we can return to them now to reclaim tantalizing glimpses of a mid-twentieth-century model that anticipates contemporary systems thinking about emergent development and planetary coevolution.

My essay takes its cue from Margulis's scientific memoir *Symbiotic Planet*. While the idea of a "symbiotic planet" addresses a primary biological dynamic of the Earth as a geobiological system, the description it proposes rests on a sustained exercise of the scientific imagination—in this case, Margulis's particular formation of a planetary imaginary. "The Planetary Imaginary: Gaian Ecologies from *Dune* to *Neuromancer*" explores science fiction, ecosystem ecology, and systems theory as these lines of literary, scientific, and philosophical production converge dramatically in the 1960s—at the same time that Waddington is exploring cybernetic and Gnostic models in relation to theoretical biology—and then extend forward into the 1970s and '80s. Frank Herbert's 1965 novel *Dune* imagines the planet Arrakis in full ecosystemic detail, while it also forecasts the ecological imaginary of a systems counterculture rising through cybernetics and psychedelics to new visions of planetary dynamics. Purveying a space-age version of systemic holism emblematic of this cultural moment, the *Whole Earth Catalog* reviews *Dune* alongside Buckminster Fuller and other cybernetic avatars of "Spaceship Earth." Its periodical successor, *CoEvolution Quarterly*, celebrates the marriage of planetary ecology and cybernetic epistemology in Gregory Bateson's *Steps to an Ecology of Mind*, published in 1972. In 1975, *CoEvolution Quarterly* is the first nontechnical journal to present Lovelock and Margulis's Gaia hypothesis to a general public, and a year later it joins a campaign promoting the construction of space colonies in high orbit. The design specs of this project for artificial environments or microplanets wind up deeply embedded within the storyworld of William Gibson's *Neuromancer*, published in 1984. From this historical set of thematic and textual echoes, "The Planetary Imaginary" gathers together key elements of an influential American systems counterculture.

In "Bringing Cell Action into Evolution," James Shapiro notes that as an advocate of positive cell action in the evolutionary process, Margulis focused her work on observing real-time interactions between cells and advocating the major role of cell fusions and symbiogenesis in rapid evolutionary change. Confirmation of her argument that the mitochondria and chloroplasts in eukaryotic cells are the descendants of well-defined prokaryotes was a major turning point away from the gradualist ideology that dominated evolutionary thinking for most of the twentieth century. Shapiro details the new molecular-biological evidence that has mounted in recent decades supporting the major evolutionary roles of cell-to-cell interactions and the intracellular control of genome structures. The now well-established phenomena of symbiosis, hybridization, horizontal DNA transfers, genome repair, and natural genetic engineering, he argues, have revolutionized our understanding of genome variation. Evolutionary changes arise not from a series of accidents randomly changing a ROM (read-only memory) heredity system, but instead from active cell processes that nonrandomly restructure a RW (read-write) genomic storage system at all biological time scales. For Shapiro, the implications of this new molecular-genetic knowledge are that living cells are cognitive systems with a large repertoire of selective genome-altering activities in response to internal or environmental contingencies, and that, as a result, life altogether is inherently self-evolutionary.

Susan Oyama moves the discussion from the molecular back to the ecological level and beyond. In a wide-ranging overview of developmental systems theory, she shows how the concept of development balances an over-emphasis on genetic coding and other deterministic schemas. Earth is what it now is because millions of millennia of ever more complicated and numerous interwoven developmental pathways have unfolded within its biosphere. Organisms develop through multiple relations to their abiotic and biotic environments, altering them as they are themselves transformed. "Sustainable Development: Living with Systems" captures a play of multiple meanings both within and outside biology. In evolution and elsewhere, development is flexible. For instance, while contingency is often associated with chance or randomness, in her treatment its operative meaning is causal dependency. Organisms are ontogenetically contingent; countless dependencies can be studied within the skin and extending out beyond it to an indefinite series of conditions and processes that are themselves contingent on other factors. Yet these densely interlocked and variable complexes can generate both the variation and the stability and recurrence needed for evolution. Narrow definitions of evolution can be expanded by adopting a contextual notion of development in which both social and ecological relations are embedded, with all the bodily and worldly involvement that entails. Oyama's discussion concludes with a consideration of sustainable development, not only in its usual contexts of agriculture and economic growth, but also in developmental and evolutionary studies. Such regularity across so many scales arises from interconnected systems of contingent influences. Genetic determinism yields before epigenetic plasticity.

Christopher Witmore's "Bovine Urbanism: The Ecological Corpulence of *Bos urbanus*" brings the focus of the volume down to a specific contemporary farrago mixing up natural resources, animal lives, and human appetites. He visits an industrial site somewhere in West Texas, what the meat industry calls a "concentrated animal feeding operation," or CAFO. Naming it Cattle City, Witmore considers this industrial system as an object of both ecological and archaeological concern. Approaching bovine living situations with an archaeological attitude, Witmore contrasts the operations of Cattle City to past pastoral practices. At the same time, his approach encompasses current ecological concerns, from the anti- and pro-biotic manipulation of the bovine microbiome to the global level of energy distributions and trade-offs. Witmore elicits these ecological ironies by positioning Cattle City at the center of a struggle for a viable combination of agriculture and livestock production, waged "against an overweight meat-industrial complex." Witmore's notion of "ecological corpulence" draws together "a raw measure of material weight and energy expenditure with the image of gluttony." Striving to understand the conditions under which the residents of such an urban habitat live out their time on Earth, Witmore observes all participants—bovine, human, or equine—and their interconnections not as insulated or immunized beings thrust into corporate relationships, but rather as actors in complex material and ecological assemblages.

Peter Westbroek begins the final essay of the collection, "Symbiotism: Earth and the Greening of Civilization," with a historical consideration of the distance between, on one hand, seventeenth-century European presumptions regarding the conquest of the Earth and its riches and, on the other, our less sanguine contemporary realization that the state of our home planet and its prospects are not finally ours to determine. Indeed, as a global culture what we must accomplish is a comprehensive change of attitude in our relationships to the Earth. For Westbroek, the insights of Earth system science allow us to glimpse "the dawn of a global metamorphosis that may lead to a new relationship between the worlds of nature and civilization." Because of a planetary perspective on geobiological history spanning more than forty-six million centuries, Earth system science is capable of inducing a new worldview, a universal, reality-congruent public orientation, essential for a proper response to global change. Westbroek leads the reader through a two-step sequence, first, to a view of social self-restraint as the motor of human civilization. However, because this first step also brings to light the anthropocentrism of Western humanity's modernistic assumptions, a second, more drastic step must still be taken, to "extend our perspective from civilization per se to the level of the entire Earth." Exercising the planetary imaginary of Earth system science, Westbroek reverses perspective and looks back at humanity's present state of ecological depredation from the Earth's point of view. It turns out that civilization is caught up in the latest throe of the Earth's own destructive self-recreation. But this also means that the human phenomenon is not locked into its present nihilistic course but can learn new ecological manners, a code of conduct he terms, in homage to Lynn Margulis, "symbiotism." With symbiotism, Westbroek's hopeful planetary plot brings the discourse of *Earth, Life, and System* to a climax, a turning point devoutly to be wished: "The anthropocentrism inherent to modernism declines as we begin to realize that we humans are ephemeral manifestations of the Earth's creative reproduction."

1.

LIFE ON A MARGULISIAN PLANET

A Son's Philosophical Reflections

DORION SAGAN

In the days, weeks, and months since her death in 2011, I have grappled with making sense, however unfinished, of the legacy of my mother, and my biology mentor and longtime writing collaborator, Lynn Margulis.[1] Listed on a popular website shortly before her death as one of the twenty most influential living scientists, she is best known for her scientific contributions to our understanding of eukaryosis—the evolution of cells with nuclei and chromosomes (such as our own, eukaryotic "animal" cells) from the symbiotic merger of bacteria and archaea—among the first cells to appear on Earth.[2] To an extent perhaps unrivaled in the history of science, my mother was steeped in the mysteries and minutiae of life's poorly fossilized—but biochemically action-packed—early evolutionary period. She expertly wedded the big and small, marrying the geological history of Earth as a chemically active planet to best-guess scenarios of the metabolic evolution of the group-living, symbiotic microbes that gave rise to all familiar macroscopic beings. She characterized us humans as walking colonies of microbes, a once-sacrilegious view now becoming increasingly popular. In archiving and continuing her work, we participate in the archiving and continuing self-regard of life's own far more than human history.

Although long invisible to evolutionary biologists focused on macroscopic organisms, especially animals in our own phylum, the transition from prokaryote (cell without nucleus) to eukaryote (cell with nucleus) was characterized by Édouard Chatton as the greatest in evolutionary history. Lynn Margulis had a file drawer labeled "The Question"—meaning eukaryosis, the details of this epic transition. She not only made the case—on the basis of comparative morphology, biochemistry, microbial ecology, cellular ultrastructure, and careful review of the international historical scientific literature (particularly in Russian)—that this huge jump in evolution had a symbiotic genesis, but she also assiduously gathered evidence until the day she died chronicling the current importance of symbiogenesis in evolution. Eukaryosis or eukaryogenesis—the ancient symbiotic events that created cells with nuclei such as those that compose the nonbacterial parts of our bodies—was a precondition for the evolution of protoctists ("protozoa," organisms like paramecia and amoebas, seaweeds, slime molds, and many others in a rather wild diversity), as well as their descendants: animals, fungi, and plants. Even her detractors, such as Richard Dawkins, who avidly defended the mainstream evolutionary biological view that evolution results primarily from the gradual accumulation of random Mendelian variations, granted that she made a great contribution in showing that symbiogenesis was crucial in early evolution, in the transition from prokaryotes to eukaryotes over two billion years ago. Here, the ideas she championed were accepted when genetic-sequence comparisons convincingly linked mitochondria (the respiring organelles processing energy in our cells) to free-living, oxygen-using bacteria; when plastids (the photosynthetic organelles of algae and plants) were similarly genetically sequenced, they showed striking similarities to the genomes of cyanobacteria ("blue-green algae").

Yet, she pushed further. For example, she coauthored a paper suggesting a symbiogenetic basis for J. B. S. Haldane's famous quip, "God has an inordinate fondness for beetles."[3] The vast variety of species of these arthropods, she argued, needed to be investigated in light of the persistent association of beetles with *Wolbachia*—a genus of mycoplasma (minimal-metabolism bacteria found also inside humans) known to genetically alter many species of insects. In her serial endosymbiotic theory (SET), our very bodies result from what, observed in a time machine, could be characterized as a pathological infection or, alternatively, terminal indigestion. Forerunners of mitochondria were invasive, respiring (as opposed to anaerobic) bacteria. Although they invaded, they did not destroy, but instead conferred an advantage upon their hosts. We have not recovered from this infection; we will never recover. We *are* this infection. "Infections" in sexually reproducing species can separate populations, causing them to speciate. When the early neo-Darwinist Theodosius Dobzhansky temperature-separated flies and found that eventually the hot-bred and cold-bred fruit flies of the same species were no longer able to interbreed, he assumed this experiment proved the biological species concept, and that new species had evolved through accumulated random mutations. It turns out, however, that one new "species" was infected with *Wolbachia* bacteria, while the other was not. As symbiosis historian Jan Sapp summarizes, "Bacteria of the genus *Wolbachia* are inherited through the eggs of at least 25–75 percent of all insect species and those of nematodes, too. And far from being slaves, they manipulate the development of their hosts, causing parthenogenesis and cytoplasmic incompatibility, and can turn functional males into functional females."[4]

Rather than argue that symbiosis (the living together of members of species for the majority of the life cycles of each) or symbiogenesis (the eventual merging and emergence of new species or other taxa due to such close interaction) were confined to founding anastomoses or fusions at the root of the "tree" of life, my mother held and tirelessly presented evidence for the idea that symbiogenesis continues to occur into present times. Conversely, though she may have overstated the case in part to balance the scales of evidentially insufficiently supported academic dogma, she argued that there is little to no evidence for the assertion that speciation occurs by the mere accumulation of random mutations. Amusingly remarking to her friend Lois Brynes that neo-Darwinists shouldn't "be blamed, because that's what they were taught," she subscribed also to the theory that karyotypic fissioning—the spontaneous rupture of chromosomes, whose numbers differ in closely related species like dogs and wolves, and in various types of deer—was a crucial part of the story of animal evolution. And for her, this too was a kind of latter-day symbiogenesis, because the chromosome ends, and the notable intracellular motility apparatus of eukaryotic cells (including mitosis and meiosis), were traceable, in her view, to another kind of ancient bacterial partner—spirochetes, corkscrewing speedsters of the microcosm, which she chronicled attached to and propelling organisms like *Mixotricha paradoxa*, side by side with congenitally attached flagella. (More careful than her colleagues, she insisted on calling these *undulipodia*, to distinguish them from the radically different propulsive appendages of bacteria.) Symbionts fed in a living environment can lose parts of themselves, switching genes and merging metabolisms with their captors. An ancient propensity to live on their own sometimes reappears, not always on synchronized schedule. This last partner in her version of SET has not been proven by genetic methods.[5] But, as with the smile of the Cheshire Cat, one can expect that the longer an organism has inhabited a living environment, the greater the chances for it to lose parts of itself, and therefore the more difficult for it to be to find smoking-gun evidence of its once-independent existence. And for her, in her later versions of SET, the spirochetes, attaching to and entering and losing parts of themselves in larger cells, were the oldest bacterial partner in the eukaryotic genome.

Nonetheless, since her death there has been, both in the scientific literature and the popular media, a flurry of notices implicitly vindicating her worldview.[6] That view is that all macroscopic beings are, evolutionarily and currently, microbial colonial composites, and that a nuanced understanding of ourselves, including advanced understandings of health as well as illness and medical treatment, needs to appropriate a "Margulisian" perspective. It is, rather suddenly, common to hear of our "microbiome," and articles of the "probiotic" contribution of prokaryotes, once derided simply as germs, to our wellness and mood can now be read in Sunday newspaper supplements.[7]

With regard to the symbiotic bacterial construction of the eukaryotic portion of our cells, there is a certain irony in these genetic proofs, since my mother's lifelong work, harking in a sense back to an earlier generation of nature-loving naturalists, did not proceed through a constricted lab- and computer-based methodology, but was intrinsically broad, born of an intimacy with the objects—subjects—she studied, with which—whom?—she was herself in intimate, one wants almost to almost say, coevolutionary or symbiotic, contact. Without this methodological broadness, she would not have gathered the interdisciplinary evidence that was later vindicated by the relatively insular modern techniques of genetic analysis. Here she resembles the novelist-lepidopterist Vladimir Nabokov, whose meticulous field observations of butterflies known as blues, as well as ten years of taxonomic microscopic studies dissecting the genitalia of this group at Harvard's Museum of Comparative Zoology, persuaded him that they had migrated across the Bering Strait from Europe into South America—a deduction based on Nabokov's intimate study of and familiarity with live organisms and specimens, which he, like she, collected in the field, not just cataloging but also communing with, engaging with, the natural and perplexing beings that provided an endless source of interest.[8]

It is difficult for me to be objective about my mother, since I helped popularize her work, became a sort of cybernetic circuit in its dissemination and reception, work which from a biospheric perspective is part of nature's self-relation, its mediation through humanity and culture, and quite possibly, an evolution that will transition beyond humanity. I find it curious that with the evolution of our communicative and recording technologies, we are becoming increasingly able to access one another instantly across great distances and increasingly better at accessing our own cultural and species histories, as well as those of life's long history. Leaving aside quantum questions of nonobjectively creating or helping create the pasts we access, she was obviously a key player in this process of empirically unveiling our remote evolutionary past. She often considered herself a passionate spokeswoman for the majority of life forms (microbes) that did not have a voice, for example, the protoctists, hundreds of thousands of species which, although our form of cell coloniality and sexual reproduction evolved in them, as well as much else we consider animal and human, continue to be all but ignored because they have no clear commercial applications.

This is part of the problem Stanislaw Lem identifies in *Summa Technologiae*: the problem of increasing specialization and limited manpower associated with a necessary slowdown in the exponential growth of our species, putting a damper on the exponential growth of science.[9] Lem argues that the amazing growth of science in the last few centuries is no more sustainable than that of the human species. The explosion of science has proceeded in sometimes arbitrary capitalistic feedback loops with technology, fueled by moneymaking opportunities and pragmatic applications rather than pure research or long-term thinking. The fields in which scientists specialize have economic benefits and may not be the right ones to ensure our survival. Judging science by the exponentially increasing number of new specialized journals, Lem identifies the growth of science itself as exponential—but insupportable, because it is ultimately dependent on the number of scientific researchers, whose growth is finite. Even with robot and computer help, it is clear we cannot sustain the growth of the pure (neither pragmatically motivated nor corporate- or state-funded) scientific research, which advances human knowledge before monetization.

Margulis (and Nabokov) gravitated toward this kind of research. She was less motivated by money or fame than a kind of spellbound admiration for the unexpected connections nature forged with itself. In one of her private letters, written at a time when she was younger than me reading, she alludes to a formative experience she had in the Midwest, an epiphany in the face of nonhuman nature. This predisposition may have been exacerbated in her trips to Mexico as a teenager, where she studied indigenous culture and was likely exposed to native conceptions (as exemplified in ahayuasca paintings) of a nature that does not keep to Aristotelian species bounds (let alone Platonic eternal categories) but flows freely, joining itself, making connections, sometimes permanent, without regard for too hastily erected human taxonomic categories. She was an adamant spokesperson too for the seven-eighths of life that preceded and prepared the way for the "Cambrian explosion," a vast swatch of time where she took a firm stand against Stephen Jay Gould's claim that nothing much happened for ever so long, arguing instead that this temporal site—our biospheric childhood—was where all of life's most important biochemical and morphological innovations, from methanogenesis to respiration to photosynthesis to eukaryogenesis, took place.

DISCOVERING INTELLIGENT LIFE ON EARTH

When I contrast her contributions with those of my far more famous father, Carl Sagan, whom she married young and who helped foster her interest in science, I find that his were mostly to democratize the scientific insights of the Enlightenment—that is, ideas no longer especially radical. Lynn's main contributions, however, centered on an idea scoffed at and then marginalized (as she points out, the American biologist who fielded a version of endosymbiosis theory, Ivan Wallin, was humiliated and left biology when he could not grow mitochondria extracted from cells on their own). The idea that complex life is the result of microbial mergers—bacteria, germs—flew in the face of complacent notions of human superiority and 1950s advertising culture. But, as discussed in our first coauthored book, *Microcosmos*, this perspective represents a deconstructive shift, provisionally upsetting life's hierarchy (a holdover from the religious Great Chain of Being), in casting microorganisms (formerly derided as mere germs), as simultaneously our ancestors and our body-mates, and as the central actors in evolutionary history.[10]

Lynn's other great contribution is to be found in her multidecade collaboration with the atmospheric chemist James E. Lovelock on the Gaia hypothesis, which is perhaps best stated in terms of planetary physiology: that our planetary biota has had a regulatory effect on the chemical composition of the atmosphere, on Earth's global mean temperature, on the salinity and pH of the oceans, and on other would-be solely physical factors of our global environment. The greatest windfall of a voyage can be the expansion of perspective upon homecoming. While my father and his colleagues sought in vain for extraterrestrial intelligence, Lovelock turned that perspective around by realizing that aliens could detect Earth life from space by studying its thermodynamic atmospheric disequilibrium. In a sense, Lovelock and Margulis set the stage for the discovery of a greater planetary intelligence right here on Earth. This more-than-human intelligence has already evolved far beyond our bipedal, weedlike species, via the machinations, natural selections, symbiogeneses, and genetic recombinations of an estimated thirty million species—and that figure may not even include the bacteria, which are so genetically fluid as to flout the macroorganism-based biological species concept.

As with initial dismissals of her microcosmic views, many of the Gaia ideas continue to be dismissed when associated with that name. Gaia, the Greek mother of the Titans and goddess of the Earth, carries scientifically unwelcome teleological, feminist, and animist connotations. Yet the ideas of Gaian science are taught the world over under the more innocuous name of Earth system science. Lovelock credits her with identifying the source—microbes and their metabolic byproducts—of Earth's atmospheric thermochemical disequilibria—whose spectrometric signature could indicate to intelligent aliens the presence of a global life form on this planet.

Helping found NASA's Planetary Biology Internship and Planetary Biology Microbial Ecology programs, she regarded the search for extraterrestrial intelligence, so-called exobiology (now astrobiology) as a highly useful propaedeutic to interdisciplinary science, but also, along with Lovelock, as slightly silly, "a science without a subject." The search for extraterrestrial intelligence, which is also of course an existential quest, a kind of replacement for religion in a secular age, is complicated by the fact that our kind of technical intelligence does not necessarily have a clear relationship to survival. "It has yet to be proven," as Arthur C. Clarke wrote, "that intelligence has any survival value."[11]

The search for life elsewhere in the cosmos continues to be a worthy scientific endeavor, but the greatest benefits to our species may lie in a renewed examination of terrestrial intelligence here at home. To give just one quick indication of the alien brilliance of our planet-mates, whose non-human intelligences we are only now beginning to plumb, consider the fact that, as plutocrats consider harrowing means to address human overpopulation—arguably the source of climate change, global pollution, totalitarian regimes, mass extinction, and war—problems of population overgrowth have long since been solved by brainless nature. The tendency for exponential growth in our bodies, beginning with stem cells, is genetically regulated. Billions of years before the exponential growth of our species of chattering mammalian weed, nature had evolved multiple elegant means of regulating exponential growth. With plentiful evidence, physicist and theoretical biologist Josh J. Mitteldorf persuasively argues that aging itself is not an accident or entropic inevitability but a genetic adaptation regulating overpopulation—which has, many times in evolution, exposed booming populations to total busts via predation, infection, and starvation. Our bodies themselves do not grow with the "imprudence" of exponential self-indulgence: apoptosis (programmed cell death), telomerase rationing (obviating unfettered growth of our tissues), and inflammation (needed for the immune system but also implicated in physiological senescence) all conspire to prevent our cells from indulging in unfettered growth and continuation.

Similarly, as we try to moderate our greenhouse emissions or geo-engineer our way out of global pollution, we may look back to our evolutionary seniors, the symbiotic cyanobacteria ancestral to plants, which, more than two billion years ago, using water for hydrogen, released oxygen—at the time, and still to many organisms today, a toxic gas—into the atmosphere, ultimately increasing its concentration there from less than one to about twenty percent. In a preview of human overgrowth, the toxicity of oxygen imperiled many organisms, most proximately the cyanobacteria themselves. And yet the greening of the biosphere (the green plastids, or chloroplasts, of all plants derive from symbiogenetic cyanobacteria) testifies to their ultimate success. It is a success of which we may rightly be green with envy, for with each laugh, sigh, and scoff we make use of the oxygen they once released as a poisonous and dangerously reactive metabolic waste.

In the long run, we may be able to environmentally overcome our thickening web of industrial toxic wastes but, if we follow the cyanobacterial model, vast numbers of humans would be destroyed before natural selection kicked in. And despite our rather inordinate confidence in our own technological prowess, biology remains much more effective than human technology in devising elegant long-term solutions to pollution and population control. In fact, we humans have never sustainably done the former, and we show no signs of being able to do the latter (although war is the top contender). Plus, cyanobacteria compose not just a single species (as we do) but also an entire multispecies "phylum [that] exhibit[s] enormous diversity in terms of their morphology, physiology and other characteristics (e.g. motility, thermophily, cell division characteristic, nitrogen fixation ability, etc.)."[12]

From a Margulisian microcosmic-biospheric perspective, we humans are fascinating sport well worth watching, but hardly capable of the sorts of biotic innovation, metabolic diversity, evolvability, time-tested hardiness, and sustainable biogeochemical recycling as our tiny derided fellows and planet-mates, one lineage of which, as Margulis pointed out, are in fact our own ancestors (thus depriving "us" of the credit for even our unique contributions!). A primate that discovered fire and used it as a kind of external stomach to metabolize food (thus relieving our kind of endless mastication, a serendipity reflected in our teeth, tiny for primates), we may be characterized as promethean pyromaniacs—not quite the gods some are wont to think.[13] On the other hand, it seems clear that nonhuman nature, having evolved shape-based parallel processing, room-temperature materials processing, pollution-free energy use, stunningly symbiotic and beautifully crowd-controlled nature—however "dumb" we in the church of latter-day apes are wont to think it—is in many ways "higher" and "more advanced" than we.[14]

DIONYSIAN NATURE: AFFECT AND INTUITION

If you knew her, you will remember her nonstop energy, her run-on sentences, her passion for nature and science. She was strangely fearless, a trait that may have helped her go boldly where few and sometimes no scientists went before. The passion, energy, and intrepidness formed a formidable triumvirate. As children, my brother and I accompanied her on nature walks, on one of which we, dressed in drab clothing, heard repeated gunshots, awaking in us a fear not mitigated by discovery of bullet-perforated beer cans; ignoring our concerns, she studied more closely some exquisite woodsy detail that captured her more than our passing apprehension of a mortal vulnerability. Such focus and behavior was typical. I remember a toddler exhausted by her; he asked if he could take a nap. She spoke of how as a child she had found adults boring; they "just sat there talking." As her Spanish colleague, Carmen Chica, wrote shortly after her death, "¿Descansar en paz? ¡Imposible! Eso no puede ser." (Rest in peace? Impossible! No way.) And it's true even now: she, who kept so many occupied while alive, has left mountains of unfinished projects in her passing; she may be gone, but the scientific and philosophical questions she opened and developed will carry on, keeping scholars and scientists busy for decades at least.

Although she is known for her intellectual investigation of nature, behind it may be an important emotional register. Recent research by neuropsychologists shows the importance of emotional orientation in attention, consciousness, and memory; we have enhanced perceptual vividness for emotionally salient things.[15] Nature as a whole seems to have been emotionally salient to her. As mentioned, her private letters refer to an early, mysterious epiphany as a girl in the Midwest. As a teenager under the auspices of the University of Illinois, she studied anthropology and shamanism in rural Mexico. Here she may have absorbed or imbibed an indigenous feeling for nature as deeply interconnected and self-acting that informed her in her scientific investigations.[16]

Also early on, as Lynn Sagan, she was an articulate advocate for the possible scientific use of psychedelic drugs, specifically LSD-25, as a means to perceive nature devoid of conceptual cultural clutter.[17] It seems possible that her early exposure to indigenous American encounters with richly biodiverse nature, perhaps attested to by the flowing colorful forms of two possibly ahayuesca-inspired shamanic artworks subtly on display in her always-busy home on Triangle Street in Amherst, Massachusetts, aroused in her a nonverbal conviction of nature as autonomous, more-than-Western, more-than-human, and agentially multipartite. Such intuitively perceived and interflowing, boundary-breaking nature might be associated with "female," or cast (in Nietzsche's schema) as Dionysian in contradistinction to the Apollonian. In retrospect, abundant evidence for lateral transfer at the base of the would-be monophyletic trunk of the "tree" of life, as well as abundant evidence for continuous genetic transfer among not just bacteria but higher taxa, confirms her suspicions of and arguments for a nature far wilder and more promiscuous than portrayed in textbooks and media.

Like the American geneticist Barbara McClintock, she had a "feeling for the organism" informed by observation of live cells and beings in contrast to academic classifications. Her penchant for Spanish culture and cross-cultural investigation brought her in contact with neglected French, Russian, and American theorizers of symbiosis, which she defined as the physical contact of members of distinct species for the majority of their respective life spans.[18] One of these was Ivan Wallin, who embraced the affective tendency of organisms for one another, which Wallin termed "prototaxis." Prototaxis, taken as local affection, tropism, and the tendency of association of some organisms for others, may be seen as a kind of biophilia, organisms' affection for, deep involvement with, one another across species barriers; some prototaxes led to our ancestors. Unlike many scientists in the wake of Cartesianism, Lynn was a natural and pan-evolutionary phenomenologist, giving all organisms the benefit of the doubt in granting them the possibility of awareness whether or not they had brains. Her graduate student Oona West brought to her attention a study that showed that foraminifera select one size and color of glass bead over another to build their agglutinative shells, which suggested something like choice by microbes. In *What Is Life?* (following Samuel Butler), we described them as having "little purposes."[19]

In *Chimeras and Consciousness*, the last volume she edited and oversaw, spawned from a conference in Bellagio, multiple authors argued in different ways that mind is more than human, indeed that it may be an emergent multispecies phenomenon, a kind of evolutionary synesthesia rather than a uniquely human let alone divinely anthropocentric trait.[20] For Lynn, our mentality, no less than our anatomy and physiology, has multiple life-form roots. The fullest extension of this tenet can be found in her essay "Speculation on Speculation":

> Speculation, I claim, is the legacy of the itching enmities of unsteady truce. . . . When they are starved, cramped, or stimulated we have in-choate feelings. . . . Here I speculate on the spirochete origin of our sensory-nervous systems. . . . Whereas in science theory is lauded, spec-ulation is ridiculed. A biologist accused in print of "speculation" is branded for the remainder of her career. . . . In a manuscript deficient in references and lacking data, field and laboratory observations, and descriptions of equipment and their correlated methodologies, I feel a huge restraint as I attempt to slacken the bonds of professionalism. . . . My inhibitions fade with opportunity to tell you what I really think! . . . What is the idea? I hypothesize that all the phenomena of mind, from perception to consciousness, originated from an unholy microscopic alliance between hungry swimming killer bacteria and their potential archaebacterial [archaean] victims. The hungry killers were extraordinarily fast-swimming, skinny bacteria called spirochetes. These active bacteria are relatives of the spirochetes of today that are associated with the venereal disease that, in prolonged and serious cases, infects the brain: the treponemes of syphilis. The fatter, slow-moving archaebacteria were quite different from the spirochetes. In resisting death the archaebacteria incorporated their fast-moving would-be killers into their bodies. The archaebacteria survived, con-tinuing to be infected by the spirochetes. The odd couple survived. The archaebacteria were changed: they were made more motile, but not killed, by their attackers.[21]

Although Lynn was not conversant with continental philosophy, we might say that the literal duplicity of the animal genome (a jointure of archaea and alpha-proteobacteria, ancestors of "our" mitochondria), the triplicity of the plant genome (this "animal" genome plus cyanobacteria), and the still more profound if not yet genetically proven multiplicity of all eukaryotic genomes when an endosymbiotically most ancient or primordial layer of spirochetoses is taken into account, are deeply consonant with the deconstruction of identity implied by Roland Barthes's and Michel Foucault's "death of the author" and similar rubrics. Humankind (and perhaps all life) operates on delusional beliefs of a discrete self, but this self is already always compromised, not just metaphysically or philosophically but also literally by multiple ongoing biological and "cultural" origins. Too-simple monotheistic stories featuring humans become diverted. The gift has no destination, as Derrida has said. Thus the Native American praxis of giving without signing, of the true gift as anonymous and thus nameless, because it really comes from the all, becomes consonant with the microbiological deconstruction of identity. As in pre-Renaissance art, according to Foucault, the artist disappears into the work and its presencing.[22] The archmetaphor of ownership is put under erasure. We do not own the Earth but owe the biogenic compounds of our bodies to a circulating biosphere out of which we come, not from a transcendental elsewhere but an immanent biogeochemophenomenological here. Ownership, agency, identity come under question. It may be a conceit to think that anything is original rather than recombination combined with a forgetting of sources. "The sun's a thief, and with his great attraction / Robs the vast sea; the moon's an arrant thief, / And her pale fire she snatches from the sun; / The sea's a thief, whose liquid surge resolves / The moon into salt tears."[23] Shakespeare's stark verse shows the complex infolding of the cosmos, relating to and taking from itself at a primordial level, one that has been empirically validated in evolutionary biology.

LONG-EVOLVED SENTIENCE

But without any requirement to pledge allegiance to such a postmodernist gloss, we can see that emotions and mental states are multiple, real-time motivators and evolved in the crucible of intense multispecies interactions. The beings in and around us affect or infect our perceptions. *Toxoplasma gondii*, a protist whose sexual reproduction takes place in the stomach of cats, is estimated to be a standard part of the French microbiogenome, in part because of that attractive, culinarily advanced populace's love of steak tartare and kindred *Toxo*-bearing foods. It contains genetic stretches thought to code in humans for the brain synthesis of dopamine, a neurotransmitter involved in thinking and pleasure. Life-loving life folds into itself. As well as speciating and replicating, it involutes, convolutes, anastomoses, and reorganizes. Colorful Lepidoptera and our taste- and fragrance- and hue-tending mammal ancestors coevolved with those natural factories of diverse alkaloids, angiosperms. Plants and fungi can sometimes alter us profoundly after we ingest them; and "simple" bacteria swim toward magnetic due north, away from and toward oxygen, away from and toward the light, aggregating into biofilms and altering the behavior, metabolism, and physiology of one another and host organisms. Prototaxis, the tendency of life forms evolving in an extremely crowded planetary environment to feel for one another, is ancient. Sensation is ancient. The sweetness and light of paradise subsist already in the microbial world, as Lynn demonstrated in one of her videos where the shadow around a circle of light projected into a bacterial medium was recognized by rapid swimmers, who turned about when they reached the edge of the light. Bacteria swim up a sugar gradient; life's sweet tooth precedes its teeth, calcium phosphate carbonate fluoride (apatite) extrusions that, evolutionarily, are modified cell waste dumps (because calcium ions are too prevalent and poisonous in ancestral marine waters) allowing organisms to get at concentrated energy gradients. Consciousness no more first emerged in humans than did life in the Cambrian era, although the paleobiological illusion persists because that is when life began to biomineralize hard parts in macroscopic abundance.

For Lynn, "our" awareness was not our own so much as an interspecies phenomenological network: "our" neuropsychology, like our physiology, is but one of many working versions of microbial ecology. She was fond of bold syntheses and what now is often referred to (in a lamentable cliché) as "thinking outside the box." She repeated James E. Lovelock's phrase "academic apartheid" to call attention to nature's independence from human classifications, Platonic abstractions, and verbal habits of labeling something and then forgetting about it. ("Label and dismiss," we called it.) She deployed Whitehead's critique of abstraction under the rubric of "misplaced concreteness." She scoured early texts in multiple languages to unveil the thoughts of previous biological thinkers that the organelles of "animal" cells had begun as bacteria—bacteria that merged in flagrant disregard for thousands of years of biologists' segregating plants and animals, of centuries of religionists' separating divine humans from animals, and of decades of medical professionals casting microbes as mere germs to be eliminated.

As science progresses, becoming more specialized, the need for synthesis grows. Terms such as "biogeochemistry," "microbial ecology," "environmental evolution" (a name of her college course), and "geophysiology" underscore that nature is a single phenomenon studied by separate sciences, divided like the elephant in the parable of the blind men each feeling a different part (the trunk, leg, the tail) and taking it for the whole. Understanding nature meant viewing it from multiple angles and seeing ourselves as inalienably embedded within it. Not only was nature of a piece, but active, sensing individuals also came together permanently to produce new, more inclusive individuals. Speciation itself occurred not just from the touted geographical separation of sexually reproducing animal populations, but also from the "hypersexual," permanent mating of wild, and wildly distinct, forms.

Nature's incestuous, Dionysian proclivities may not have been appreciated by academic typologists with a penchant for erecting eternal platonic categories, but changing nature didn't care, and neither did she. Her personal life itself—where she cc'd multiple parties, of dubious appropriateness to the matter at hand, on emails; where she took her lovers and husbands/long-term partners (sequentially: astronomer, crystallographer, microbiologist) from the sciences; where she employed and collaborated with her children and treated her graduate students as family; and where she developed a community ethic of intellectual ferment in the extended family of her homestead—attest to a pragmatic instantiation of the multifarious connections she saw everywhere, and lovingly and painstakingly chronicled, in polymorphous, autoaffective nature. In retrospect we can surmise that, conscious or not, her *practice* of the connectionism she gleaned to be at the core of the inner workings of nature was itself a kind of biomimicry; human values such as social and ecological connectionism, diversity and biodiversity, repeat a basic operation of overwhelming importance to the appearance of "higher" life forms, evolution by association.[24]

BACKGROUND TO SYMBIOSIS AND GAIA:
THERMODYNAMIC NATURE

Lynn's agile mind did not ossify but continued until she died to welcome new ideas and be open to changing and syncretive worldviews. She welcomed my work with Eric D. Schneider, who presented ecology in a thermodynamic context, developing the work of Howard and Eugene Odum and others to show life as a collective energy phenomenon showing growth and cycling regularities and obeying thermodynamic laws. Together we discovered and repeated physicist David Bohm's line that science is about finding the truth, whether we like what we find or not. This is deeply dissonant with certain strains in social studies of science, epistemological relativism, Bruno Latour's critique of mononaturalism (the idea that there is "one nature," defined for example by Western science) and other culturally compensating critiques of apodictic science. But the pendulum has swung perhaps too far in the opposite direction. Isabelle Stengers speaks of moving from the relativity of truth to the truth of relativity and also of "the curse of tolerance." At the present juncture we must navigate between language games, postmodernism, and the fashionable academic critique of power on one hand and the revalorization of truth in a time of increasing propaganda and corporatocracy on the other. Behavioral finance's ownership effect applied to ideas already prejudices people in favor of their previous opinion at the personal level; institutional inertia, career security, and desire for social approbation dissuade minority and novel views, and when corporate funding and government politics added to the mix, the "truth" has even more formidable foes. It is a tough row to hoe.

Before moving on to discuss Lynn's advocacy of minority positions—by which I mean viewpoints, not ethnicities—on science, I would like to touch briefly on what she identified as the largest framework for observed biological process. This was a view of chemistry and energetics. Here she was Vernadskian, as well as Lovelockian or Gaian. If Vernadsky in his Soviet society disdainful of religious superstition could accurately portray living matter as an energetic phenomenon feeding on available sunlight, whose energy was transmuted into life at Earth's surface (a view that influenced Bataille, and through him postmodernism), so Lovelock, in the undeconstructed British empire, in the relatively Christianized West, could portray Earth as an organism with true physiology, cybernetic feedbacks, and a female name. Both were seeing life as a thermodynamic system, that is, as a transducer of energy. Gaia was explicitly discovered on the basis of spectrographic investigations revealing the planetary biota's chemical signature in the atmosphere, a chemical signature that could theoretically be recognized from afar by alien scientists—just as Earth scientists, including NASA in its protocols, could search for life by looking for atmospheric chemical and thermodynamic disequilibria. Lovelockian planets would be like Earth, open thermodynamic systems away from equilibrium. Lynn's friend Lewis Thomas, the chain-smoking chief of the Sloan-Kettering Institute, popularized her cell symbiosis theory, compared Earth to a cell, and expressed disbelief that such a being would not be sentient at its own level.

Sentient or not, organisms and the cells they contained were social beings, merging with members of the same species (or not always detaching from them after reproduction). Individuality arose and continues to arise at multiple levels via life's favored module, cells—not genes: Eubacteria and archaea merged to make eukarya (protists); protists merged to become protoctists that evolved and diverged into the great plant, animal, and fungal kingdoms. Eusocial insects and mammals evolved still more inclusive levels of organization, and populations were regulated into biofilms, communities, and ecosystems. *Ophrydium, Mixotricha paradoxa*, the swarming hindguts of termites digesting wood from inside their mobile mandibular containers, cows doing likewise with the methanogens in their rumens adding oxygen-stabilizing methane to the atmosphere, kefir carbonated beverages with curds composed of multifarious microbial communities, seaside microbial mats in Baja California, Massachusetts, and elsewhere representing living fossils of early life, stromatolites in Australia as their lithifying forms, manganese nodules infused with bacteria beneath the ocean's surface, biomineralizing microbes shaping the entire crust of an Earth that was not infested by but composed of living beings—there was no place on the planet that was not biogenic, the reworking of ancestral breaths and bones, the result of vast hordes of chemically specific, mixed, environment-changing beings of which humanity was only one instantiation, no more evolved, in terms of hereditary tenure on Earth's surface, than any other extant organism.

Gaia—she liked the gloss, which she took from her graduate student Gregory Hinkle—was just "symbiosis seen from space." Hers was not the world of academic apartheid or changeless Platonic abstractions, filtered through the misplaced concreteness of Linnean nomenclature. She loved and knew nature's names and the detailed exploration of species differences, but also that beings could change quite quickly by prototaxis and fortuitous congress. Evolution is Heraclitean. Organisms are thermodynamically open systems. Gaia itself is thermodynamically open to energy from the sun, on which it mainly feeds, incrementally dissipating concentrated energy to space after capturing photons to make the foodstuffs that feed animals and their sensing but not sacrosanct minds.

And it is in this world of open, flowing, genetically and thermodynamically connected forms that sex and death evolve from sexless microbes that may be killed but do not die, not as absolute categories or out of biblical necessity, but because of historical cell interactions. Marching to the beat of her own theoretical drum, Lynn agreed with Stephen Jay Gould's critique of "Darwinian fundamentalists" (or "ultradarwinists," i.e., neo-Darwinists): that they were far too quick to indulge in a misplaced concreteness of mathematical explanations based on untenable assumptions, and to assume that extant traits had an adaptive advantage.[25] But while respecting Gould, she also disavowed his typical zoocentrism, as in his comments to the effect that nothing much happened in the three billion years before the Cambrian explosion.[26]

Here was a new, temporally deeper, less humanist and anthropocentric perspective. In *Origins of Sex*, we developed a story of how meiotic sex in eukaryotes evolved based not on genetic advantage but on contingent histories of symbiosis-like mergers among hungry cells that, in the attempt to eat one another, eventually fused cell membranes. Harvard biologist L. R. Cleveland had chronicled the engulfing of one cell by another in times of hunger, the merger of cell and nuclear membranes, and the doubling of chromosome numbers to make bloated diploid monsters. Lynn's intimate knowledge of protoctistan life alerted her that such doubled cells existed in lineages that did not usually reproduce sexually, but only when deprived of nutrients such as nitrogen, at which point they would merge to form hardy propagules. Reproductive sex followed by death was just one avenue taken by protoctist evolution. Her focus on the realities of metabolism, energy use, and chemistry contrasted with the mathematical abstractions of the neo-Darwinists who were often ignorant of both chemical and microbial ecological details.

Citing Simon Robson, she contended that mainstream theorization about reproductive (meiotic) sex's adaptive advantage was based on a data set billions of years out of date.

Lynn believed that reproductive sex evolved multiple times and was not directly selected for so much as embedded in lineages that began to alternate between haploid and diploid generations, cyclically "relieving" their diploidy (status of having two sets of chromosomes) by producing protist-like sex cells (as in humans sperm and eggs). The typical animal species has no option to "lose" its sexuality and reproduce asexually, as neo-Darwinist models assumed; even all-female populations such as whip-tail lizards appeared to have evolved from sexually reproducing ances-tors, and it was not clear that they had evolutionary staying power. There was, moreover, no evidence that such "asexual" (though they mounted one another and effectively fertilized themselves) organisms were less ge-netically variable than their reproductive relatives. This checking of neo-Darwinism against the facts, and comparing the morphology and behavior of multiple species in the field, as well as investigating questions using the natural laboratory of microorganisms with their astounding variety in traits that have become more fixed as animals evolved from their micro-bial ancestors, was typical of her investigations and critique of mainstream biology.

Death was for her an evolved trait that was not so prevalent or formal-ized at life's origin. It could be imposed by predation, disease, accident, or cellular malfunction, but in principle most microbes can reproduce in-definitely. Death as the natural end of multicellular organisms evolved in clonal communities with complex tissue differentiation. The doubled monsters that grew but had to return to their single-celled protist-like state in order to reproduce, recapturing in the fertilization event the cannibal-ism that had stochastically saved them, were subject to aging and discrete life spans associated with their highly developed forms. Moreover, Lynn's emphasis on symbiotic mergers in contrast to speciation as extrapolated from human breeding depended on seeing living beings as eminently prone to entanglement, coming together not just temporarily in sex but also permanently in symbiosis. She stated to me her agreement that a top-down view focusing on ecosystem energetics, the thermodynamics of ecol-ogy, was bigger than Gaia. Living or nonliving, the most interesting complex systems—from Bénard cells to the hexagonal vortex of Titan, from Belouzov-Zhabotinski chemical clocks to the multicolored seaside expanses called microbial mats—are *open*: exchanging energy, matter, and, in the case of life, genetic information, they share a naturalistic purposiveness, to process available energy, which becomes more effective in the long term if they do so in a stable fashion. Gaia itself, as stated, had its genesis in the realization that Earth's atmosphere was away from thermodynamic equilibrium—that Earth's atmosphere was an extension of the biosphere, organized as a system the better to capture and degrade solar and chemical energy. Genetically open systems with semipermeable membranes can merge in surprising ways; bacteria trade genes before, after, and during reproduction without any need to do so to reproduce. This openness and malleability of always already thermodynamically open living forms upsets zoocentric notions of individuality. We are multiple not only in the symbiotic bacteria that compose us but also in the differentiated clones that are these symbiotically derived eukaryotic cells working as bodies organized into tissues. And since the work cells do together to form colonies is by turns slippery and agglutinative, open and bordered, the individuality of organisms is labile in ways that upset neo-Darwinism's contradictory focus/fetish on genes and individual animals as "the" unit of selection. She was a de facto supporter of the group selection thesis in evolutionary theory that had been cattily and prematurely put to rest ("labeled and dismissed") by mainstream mathematical biology out of touch with natural history's facts, insufficiently attendant to nature's ability to produce individuality at various levels over deep time.

As I quipped and she repeated to Richard Dawkins at the "Battle of Balliol,"[27] genes cannot be selfish because they have no selves. Modern evolutionary biology tends to presume for the sake of mathematical convenience the false notion that genes in organisms act independently; this is indicated by the use of a special term, "epistasis," to refer to correlated gene activity. More unrealistically, neo-Darwinism, again for the sake of mathematical convenience in opposition to empirical natural history, has long assumed populations to be stable—flying in the face of formative boom-bust cycles, not to mention our own economic experience and global population doubling over the last fifty years. Powerful selective forces, however, just these boom-bust cycles can destroy populations and species that have not evolved means to regulate population size so that, for example, the entire group doesn't "shortsightedly" strip an entire food or resource base. Aging, which cannot be attributed just to antagonistic pleiotropy or thermodynamic wear and tear, appears to be "an inside job." Programmed death at the cell and animal level (and nothing could be worse for a selfish gene) offers evolutionary advantage to populations (despite being disallowed in traditional mathematical neo-Darwinism and selfish-gene rhetoric). But aging as a genetically accelerated onset of death begins to makes sense when we consider the dire consequences to interconnected populations that grow too fast, exposing them to complete wipeout from predation, starvation, and infection. The increasing medical interest in the human microbiome, which includes metabolic, behavioral, and psychological influences from our bacteria and other microbes, belies the still common dismissal of group selection in genocentric microbiology. We are multiple, not only in being constituted of multiple species but also socially in our dependence upon one another, and on long-evolved genetic programs like aging that attack individual fitness yet insure the survival of the groups in which individuals necessarily evolve. Mitteldorf and Pepper give a sophisticated mathematical treatment of such group selection that is more tethered to life's "stubborn facts" than neo-Darwinism's mathematically convenient abstractions.[28]

One might say that group selection (my mother was amused at my surprise that it was prohibited in mainstream Darwinism) works great in practice but does not work in theory. Ironically, now it does. Martin Nowak at Harvard (among others) has legitimized or at least widened the acceptability of group selection by giving it a mathematical treatment. And E. O. Wilson, founder of sociobiology, replaces selfish gene–based kin selection theories with group selection theories in *The Social Conquest of Earth*.[29]

AUTOPOIESIS AND GAIA

Lynn's system-based biology studies wholes without losing sight of detail, not just the parts as building blocks sufficiently explanatory of the whole, but also wholes considered on their own terms. She liked the concept, introduced in 1972 by Chilean biologists Humberto Maturana and Francisco Varela, of autopoiesis ("self-making"), which refers not to the substance composing a system but to the process by which it reproduces its own processes: "Consider, for example, the case of a cell: it is a network of chemical reactions which produce molecules such that . . . through their interactions [they] generate and participate recursively in the same network of reactions which produced them."[30] What is the difference between a living system and a dead one? Chemically they may be all but identical, but the functional organization of the latter has departed. It no longer reproduces its own living processes. Biological autopoiesis points up the importance of metabolism in life. Life is not just a replicative phenomenon; replication can take place only within the context of energy degradation. It is the lighter, the match, that keeps aflame the naturally cycling system whose phenomenological inside is attuned, both unconsciously and consciously, to searching and reducing ambient gradients, concentrated as yet unexpended energy reserves that are the object of living activity, and without which neither life nor its intrinsically telic character would be comprehensible.

Metabolism increases the production of entropy, the delocalization of concentrated energy mandated by the second law of thermodynamics. We hear this even in the common language. "To *live*" rather than merely exist means to be active. The live and the dead cat consist of the same compounds, but the dead one no longer coheres as a functionally recursive chemical system. It no longer locates the material substrates and energy sources it needs to maintain itself. It no longer breathes or defecates, adding heat and thermodynamically equilibrated carbon dioxide and oxygen-reactive methane to the environment. Its natural functionality has ceased. It has the same genes but is no longer autopoietic, either physically or cognitively in the sense of making sense of the world in terms of its needs. It no longer anticipates its next activity. Imagine the early planet strewn with those all-important, replicating DNA molecules or, alternatively, with RNA molecules. But if replicating nucleotides preceded the first cells, what would be their advantage in slowing up their naturally selected replication process by evolving cells? This would be like an Olympic runner voluntarily handicapping himself by wearing a potato bag over his feet. This thought experiment (from Jeffrey Wicken) shows the natural primacy of metabolism, cybernetically continuous open-thermodynamic cycling. A systems approach thus suggests that metabolism is at least as basic, and likely more basic, than the genomic "mechanisms" that insure it.

A viewpoint from the history of science reveals what might be termed the return of the scientific repressed.[31] What returns—whether it be the experimental apparatus in quantum mechanics, the events observed in thermal exclusion that then become applied to the projected "death" (stasis) of the entire universe (so-called heat death) in thermodynamics, or the vibrancy of the body once it is laid out as cadaver for anatomical dissection—is what has been excluded from an experiment for purposes of convenience and observation. As Nabokov might say, the shadow of the instrument falls over the specimen. In the case of the modern evolutionary prejudice for replicative investigations and explanations over metabolic inquiries, Freeman Dyson points out that Max Delbruck, an early pioneer of molecular biology, simplified the experimental realm by studying bacteriophages, viruses that feed on bacteria but cannot replicate on their own. Removing genetic elements from their autopoietic, metabolic context belies their partiality, making them seem more competent and complete then they are. For Lynn, the minimal unit of life was not the gene but the cell. Lynn's embrace of autopoiesis helped her in her battle against the theoretical excesses taught as Darwinism but perhaps better parsed as neo-Darwinism. She considered Gaia a selective force and the cell the unit of selection that could merge to form patinas, quorums, microbial mats, stromatolites, eukaryotic cells, and animals grown and evolved of eukaryotic cells—systems whose innards are not clockwork springs but bacteria.[32] The biosphere displays physiological behavior in its modulation of atmospheric chemistry, global mean temperature, marine salinity and pH, and other variables. For her Earth's surface was autopoietic. Autopoiesis offers conceptual help to the idea of a physiological Earth, for which Lovelock showed there's evidence, but which is anathema if natural selection alone is required to explain biotic complexity. Organismic and ecosystemic, Gaia is not an organism precisely but the autopoietic system that arises from the sum of biological autopoiesis coupled to its Earthly and cosmic, material, and energetic conditions.[33]

MENTOR TO SCIENTIFIC UNDERDOGS

Lynn's view of life as bound up with an open-thermodynamic and autopoietic system called Gaia, both more and less than an organism, her nononsense view of life not as an animal zoo but as a microbiogeochemical planetary phenomenon providing the evolutionary and ecological context for humanity, led her to be far more accepting than most of her peers of other minority scientific views.[34] Like hers, these too might prove correct; like those hers received, remonstrative remarks and defensive overreactions might signal not the expertise of the dismisser but the potential worthiness of the too-hastily dismissed intellectual problem or potential solution.

She was especially receptive to theories that were dismissed because they elicited knee-jerk reactions against improbable or, better, prohibited unions. For Lynn, the biological species concept central to neo-Darwinism was not sufficient, since it was predicated on a zoocentric viewpoint that extrapolated from animals, a specific kingdom of recently evolved organisms, to all of life. To extrapolate from animal speciation to the whole of life is to explain the past in terms of the present. But animals are far more microbial communities than "protozoans" (protists) are tiny animals. Relative to microbes, we animals are metabolic midgets that have the truncated the richness, the full chemical flowering of early life. She embraced the position of Sorin Sonea and Maurice Panisset in *A New Bacteriology*. These authors emphasize the promiscuous exchange among, in principle, all bacteria on Earth, of genetic elements without regard to morphological species borders. Prokaryotes practice natural genetic engineering; they make up a kind of global supercomputer or primordial genetic Internet. I have heard there can be up to 50 percent genetic difference between two cells of *E. coli*—so much difference that the term "pan-genome" has been coined so that individuals within particular bacterial "species" can possess genes that are not, in fact, in their own bodies—unless those bodies are distributed in a way that ours are not.[35]

Genetic exchange is ancient and wild in nature. The movie *The Amazing Spider-Man* (2012) imagines "cross-species genetics" as if it were something futuristic; but for the nature from which our technical intelligence is hewn, it is old hat. A paper from the year before *The Amazing Spider-Man* shows that microRNAs acquired from food such as rice or corn can alter your own gene expression.[36] As lead researcher Chen-Yu Zhang puts it, "It would indicate that in addition to eating 'materials' (in the form of carbohydrates, proteins, etc.), you are also eating information."[37] James Shapiro chronicles multiple examples of real-time genetic machination he calls "natural genetic engineering." (Carl Woese, universally respected for his RNA phylogenetics, and like Lynn one of the great biologists of the twentieth century, praised Shapiro's account.) Thermodynamically open, prototactically associating, group-living and symbiogenetic organisms have multiple modes of response, change, and evolution: They transfer genes and cytoplasmic proteins to the nucleus, coordinate cell cycles after symbiotic mergers, proofread DNA during and after replication, use episomes to repair DNA damage, and exchange distinct mobile genetic elements with names like plasmids, integrons, superintegrons, shufflons, retroviruses, and transposons. Two putative rapid key duplications of the entire genome are deduced to have occurred in our lineage, one at the root of vertebrates and then again in the transition from jawless to jawed vertebrates. Nature indeed seems more "saltational," more quick to evolve and respond in intelligent or sophisticated ways in real time, than dreamed of by classical Darwinians.[38]

Life is Dionysian, theory Apollonian: Lynn's dancing green eyes were always on the lookout for illicit liaisons, prohibited in theory but rampant in practice. Her own experience with the "old boy's network" of traditional biology, initially condescendingly dismissive of the evolution by symbiogenesis that it now accepts, but still tends to marginalize, led her to study the history of her illustrious predecessors in Russia, France, and the United States; and it prompted her to offer her support to those scientists who were dismissed without due consideration of evidence or testing just because they were working at the margins of orthodoxy. She was attracted to Peter R. Atsatt's theory that land plants arose from a nutritional hybrid between photosynthetic algae and degradative-absorptive fungi.[39] She was interested in the strange theory that fruits may have begun as arthropod-induced fungal galls, which would be another example of a disease structure becoming a part of the body.[40] And she was extremely supportive of Donald Williamson's detailed work and experimental evidence that seemed strongly to support his idea that biological metamorphosis, both in marine invertebrates and Lepidoptera, were the result of verboten fertilizations between genomes of organisms belonging to completely different phyla.[41]

Williamson's theses would account for the observed radical differences between juvenile and adult stages of metamorphosing beings, the juvenile stages resembling the adult organisms of distinct species more than the alternative life cycle stage of their own species. The resistance this marine biologist and native of the Isle of Man encountered in exploring his ideas no doubt reminded her of her own difficulties in melting prejudices against the empirically backed notion of the symbiotic origin of larger cells. As she described it,

> Williamson's radical proposed explanation sounds less fantastic if you realize that many nonmammalian animals fertilize eggs outside their bodies. For example, it is completely within the realm of possibility that flying insects in the late Paleozoic swampy tropics dropped some of their sperm on the egg masses of velvet worms. Velvet worms, called *Onychophora*, abound in tropical leaf litter on the forest floor. . . . Velvet worms are tropical, crawling, many-legged beings who would look like caterpillars to your mother. Most of the time, such inappropriate sperm drops by an oversexed impatient flutterer would have had no effects, but not always. Sometimes sperm and egg unions of very distantly related animals are neither infertile nor aborted.[42]

Whether or not Williamson's ideas become accepted, it does seem that nature is more supple than scientific orthodoxy, and that her tendency for change and forging new connections has been in place since the beginning, and is not in the least hampered by academic rigidity. "Becoming," wrote Paul Klee, "is superior to being." Or as Lynn's favorite philosopher, Alfred North Whitehead, wrote in his 1939 essay, "John Dewey and His Influence,"

> The human race consists of a small group of animals which for a small time has barely differentiated itself from the mass of animal life on a small planet circling round a small sun. The universe is vast. Nothing is more curious than the self-satisfied dogma with which mankind at each period of its history cherishes the delusion of the finality of its existing modes of knowledge. Skeptics and believers are all alike. At this moment scientists and skeptics are the leading dogmatists. Advance in details is admitted: fundamental novelty is barred. This dogmatic common sense is the death of philosophic adventure. The Universe is vast.[43]

That would include the biological universe, which we are still just beginning to explore and which, as we do so, reveals hitherto unseen aspects of our own being and potential.

2.

THE RNA/PROTEIN WORLD AND THE ENDOPREBIOTIC ORIGIN OF LIFE

SANKAR CHATTERJEE

When the Earth formed some 4.6 billion years ago, it was lifeless and inhospitable to life, a cauldron of erupting volcanoes, raining meteors, and hot noxious gases.[1] One billion years later, it was a placid watery planet teeming with microbial life, the ancestors to all other living things (Figure 2-1).[2] How could this have happened? The problem of the origin of life on this planet has engaged the attention of some of the keenest minds in philosophy and science since the time of Aristotle. For the past century, the debate on the origin of life was centered entirely on the chemical evolution of cells from organic molecules by a long succession of chemical processes. However, the origin of life was not an isolated prebiotic chemical event. I argue it included four hierarchical interconnected stages—cosmic, geological, chemical, and biological. Additionally, in recent times, with the exploration of space, the study of the origin of life has shifted to a broader perspective including planetary beginnings and exobiology, or the question of life on other planets. Current analysis of surface material by the Curiosity rover may provide crucial evidence whether Mars could ever have supported life. By studying the inner planets and their moons, we have begun to reconstruct the early history of our own planet when life might have emerged. The universal and uniform architecture of bacterial cells and the genetic code point to unity of type and

FIGURE 2-1. Earth about four billion years ago, once it cooled down to form a solid crust during the early Archean period, becoming a placid planet with immense oceans. The Moon was almost twice as close as it is now and would fill much of the night sky.

singularity of origin. Both molecular phylogeny and the microfossils provide a rare glimpse about the nature of the earliest forms of life.

The moment of the origin of life is considered to have been a singular, nonlinear, chaotic event of historical contingency in time and space, an outcome of long chains of unpredictable antecedent states.[3] To begin with, substances found in meteorites support the thesis that comets bombarding early Earth carried and deposited key ingredients necessary for life to emerge.[4] At the end of the Late Heavy Bombardment Period (LHB ~3.9 Ga), meteorites delivered carbon-based molecules and, more important, created suitable sites that became crucibles for prebiotic chemistry and the emergence of life. Impact cratering evidence from the pockmarked crater surfaces of Moon and inner plants suggests a violent environment during the early history of our young planet. Two distinct processes then promoted the origin of life: the exogenous delivery of carbon-based molecules from space and the endogenous synthesis of life on Earth.[5] In my view, both comets and asteroids played two distinct roles in the prebiotic period. Comets delivered water and building blocks of life to this planet, which were then accumulated in asteroid-induced crater basins, heated by hydrothermal vent systems. Impact crater basins of up to 500 kilometers in diameter offer a number of possible sites for prebiotic chemistry including: impact-generated hydrothermal vent systems similar to present-day niches of thermophilic microorganisms; prolonged convective circulation of heated water; sequestered sedimentary basins; and nanopores and pockets in mineral substrates for the concentration of biomolecules.[6]

Let us suppose that life arose in a sequence of steps in hierarchical fashion. First came the cometary building blocks. Next came the concentration, catalysis, chirality, and polymerization of these biomolecules in crater vents of the young Earth. Eventually, an evolving, self-replicating collection of macromolecules emerged on mineral substrates of the crater vents; these polynucleotide and polypeptide molecules were eventually encapsulated—that is, they developed semipermeable envelopes. They began to evolve through interactions that shared molecules fit for coding, and finally, to replicate to form the first living cells. I propose a view of the origin of life that integrates these four hierarchical stages—cosmic, geological, chemical, and biological. Such multilayered hierarchy provided a kind of quality control at each assembly level. Each stage gave birth to a new sequence of biomolecules with new characteristics and increasing complexity. Once a sequence is established, it is hard to destroy. This is a *bottom-up* approach that focuses on assembling a minimum protocell from cosmic ingredients of biomolecules in hydrothermal vent environments and on the process of self-assembly.

During the prebiotic evolution, which came first—the DNA or the protein? In a living cell, DNA depends heavily on catalytic proteins, or enzymes for functioning, yet protein relies upon DNA for correct sequencing. In short, proteins cannot form without DNA, but neither can DNA form without proteins. However, the paradox would disappear if RNA serves as precursors to both DNA and proteins. In the RNA worldview, RNA appeared before proteins and DNA. RNA has the capacity to store information (like DNA) and some RNA molecules such as ribozymes can catalyze reactions (like enzymes). The dual function of RNA makes it a versatile molecule. In the RNA world there is no need for the complex translational apparatus required for the production of coded proteins. But this does not remove the riddle entirely. RNA still has to be decoded by very specific proteins that are themselves coded for by the information contained in the RNA. As researchers continue to examine the RNA world concept closely, more problems emerge. RNA and its components are difficult to synthesize in a laboratory under the best conditions. Yet, no one has been able to demonstrate how RNA could have formed on early Earth in the absence of living cells. Besides, RNA molecules are very unstable and fragile. The origin of RNA is one of the most formidable problems facing prebiotic chemists.

Although the RNA-world hypothesis is widely popular in the origin of life debate, we prefer an alternative hypothesis: the coevolution of RNAs and proteins may be a better evolutionary model. If the earliest ribosomes had proteins as well as RNA, how were the RNA and proteins produced? Most likely, the proteins must have been produced by a non-ribosomal process. Amino acids, the monomers of proteins, are abundant in chondritic meteorites and are readily formed in the laboratory. Thus, the origin of prebiotic proteins from cosmic ingredients is less problematic than RNA. Once enclosed in protective cell membranes, RNA and protein molecules increased their chances of survival and began to interact to form the hybrid ribosomes. Eventually RNA would give rise to custom-made protein molecules inside the protocells via ribosomes. In presence of proteins, RNA would modify and replicate to form a self-amplifying feedback loop. The dual origin of "RNA/protein world" is more parsimonious than the widely embraced "RNA world" from the constraints of hydrothermal vent environments and the phylogeny of ribosomes. This view of origin of life is analogous to the endosymbiotic theory of Lynn Margulis for the origin of eukaryote cells. Margulis suggests that several key organelles of eukaryotes such as chloroplasts and mitochondria originated as symbiosis between separate free-living bacteria that were taken inside another cell as an endosymbiont. Following the footsteps of Margulis, I propose that symbiosis and cooperation between informational molecules such as RNA and proteins operated in the prebiotic state inside the lipid membranes, a process called here "serial endoprebiosis" that led to the emergence of the first protocells. Symbiosis and cooperation are strong evolutionary forces of nascent life, as Margulis has championed, and in fact may have been the defining element of the chemistry to biology transition. Modern viruses and prions may represent the evolutionary legacy of the original components of emerging life in the prebiotic world, such as RNA and protein molecules. In sum, I offer here an integrated and parsimonious model that gives sequence and shape to the mystery of the origin of life.

Impact cratering affected the geologic and biologic evolution of Earth, from the earliest stages of heavy bombardment period to the present. The Late Heavy Bombardment (LHB) was a cosmic event around 3.9 billion years ago (Ga), when the intense onslaught of comets and asteroids produced large numbers of impact craters on the Moon, Mercury, Venus, and Mars. The evidence for this intense impact event comes from the isotopic dating of shocked minerals and glasses from lunar samples.[7] The lunar surface is pocked with thousands of pristine impact craters preserved for billions of years because of the Moon's lack of water, atmosphere, erosion, or plate tectonics. In all likelihood, because the larger Earth has more gravity, in its early period it experienced more impacts than the Moon; the number of impacts occurring on Earth would have been an order of magnitude larger, implying more than ten thousand large impact events. However, the record of these events is all but lost. Our planet's plate tectonics and active erosion ensure that its surface is constantly renewed and recycled. Early in our planet's history, for the first 600 million years, large impacts must have liquefied Earth's lithosphere and mantle repeatedly, making it a fiery ball of molten magma.[8] Eventually, as the bombardment slowed down, Earth cooled, clouds formed, torrential rains created large expanses of oceans, and the crust began to harden, forming a terra firma. Clues to the condition of early Earth and the emergent biospheres are recorded in the oldest known sedimentary rocks.[9] However, setting the stage for the rise of life, the young crust was punctured by the last phase of heavy bombardment, which created large crater basins with hydrothermal vent systems, forming the ideal crucibles for life synthesis. In spite of crustal recycling, a few of these Archean craters may have survived in the Greenstone belts of Greenland, Australia, and South Africa that contain the oldest sedimentary sequences and the earliest signs of life.

Greenstone Belts and Crater Basins

Once the meteoritic bombardment ended, impact crater basins up to 500 kilometers in diameter offered a number of possible niches for prebiotic chemistry: impact-generated hydrothermal vent systems similar to present-day niches of thermophiles or heat-loving bacteria; prolonged

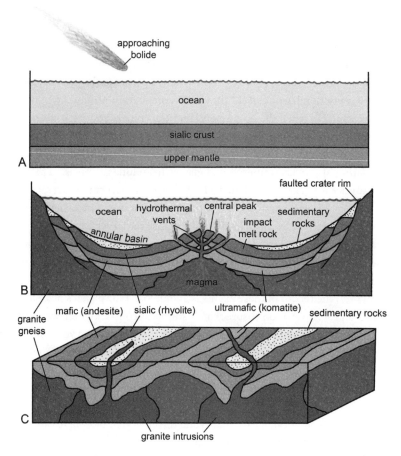

FIGURE 2-2. Development of an early Archean greenstone belt by a large bolide impact. (A) Approaching bolide in shallow marine setting. (B) Formation of a large complex crater with hydrothermal vent system sequestered and separated from the ocean basin; the crater basin was covered with impact-triggered lava flows as terrestrial equivalents of lunar maria. (C) Formation of the greenstone belt when the crater was folded, metamorphosed, uplifted, and eroded.

convective circulation of heated water with temperature gradients; seques-tered sedimentary basins where cosmic ingredients could be concentrated; and nanopores and pockets in mineral substrates for the concentration of polymers such as primitive RNA and protein molecules (Figure 2-2a, b).

Since numerous impact craters are preserved on the surfaces of the Moon and Mars, most likely Earth had a similar distribution of crater systems on its nascent crust. Unlike the Moon and Mars, however, the craters on early Earth were covered by water and developed hydrothermal vent systems driving high-temperature fluid flow in the crater floor. These crater basins were vessels for prebiotic chemistry and habitats for the early evolution of life.[10] Surface geological processes have erased the traces of most of the impacts on early Earth, but a few early craters may have survived. Greenstone belts preserve the oldest systems of this sort on Earth and contain the earliest evidence of life.[11] Although highly deformed, altered, and difficult to recognize, some of the oldest Greenstone belts in southern Greenland, western Australia, and South Africa may be the relics of these ancient primordial craters, where the oldest signs of life have been detected.[12] Fallout from these ancient impacts lies in the spherules and iridium layers of the Kaapvaal craton in South Africa, Pilbara craton in western Australia, and Isua Supergroup in Greenland that indicate the impact origin of these basins. The algoma-type banded iron formation and organism-bearing chert in Greenstone belts suggest hydrothermal vent environments. The volcanic-sedimentary sequences of Greenstone belts were deposited on shallow marine platforms when protocontinents were beginning to form.

Greenstone belts represent the core of the Archean protocontinents. An idealized Greenstone belt consists of metamorphosed, troughlike folded synclinal sequences associated with three major volcanic rock units, ranging from ultramafic (komatite) near the bottom, to mafic (andesite) at the middle, to sialic (rhyolite) near the top. Komatite is a mantle-derived volcanic rock with a high magnesium content that reflects a hotter mantle during the early Archean period.[13] Above the volcanic complexes, the upper sedimentary unit in Early Archean Greenstone belt consists of cherts, hydrothermal Algoma-type Banded Iron Formation, volcanic sediments, and impact-induced spherule layers. These cherts contain some of the oldest known microfossils. Additionally, pillow lavas associated with igneous rocks in Greenstone belts suggest underwater volcanism in hydrothermal vent environment. Shallow water and subaerial eruptions are indicated by pyroclastics, and in some areas, large volcanic centers built up above sea level. A late event in the history of the Greenstone belt is the folding, metamorphism, and intrusion of granites about 2.6 billion years ago, followed by uplift and erosion (Figure 2-2c).

Impact Origin of the Greenstone Belts

How a large meteorite impact could create a Greenstone belt is reconstructed here. The crust was thin and thermally active, buoyant, and ductile during the early Archean time and was covered by the universal ocean.[14] A large asteroid, about the size of the Near-Earth Object Ganymede (~40 km across), slamming into submerged crust could have excavated an enormous crater, about 500 kilometers in diameter, about the size of these Greenstone belts.[15] Impact would shatter the thin lithosphere to trigger magma to erupt to the surface as ultrabasic-basic volcanism and create an active underwater hydrothermal vent system along the diameter of the crater. Some of the largest craters on Earth, such as Sadbury, Chicxulub, and Shiva, provide terrestrial evidence for impact-generated thermal activity and volcanism.[16] Similarly, craters filled with impact-generated lava are the most common features on the surface of the Moon.

The large bolide impact scenario could nicely explain the magmatic and tectonic episodes of the Greenstone belts. Most likely the target rocks were thin sialic crust submerged under shallow sea (Figure 2-2a). For a high-angle impact by a 40-kilometer-diameter asteroid, a transient crater of 50 kilometers in depth would form, penetrating the thin crust and the upper mantle, followed by a rebound of the central peak ensuing in a complex crater of 500 kilometers in diameter. Because of large bolide impact, melting within the crater took place in two modes. The first stage was shock melting from the energy of impact. Soon after the impact, the crater basin was filled with shock melt of sialic crust such as rhyolite. The second mode of melting was decompression of the crust and upper mantle.[17] Excavation of the crater reduces pressure beneath the crater that produced volcanism from the mantle, filling the crater basin with mafic (andesite) and ultramafic (komatite) lavas below the sialic volcanics, causing isostatic uplift of the central peak. Komatites crystallize at relatively high temperatures greater than about 1600° C and were a prominent component of early Earth's crust billions of years ago. Today, Earth's interior is too cool to produce this particular rock. It may be that the superheating of the mantle due to large bolide impact in Archean times formed the komatite lava.

The architecture of the Greenstone basin mimics the large complex crater basin.[18] In cross-section, the central part of a Greenstone basin shows an anticlinal structure, which is flanked on either side by two synclinoriums. Its bilateral symmetry, like that of a complex crater, is unusual. Perhaps, the central part of the basin represents the central peak of the immense crater, which is surrounded by deep annular basins where younger sediments were deposited, followed by raised and faulted crater rims (Figure 2-2b, c). Today, these ancient impact scars are highly eroded, uplifted, and deformed, retaining only the roots of the craters.

Two additional lines of evidence support the impact origin of the Greenstone belts. The first is the occurrence of impact spherule and ejecta layers in Greenstone sediments that contains tiny balls of melt rock vaporized by massive impacts.[19] Spherule layers, which formed when target rocks vaporized by impact forces condensed into molten droplets, have been reported from the belts in Greenland, South Africa, and Australia, indicating that impacts might be linked to the formation of these oldest basins. The second is the discovery of an iridium anomaly, high concentration of the element iridium of meteoritic origin.[20] Iridium is rare in the crust but abundant in certain asteroids. These impact layers have been discovered from the Pilbara and Kaapvaal cratons between 3.5 to 3.2 billion years ago.

THE OLDEST EVIDENCE OF LIFE: ISOTOPIC SIGNATURES AND MICROFOSSILS

We know that Earth is 4.6 billion years old, but what is less clear is precisely when it first became the only known home of life. A grasp of the timing and environmental setting of early life on Earth is important for understanding its past ecosystems. However, locating the first evidence for life, a fossil about the size of a bacterium, is a question wrapped in considerable controversy. Efforts to find the earliest life in the Greenstone belts of Greenland, Australia, and South Africa have been focused on metasedimentary rocks such as cherts, banded iron formations, and hydrothermal vent rocks (Figure 2-3). The presence of the earliest forms of life—single-celled prokaryotes, such as bacteria or archaea—may be surmised from their isotopic signatures.

The Akilia and Isua belts of Greenland are some of the rare places that provide indirect evidence for the oldest traces of life on Earth. Unfortunately, these rocks have been subjected to intense metamorphism, so that any microfossils, if any ever existed, have been destroyed by heat and

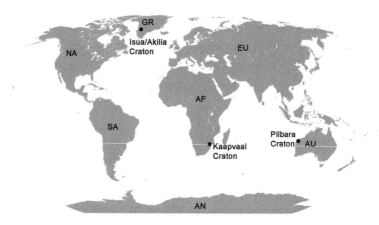

FIGURE 2-3. Location of three early Archean greenstone belts where signs of earliest life have been found: Isua/Akilia Craton in Greenland, Kaapvaal Craton in South Africa, and Pilbara Craton in Australia. These greenstone belts may be the relics of ancient impact scars.

pressure. The presence of life in these rocks is faint, based solely on biochemical signatures such as light carbon (carbon-12) in sedimentary rocks, the probable chemical remains of disintegrated microbes.[21] These sediments were laid down in Greenstone basins associated with hydrothermal vent volcanism. However, others have challenged this microbial activity in Akilia rocks, suggesting that biogenic carbon usually occurs as kerogen—coaly organic matter—not as the crystalline graphite that is present here.[22] A more convincing hint of early life has been documented about 180 kilometers northeast of Akilia in contemporary sediments (~3.7 Ga) of the Isua Group of Greenland, in the form of light carbon tied up in the thick layers of clay.[23]

Direct evidence of very early cellular life comes from the ~3.5 billion-year-old Apex Chert and Strelley Pool Formation in Pilbara craton, the oldest continental remnant in Australia, and the Swaziland Supergroup in Kaapvaal craton in South Africa.[24] These microfossils are most likely chemosynthetic bacteria similar to those of modern heat-loving thermophiles, thriving around the hydrothermal vents that survived the violent and extreme environments of early Earth.[25] These findings suggest that life emerged soon after the cessation of the Late Heavy Bombardment.

THE CRUCIBLE OF LIFE:
HYDROTHERMAL CRATER VENTS AND THERMOPHILES

The most likely site for the origin of life was not the open ocean or beaches, where the biomolecules would be diluted, dried, and dispersed. Instead, the most plausible incubators for the assembly of life molecules were places where liquid water and cosmic ingredients were sequestered, churned, concentrated, and synthesized at different interfaces of solid, liquid, and gaseous phases. Such an ideal crucible for the origin of life appears to be primordial crater basins with hydrothermal vent systems driving hot fluid flow by convection currents for concentrating and synthesizing life components (Figure 2-4).[26] Hydrothermal systems have prevailed throughout geological history on Earth, and ancient Archean hydrothermal deposits in Greenstone belts provide clues to understanding Earth's earliest environment for incubating and supporting life.[27]

Large crater basins with hydrothermal vent systems are appealing as a site for the origin of life because they incorporate a number of different aspects of biogenesis into a single hypothesis. These include:

1. an early crust bombarded by meteorites and leaving some telltale signs in the Greenstone belts;
2. crater basins with hydrothermal vent systems providing many opportunities for key geological events to take place;
3. high crater rims sequestering cometary biomolecules for concentration;[28]
4. geochemically reactive and redox environments of hydrothermal vents and prolonged convective circulation of heated water suitable for the chemical reactions leading to the establishment of life;[29]
5. nanopores and pockets in mineral substrates of clays and pyrites in crater basins ideal for catalysis and polymerization of biomolecules;[30]
6. crater lakes shielding early life from meteoritic impacts and the subsequent partial vaporization of oceans[31]; and
7. thermophiles that presently inhabit submarine hydrothermal vent environments, among the most primal organisms known.[32]

Most likely, the first cells had metabolisms similar to those of thermophiles. Such microbes, thriving in modern high-temperature deep-sea

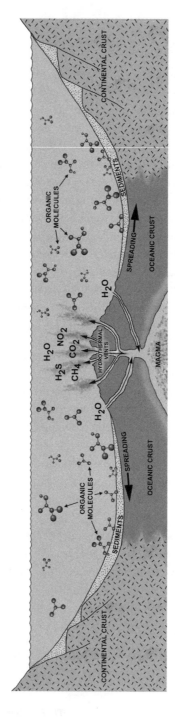

FIGURE 2-4. The crucible of life: large crater basins during the early Archean period. Submarine hydrothermal vent environments would have offered a protected setting for the synthesis and emergence of life. The water in the basin was rich with organic molecules delivered by comets. Heat and gases released from the hydrothermal vent brewed the vital components of life that began to accumulate at the mineral substrate at the bottom of the basin, where they were polymerized by catalytic action.

hydrothermal systems, have stimulated new theories of life's origins. It is an environment that seems to have all the necessary ingredients to spark chemical and biological reactions that could create the first cells. The primary energy to power this ecosystem comes from the dissolved chemicals of the vents such as methane and hydrogen sulfide. Among various chemical components released by the vent systems are: H_2O, H_2S, HCl, HF, CO_2, CH_4, SO_2, H_2, and various mineral and rock components.[33] The chemically enriched, superheated seawater rises through vent conduits, and picks up iron, copper, and zinc sulfides, forming a chimney-like structure on the seafloor, and supporting an exotic oasis of life, where thermophiles are at the base of the food chain.

We can improve our view about life's origins by working backwards from extant life. Molecular studies have confirmed that all living organisms trace their ancestry back to a single common ancestor, named LUA (for last universal ancestor); thermophiles are considered to be the oldest forms of life on Earth, thus the elusive LUA.[34] Recently, paleo-temperatures inferred from resurrected proteins reinforce the prevailing view that early life indeed lived in hot, hydrothermal environments. Proteins from ancient bacteria and their heat stability suggest that the last common ancestor was thermophilic and lived in oceans at temperatures of 65–73° centigrade.[35] Thermophiles originating in hydrothermal vent environments may be the oldest forms of life.

THE HIERARCHICAL ORIGIN AND EMERGENCE OF LIFE

We have no way of elucidating precisely the chain of events that led to the first life on Earth, but from our multidisciplinary approach we may speculate upon the plausible mechanisms and likely stages. Here I identify four distinct stages for the synthesis of the very first cells in the hot and dark environments of the hydrothermal crater vents (Figure 2-5).

1. The cosmic stage, representing the import of cometary molecules to the early Earth.
2. The geological stage, providing special environments for the potential incubators, the crater basins with hydrothermal vent systems, where segregation and concentration of these exogenous

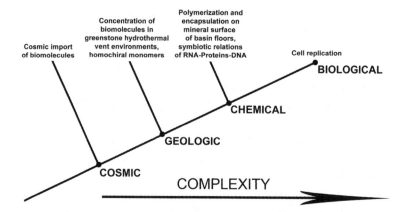

FIGURE 2-5. The origin of life in four hierarchical stages: cosmic, geologic, chemical, and biological. Each of these stages took place in the hot and dark environments of the crater basin, powered by the hydrothermal vents, which may represent Earth's oldest ecology. Increasing complexity of biomolecules in the prebiotic world led to the development of the first cells.

molecules (monomers) took place by convective currents; primal cell membranes probably appeared at this stage.

3. The chemical stage, representing the concentration of chiral monomers, such as the left-handed amino acids and right-handed sugar molecules, at the nanopores and pockets in the mineral substrates that surrounded the hydrothermal vents; simultaneously at the mineral substrates of the crater basins, these monomers joined together to form complex polymers, such as RNA and proteins, which were subsequently enclosed within cell membranes; development of prebiotic relationships between nucleic acids and proteins to store and process pregenetic information. Serial endoprebiosis led to the formation of the plasma membrane, RNA, ribosomes, customized proteins, and finally, DNA and the development of the genetic code.

4. The biological stage, representing the origin of reproducing protocells as they began to store, process, and transmit genetic information to their daughter cells.

A serial scenario for the hierarchical assembly of the building blocks of life from simple biomolecules to the first cells provides several advantages for a theory about the emergence of life. It provides "quality control" that can be exerted at each level of assembly, allowing natural selection to operate so that defective molecules can be discarded at an early stage rather than built into a more complex structure. The overall strategy of hierarchical assembly from subunits has the dual advantage of chemical simplicity and efficiency of assembly.

What Is Life?

Before we can tackle the question of biogenesis, we should have criteria for distinguishing living from nonliving matter. All living things are characterized by cellular organization, metabolism, growth, reproduction, and heredity. All organisms depend on external sources of energy to fuel their chemical reactions. Metabolism is the sum of all chemical reactions taking place in an organism. As a result of metabolic activities, organisms may increase the number of molecules of which they are composed—that is, they grow. Finally, all organisms replicate themselves, but there are many modes of reproduction. In the origin of life question, we are concerned with simple form of reproduction, where the parent cell divides into two identical daughter cells, as seen in bacteria.

Life has a unique chemistry, based on six chemical elements: C, H, O, N, S, and P, of which carbon is the key element. The simplest living cell is enormously complex, in which several hundred genes and their expressed proteins control and catalyze hundreds of reactions simultaneously within the same tiny compartment. As we wonder how the first living cell might have looked and how it worked, we must try to strip away these complexities to a minimal design. Minimally, a cell has to have three parts to function as a living entity.[36] First, it has to have a membrane as a boundary between itself and the rest of its environment. Second, there has to be some genetic material such as nucleic acid (DNA and RNA) for inheritance and reproduction, and for "information" that specifies the assembly of cell parts. Nucleic acids contain the information necessary to code for proteins. The synthesis of protein takes place on ribosomes, constructed not only with a large set of protein molecules but also with several molecules of RNA. Third, the cell needs metabolism or nourishment, self-maintenance mediated by its proteins. Proteins are involved virtually in all cell functions. The overall process is DNA → RNA → protein, where the arrows show the direction in which the sequence of information is transferred. If we can design a minimal cell with a membrane, nucleic acids, and proteins, making them interactive and capable of reproduction, we can speculate more precisely how the first life might have originated.

THE COSMIC STAGE: THE IMPORTATION OF BIOMOLECULES

Perhaps part of the chemistry needed for life was not produced on Earth but came to Earth from space. Our seawater and present atmosphere bear some tantalizing chemical evidence for the massive cometary bombardments that occurred four billion years ago.[37] A rain of comets delivered the early Earth its water, its biomolecules, and its first atmosphere—key ingredients for life's beginnings—whereas a barrage of asteroids created the hydrothermal crater vents, where prebiotic synthesis took place.[38] Our starting point in the chronicle of the genesis of life is the importation of prebiotic substances by comets and asteroids to Earth, where they were finally assembled. The Murchison meteorite that crashed on Australia in 1969 provided a clue as to how these biomolecules could be delivered from space.[39] Some of the building blocks from the Murchison meteorite include nucleotides of DNA and RNA, of left-handed amino acids and cell membranes, and other ingredients. Comets also contain prodigious quantities of amino acids, adenine, ketones, quinones, carboxylic acids, hydrogen cyanide, polycyclic aromatic hydrocarbons, thioformaldehyde, acetaldehyde, sugars, cyanogens, cyanide, methanol, and ethanol.[40] These findings demonstrate that many organic compounds, key components of life on Earth, are already present in the early Solar System. They must have played a key role in life's origin, providing crucial parts of the cell, such as the sugar and phosphate backbone of DNA, its information-bearing nucleotide bases for making nucleic acids, amino acids for making proteins, and lipids and fatty acids for making membranes. Just as significantly, further building the framework of life, many of the extraterrestrial cargoes delivered vast quantities of water to cool the Earth and make it habitable.

THE GEOLOGICAL STAGE: A SIMMERING CAULDRON

The geologic stage provides special dark, hot, and isolated environments of the crater basins with volcanically driven hydrothermal vent systems that served as the potential incubators of life, where segregation and concentration of monomers took place by convective currents of the vents spewing chemicals and gases to form a dynamic geochemical setting. The large, sequestered crater basin with hydrothermal vent systems is unique in geologic history and is very different from black smokers, which we encounter today along the spreading ridge of the ocean floor. The hydrothermal vents in large crater basins of early Earth were the likely sites of life's origin.[41] They would have provided energy and opportunity, and the

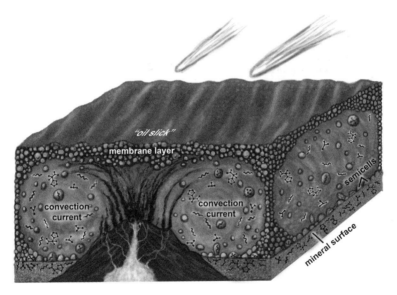

FIGURE 2-6. Three-dimensional cross-section of the hydrothermal vent environment of the crater basin. The vent acted as a giant cauldron to mix cosmic biomolecules. At the water surface, primitive lipid membranes and other hydrocarbons float as an oil slick. The mineral surface at the floor of the basin might have acted as a catalytic surface for concentration and polymerization of monomers. Some of the membranes and other organic molecules were mixed by convection currents, brought back to the surface, and concentrated. Membranes attach to the mineral surface to form blister-like semicells ready for encapsulation of polypeptides and polynucleotides.

minerals dissolved in water flowing from such vents would have provided a constant source of chemical nourishment (Figure 2-6). The crater basins where cometary biomolecules began to concentrate, polymerize, and transform to more complex molecules were sheltered from the vast stretches of oceans by their raised rims. Most of the volcanic activity was underwater, and hydrothermal vents provided thermal energy, inorganic resources, and various minerals that acted as catalysts (Figure 2-4). The temperature of black smokers in hydrothermal vents could be as high as 400° C. Basin water controlled the violence of the slow constant seepage of molten lava from the vents. Like a simmering cauldron, heat from the hydrothermal vents churned the basin water, mixing and combining biomolecules at different layers to form the thick primordial soup, and facilitated rapid chemical reactions. As the basin became rich in organic molecules by continued accumulation and concentration, they joined to form more complex molecules. During the geological stage, several key components of life, such as lipid membranes, amino acids, sugar and phosphate molecules, and nucleotides, began to assemble in the hydrothermal vent settings.

Formation of Lipid Membranes

The earliest components of precellular life were the lipid membranes that must have enclosed complex biomolecules, preventing them from getting out while letting smaller nutrient molecules in.[42] Meteorites may have brought this fatty lipid material to early Earth. The hollow spheres extracted from the Murchison meteorite mimic cell membranes and provide some ideas about the origin of primordial cell membranes. These spheres are not true cells; they are empty, but they do have lipid molecules on their surface. They have enough mechanical strength, both to have contained living matter and to have given it some protection from outside forces.[43] These lipid molecules began to concentrate on the water surface to form a thick scum of bubbles.[44] Evaporation thickened the bubble slick that served as a shield for underwater monomers, protecting them from ultraviolet radiation, making the hydrothermal vent environment pitch black. Most likely, the primitive, monolayer membranes that appeared during the geologic stage floated on top of

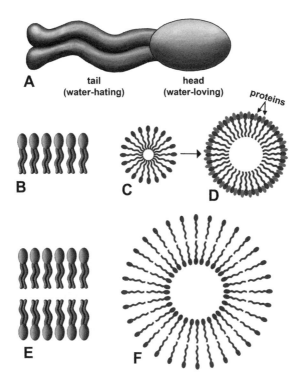

FIGURE 2-7. Formation of a primitive cell membrane. (A) A generalized phospholipid molecule showing the head (water-loving) and tail (water-hating). When mixed in water, these phospholipid molecules form spherical structures, which may be a monolayer (B–D) as in thermophiles, or a bilayer (E–F), as in most other cells. The simple monolayer membrane (C) was probably formed at an early geologic stage as a fatty bubble on the surface of the greenstone basin. However, more sophisticated plasma membranes, strengthened by proteins, would form only when proteins were encapsulated (D). Bilayered membranes (E–F) probably formed during the emergence of cyanobacteria.

the water surface of crater basins but moved to the bottom by convection currents (Figure 2-7).

Lipid molecules such as fatty acids are amphiphilic, which means that one end of the molecule tends to dissolve in water (the polar, or hydrophilic group), while the other end tends to stay away from water (the oily hydrophobic group). Lipids can be arranged in monolayers and bilayers, with heads and tails always facing in opposite directions. The "head" loves water, whereas the "tail" shuns water. In a typical cell, the water loving heads contact water both inside and outside while the water-repellent tails connect with one another to form a waterproof container. The monolayer micelles (such as soap bubbles) are roughly spherical aggregates of lipid molecules with the polar heads oriented toward the outer surface so that they interact with water, while the nonpolar tails are directed toward the center. In thermophiles, the cell membrane is a monolayer where the tails are glued together to hold the cell together to withstand intense heat. Spheroidal micelles may be the earliest forms of cell membranes to house thermophile-like protocells; bilayer membranes evolved later for those microbes living at normal temperatures.[45] Some of the lipid molecules were recycled continuously from the surface to the crater floor by the convection current of the volcanic vents; these bubbles were stuck on the mineral substrate at the basin floor for encapsulation, where RNA and protein molecules were catalyzed and accumulated.[46] Most likely, as seen in thermophiles, the cell membranes of early life were monolayer. The double-membrane cell appeared later during the origin of cyanobacteria around 2.7 Ga.

The Concentration of Monomers in Nanopores of Minerals

The basic molecular components of cells are small organic molecules called monomers. About thirty small monomers, called the "alphabet of biochemistry," constitute the cells. This alphabet includes the twenty amino acids found in proteins, the five nucleotide bases present in nucleic acids (adenine, cytosine, guanine, thymine, and uracil), the two sugars (ribose and deoxyribose) in nucleic acids, and three lipid molecules in membranes.[47] Most of the monomers may have had a cometary origin.[48] They began to concentrate in the pore spaces of the mineral substrates of the vent systems. Both hard and soft substrates were available. The condition, permeability, chemistry, crystal lattice, and electrical charges at the floor of the crater basin were important for the concentration of the monomers.[49] Intricate networks of pores, tunnels, and cavities of mineral substrate trapped and concentrated monomers, then protected and prevented them from degradation. The hard substrate consisted of minerals and rocks, such as pyrite, borates, and zeolites that crystallized out of the molten basaltic lava from the vent. The nanopores of pyrite substrates might have encouraged the concentration of amino acids and acted as catalytic surfaces to enhance chemical reactions, linking these biomolecules into complex forms of proteins.[50] The soft substrate was essentially a sedimentary layer, such as clay at the floor of the crater basin. The complex architecture of nanopores of clay minerals and their stacking sequence might have formed the rudimentary scaffolds for the monomers of polynucleotides such as sugars, phosphates, as well as the nucleotide bases.[51] Borate mineral, such as colemanite, might have acted as a catalyst in synthesizing key ingredients of nucleic acids, such as ribose and nucleotides.[52]

THE CHEMICAL STAGE: HOMOCHIRAL MOLECULES AND POLYMERS

The geological stage grades smoothly into the chemical stage, when complex polymers such as nucleic acids and proteins were catalyzed and assembled from their monomers. The heat churning the water column in the crater basin mixed chemicals together and caused simple chemicals to grow into larger and more complex ones by self-assembly. Three critical steps took place at the chemical stage:

1. Homochiral monomers of amino acids and sugars became segregated and selected as left-handed and right-handed molecules respectively;
2. Homochiral monomers were catalyzed into complex polymers such as proteins and RNA; and
3. RNA and protein molecules were encapsulated by lipid membranes to prevent disintegration and initiate serial endoprebiosis.

Homochiral Monomers

Most organic molecules occur in two chemically identical forms that are left-handed and right-handed mirror-reversals around an asymmetric carbon group. These mirror images are known as chiral molecules. While these two chiral forms are chemically identical, their chemical actions and biological effects surprisingly are quite different. The single-handedness of biological molecules is called homochirality. Homochirality is a signature of life on Earth.[53] Right-handed sugars in nucleic acids (D-deoxyribose

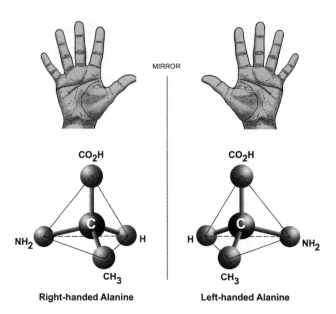

MIRROR

CO$_2$H

C

NH$_2$ H

CH$_3$

Right-handed Alanine

CO$_2$H

C

H   NH$_2$

CH$_3$

Left-handed Alanine

FIGURE 2-8. Enantiomers. An important property of carbon allows the formation of two molecules with identical composition but different properties: just as the left hand is the mirror image of the right hand, one form of the molecule (called homochiral) is the mirror image of the other. Living cells have left-handed amino acids but right-handed sugars. The selection of homochiral monomers took place in prebiotic environments, probably mediated by certain minerals.

in DNA and D-ribose in RNA) and left-handed amino acids in proteins (L-amino acids) exist exclusively in living organisms. Yet, if these components are synthesized in the laboratory, they come in a racemic mixture with equal amounts of left- and right-handed enantiomers of chiral molecules (Figure 2-8). The selection and separation of homochiral monomers in the beginning of life is not fully understood, but some mineral filters, biologic processes, or cosmic mechanisms may have contributed to their formation. For example, calcite crystals can separate and concentrate left-handed amino acids.[54] In the crater environment, one particular face of calcite crystals might be preferentially selected, thus sorting and concentrating the left-handed forms of amino acid for making proteins. It seems likely that certain minerals in addition to calcite, which were available in the hydrothermal environment, selected and sorted chiral molecules such as left-handed amino acids and right-handed sugars.

A biological view suggests that the chiral selection of specific monomers occurred in one or more self-assembly processes and was amplified then through an autocatalytic or self-sustaining feedback loop.[55] Certain organic molecules have the capacity to react with each another to form a self-sustaining chain reaction. Some of these reactions can spontaneously select one homochiral form over another by positive feedback loops that gradually amplify and concentrate it. Recently scientists have found that left-handed amino acids can catalyze and encourage the formation of right-handed molecules of sugars.[56] Somehow, the L-amino acid interacts with the components of the reaction and guides the chirality of D-sugars. This could explain how carbohydrates originated and why the left-handed amino acids and right-handed molecules of sugars are exclusive to all life forms.

However, chiral molecules might have cosmic as well as terrestrial origins, because some meteorites show a slight surplus of left-handed amino acids.[57] Ultraviolet irradiation of interstellar ice analogs mimicking the space environments with circular polarized light has produced left-handed amino acids, suggesting that the origin of homochirality is prebiotic and cosmic.[58] Astronomers speculate that neutron stars throw out electromagnetic radiation in chiral forms that may have selected more left-handed amino acids in space. Likely these left-handed forms were seeded to the young Earth, where they began to concentrate. Whatever may be the cause for the origin of homochirality, left-handed amino acids and right-handed sugars were selected and concentrated in the hydrothermal vent environments before they turned into more complex polymers such as proteins and nucleic acids respectively.

The Formation of Polymers: Polynucleotides and Polypeptides

Both proteins (polypeptides) and nucleic acids (polynucleotides) are polymers, and would be synthesized from their monomers spontaneously in vent environments by catalytic processes. Polymerization to form large complex macromolecules from simpler ones is a crucial step in the chemical evolution of life.[59] Mineral substrates in the crater floor would have been ideal settings for the synthesis of prebiotic polymers by repeated cycles of dehydration-rehydration reactions and concentration of substances in evaporating waters. Several minerals in hydrothermal vents such as pyrite and clay could act as catalysts for the polymerization of proteins and nucleic acids.[60] Proteins are the most abundant polymers in living cells, and their monomer components are amino acids. To make a protein, the ribosome strings amino acids together into a chain, called a polypeptide, a single linear polymer chain of amino acids. Condensation or loss of water (dehydration) between two monomer molecules facilitates linking them into long chains of polymers. Amino acids build proteins by this method by rearranging covalent bonds in such a way that two amino acids are connected and water is removed. Proteins consist of one or more polypeptides, typically folded into a specific three-dimensional shape (Figure 2-9).

Like the amino acids in proteins, nucleotides can be assembled end to end into linear molecules called polynucleotides. Each nucleotide is made of a phosphate group, a right-handed sugar molecule, and nitrogen-containing organic bases. The basic constituents of RNA molecules such as D-ribose, and the four bases, adenine, guanine, cytosine, and uracil may well have been delivered to hydrothermal vents by meteorites, but they would have been assembled and polymerized in crater basins.[61] Polynucleotides have more limited capabilities as catalysts, but they can directly guide the formation of exact copies of their own sequence. Replication of a large nucleotide sequence requires polypeptide enzymes to form a complementary sequence. Additionally, the mineral surface on the crater floor, functioning as a catalytic scaffold, could catalyze monomers into polymers and jumpstart critical chemical reactions.[62] Convection currents would continuously supply assortments of monomers including amino acids, nucleotides, phosphate, and sugar molecules into the mineral pore spaces at the floor of the crater basins, condensing them into proteins and nucleic acid chains. Several natural catalysts such as clays, pyrites, and zeolites at the ocean floor might have been important resources for polymerization. Pyrite could have catalyzed the polymerization of proteins and served as the template and energy source that drove the formation of the polymers.[63]

Zeolites are complex silicates with repetitive structures like clays and surfaces that can catalyze organic reactions. They could have absorbed organic molecules out of the water and concentrated them through intricate networks of porous channels. Their cavities could have trapped amino

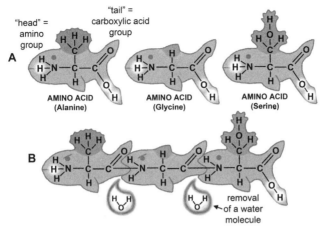

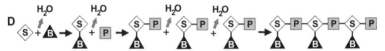

FORMATION OF NUCLEIC ACID BY POLYMERIZATION

FIGURE 2-9. Formation of polymers such as proteins and nucleic acids from monomers (amino acids and nucleotides) through dehydration and polymerization in vent environments. (A) Imagining the amino acid as a fish helps us visualize several important parts of the molecule; different kinds of amino acids are nearly identical in possessing an amino group (the "head") and a carboxylic acid group (the "tail"); they differ only in the "fin" on the fish back, which determines the kind of amino acid. (B) The amino acid "fish" can form a protein by linking up in long chains, combining the amino and carboxylic acid groups of the "head" and "tail;" in the process, a water molecule is formed by combining a hydrogen (H+) from the amino group and a hydroxyl group (OH–) (these are shown highlighted in A). (C) Formation of longer and longer chains of polypeptides by removing water molecules from each link. (D) Formation of nucleic acids (RNA) by linking sugars (S) and bases (B) (which form a nucleotide) with a phosphorous (P) backbone. This also involves removing a water molecule.

acids and protected them from harsh surroundings. As catalysts, zeolites could have polymerized into proteins these trapped amino acids in the pore spaces.[64] Due to their interconnected architecture, clay minerals exhibit enhanced flexibility over crystal networks and probably functioned as scaffolds to synthesize and polymerize amino acids into simpler proteins that promoted correct folding. Clay minerals also have a small surface of electrical charge that quite efficiently and selectively attracts and adsorbs organic molecules, such as amino acids, nucleotide bases, and sugars. They could have immobilized these monomers on their surfaces, catalyzed and concentrated them, thus assisting the formation of chains of proteins and nucleic acids. The long, straight cleavage planes of clay minerals would have encouraged monomers to arrange themselves in lines to form complex chains of polymers.[65]

THE EMERGENCE OF SELF-REPLICATING SYSTEMS

Life is associated with three classes of macromolecules—DNA, RNA, and proteins, enclosed in phospholipid membranes. Among these molecules, which came first? Did they occur sequentially or concurrently? Most scientists agree that DNA-based life probably emerged from a much simpler life form. Judging from the roles that RNA has in modern cells, it seems likely that RNA appeared before DNA.[66] Different views about the origin of life give different accounts of the order of appearance of the three major components of cells—membranes, proteins, and RNAs. To most people who search for life's origins, either proteins or genes are everything; however, without a container or membrane to place a boundary around the operations that involve genes and proteins, there can be no life.[67] In light of this, I see the membrane as the first cell component to emerge.

The Membrane World

Earlier forms of life definitely needed a permeable membrane compartment to keep the polymers such as polypeptides and polynucleotides safe and separate from their environments. Lipid cell membranes are easy to synthesize and have been detected in meteorites.[68] Most probably, membranes first appeared during the geological stage. An outer membrane encapsulating cooperative molecules of RNA and proteins would bring them close together for interactions. Without the interaction and physical proximity of polypeptide molecules, a RNA molecule would not replicate to another RNA molecule; it would be dispersed in the vent environments. Some containment of RNA and protein molecules is essential to the first synthesis of life. Encapsulation forced primitive RNA and protein molecules to stay together in close proximity and evolve together by what I have called *serial endoprebiosis*.

The RNA World

The way we think about the origin of life changed once it was discovered that certain RNA enzymes, called ribozymes, could act, like proteins, as active catalysts for their own replication. Many in the scientific community accept the RNA-world hypothesis for the origin of life. This hypothesis states that self-replicating RNA molecules were precursors to current life; in other words, the RNA world preceded the DNA/RNA/protein world.[69] In such a definition, the RNA world is to be considered either as living or as prebiotic. RNA has the ability to act both like a gene (like DNA) and like an enzyme (like proteins), and can duplicate itself and catalyze reactions for protein synthesis.[70] RNA was regarded as the first self-replicating molecule that could perform two primary cell functions.

Just like DNA does, RNA could store and transmit genetic information. The RNA world began to conjugate amino acids and synthesize proteins, which in turn accelerated the production of more RNA and more proteins. The structures that translate RNA into proteins, called ribosomes, are hybrid RNA-protein machines, and it is the RNA in them that does the catalytic work.[71] A positive feedback loop or hypercycle was established that amplified the production of both RNA and proteins. Furthermore, in reverse of the normal process of transcription, RNA can be transcribed into DNA. These facts are reasons to consider that the RNA world could be the original pathway to cells. As their genetic systems became more and more complex via the feedback loop of RNA and proteins, the storing of genetic information by a separate genetic molecule became essential. This is when double strands of RNA molecules zipped together and evolved into DNA molecules. Over time, DNA took over information-storage functions, and proteins took over structural functions. RNA does the lion's share of protein synthesis, but DNA provides the template.

The idea that life began as a solitary self-replicating molecule of RNA without protein or membrane is appealing because of its simplicity and versatility.

Although the RNA-world hypothesis is attractive, many scientists question its having existed.[72] They argue that it is not chemically realistic for the naked RNA world to give rise to the first cell. RNA molecules are not very stable and in vent environments decompose easily into their monomer constituents.[73] Most workers now believe that to perform its many tasks RNAs could not be free molecules but must have been encapsulated by vesicles.[74] This would imply, once more, that membranes arose earlier than RNA.[75]

The basic premise of the RNA-world hypothesis is a protein-free RNA world. In the RNA world, RNA is the pioneer molecule that plays a central part in directing the synthesis of polypeptide molecules via ribozymes. In this view, polypeptide molecules did not exist before the emergence of RNA molecules; RNA molecules created all the protein molecules by translation and made them available when needed. Instead of protein molecules, the ribozyme, the RNA catalyst, is believed to function as the first catalyst for a wide variety of different reactions, including the polymerization of amino acids.[76] Ribozymes have some ability to perform as catalysts, but in other crucial respects, RNA cannot duplicate the functions of proteins. For example, RNA cannot act as an integral membrane protein, and therefore cannot be involved in complex cellular functions such as nutrient and ion uptake and the discharge of waste products.[77] How could a rudimentary RNA molecule, lacking protein machinery, carry out these crucial tasks of construction and repair of the membrane and pump molecules in and out of the cell? Most researchers now believe that for primitive RNA to replicate, other catalysts of some sort, such as peptide-like short chains of proteins, were necessary.[78] RNA molecules are inherently fragile, and they can easily be broken down into their constituent nucleotides through hydrolysis. Moreover, catalysis is a relatively rare property of long RNA sequences only.

Given the Miller/Urey experiment's demonstration of the ease with which amino acids form and polymerize, there is no geochemical reason to doubt that primordial polypeptide molecules could be synthesized in the prebiotic world of the vent communities. Polypeptides are simpler to create than polynucleotides, and from the meteorite input amino acids were more readily available than nucleotides.[79] As suggested by the recent phylogeny of the ribosome, a ribonucleoprotein machine for making specific customized proteins by translation from the RNA template, proteins were already on the scene.[80] In our view, primal polypeptide molecules created in hydrothermal vents provided the crucial catalytic help to RNA molecules, which in turn began to replicate and produce more intricate protein molecules inside a protective membrane via a rudimentary ribosome, a hybrid of proteins and RNA. RNA could not produce life synthesis in the absence of simple forms of proteins. Protein enzymes always outperform RNA catalysts. Once we accept the availability of polypeptides along with polynucleotides in the vent communities, many of the hurdles of the RNA-world hypothesis disappear as a simpler origin for life emerges via the hypothesis of an RNA/protein world. It was not merely an RNA world, a peptide world, or a lipid world. The life system emerged only because all its elements were connected.

The RNA/Protein World

All prokaryotic cells have, without exception, a plasma membrane, a nucleoid, and a cytoplasm filled with ribosomes. Ribosomes are required for translation of the genetic information into polypeptide chains. Recent phylogenetic work on ribosomal history challenges the RNA-world hypothesis.[81] According to this new analysis, even before the ribosome's many working parts were recruited for protein synthesis, proteins were interacting with RNA. RNA and protein molecules began to communicate by means of chemical and molecular language as soon as they were encapsulated to form the hybrid ribosomes. The model suggests that the ribosome is an ancient ribonucleoprotein (RNP) complex, implying that both proteins and RNA contributed to its formation during the proto-cellular period. The primordial ribosome was not made entirely of RNA molecules; proteins contributed the lion's share. Molecular phylogeny suggests that proteins appeared before the ribosomal RNAs.[82] Thus, ribozymes are not ancestral ribosomes—a view that contradicts the strongest argument of the RNA-world hypothesis. Ribosomal proteins are not passive contributors to ribosome function but play important roles in translation along with rRNA.[83] Both proteins and RNA are essential for ribosomal functions. Ribosomes fill the need for a site where a crucial cellular activity—protein

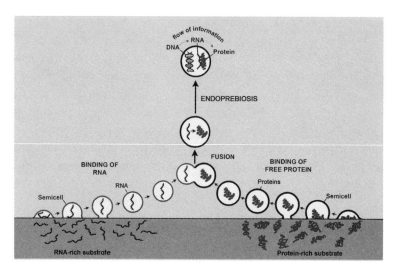

FIGURE 2-10. Proposed model for the encapsulation of RNA and protein molecules by phospholipid membranes from clay and pyrite mineral substrates respectively. Once these molecules of polynucleotides and polypeptides were sheltered in membranes from the outside world, they began to fuse and interact; endoprebiosis led to the formation of DNA and the genetic code.

synthesis—can take place. Polynucleotide and polypeptide molecules must have coexisted in vent communities long before the first cells arose. It was, instead, a ribonucleoprotein (RNA/protein) world. Such a prebiotic dual origin of rudimentary metabolism and replication offers a simpler and more parsimonious view of the origin of life. These two polymers—the polynucleotides and polypeptides—emerged simultaneously on the mineral substrates and were subsequently encapsulated to initiate prebiotic interactions (Figure 2-10). Both the phylogeny of ribosomes and the environmental constraints of hydrothermal vents suggest that the "RNA/protein world" is a better precursor for the assembly of life than the popular "RNA world" model.

Protein and RNA molecules produced at mineral surfaces were likely small in size, simple in shape, and relatively random in sequence, creating different chemical species and assortments. Certain peptides engulfed nearby iron-sulfur clusters of the vents to yield the ferredoxin, the oldest protein known in thermophiles. A simple and ancient thermophilic protein with an iron-sulfur cluster, ferredoxin uses mainly simpler amino acids, which suggests that it predates the complete emergence of the genetic code.[84]

SERIAL ENDOPREBIOSIS

Lynn Margulis championed the idea that the eukaryotic cell arose by symbiotic merger, the union of at least two kinds of bacteria with a host cell, in a process called endosymbiosis (living together internally).[85] Here I propose that something like endosymbiosis operated even at the lowest hierarchical level during the emergence of life. However, since it was a prebiotic world, instead of "endosymbiosis" I coin a neutral term, "endoprebiosis," where the membrane was the host, and polypeptides and polynucleotides were prebiont molecules that mutually benefited—in the sense of supporting each other's faithful replication—while working together within the protected environments of prebiotic membranes. I speculate that a serial endoprebiosis occurred just above the mineral substrates as the lipid membranes randomly began to encapsulate newly synthesized polynucleotides and polypeptides from the crater floor. In the process of random encapsulation, RNA and protein molecules were enclosed separately and randomly from two adjacent niches of mineral substrates; each membrane contained either RNA or protein molecules, which would fuse to form a larger cell (Figure 2-10). Encapsulated molecules began to interact with one another—protein as a catalyst and RNA as a primitive gene to create a hybrid ribosome. Eventually these molecules began to foster partnerships among their components, and to communicate with each other, circulating information in complex feedback loops.

I identify seven hypothetical stages for the emergence of life by serial endoprebiosis:

1. the encapsulation of protein molecules and the origin of a plasma membrane;
2. the encapsulation of RNA molecules;
3. the fusion of two protocells containing both RNA and protein molecules;
4. the origin of ribosomes;
5. the origin of retroviruses;

6. the origin and replication of DNA; and

7. the reproduction of the first cell.

The crater vent systems could have generated large numbers of varied complex molecules. It may be that a random collection of RNA and protein molecules of different sizes and properties emerged simultaneously at the chemical stage and were encapsulated by membranes to start the RNA/protein world along with a primordial soup rich in protocell nutrients. It was a combinatorial chemistry in a microworld inside billions of membranes. The assembly of cell components occurred through trial and error over a long period. Infinite combinations must have taken place, and a countless number must have failed to function before the secret was broken and the proper selection occurred. The first replicating cells on Earth emerged from a nonlinear chaotic system of historical contingency in which functional success provided the direction in each stage. Some of the serial endoprebiotic stages and the emergence of critical cellular parts and their functionalities are described in the following section.

Origin of the Plasma Membrane

Today, a plasma membrane contains by weight about 50 percent phospholipid and almost 50 percent protein.[86] As suggested by meteoritic evidence, the primal membranes were simple phospholipid monolayer sacs.[87] A phospholipid is a lipid with a phosphorous-containing polar group at one end and a nonpolar fatty acid group at the other (Figure 2-7c). As phospholipid membranes encapsulated proteins, they provided a haven for them; proteins, in turn, contributed some of their molecules to phospholipid membranes to strengthen the cell wall and make it more efficient for multitasking, including the passage of sodium and potassium ions and other nutrients through the membranes (Figure 2-7d). Proteins embedded in the membrane acted as gatekeepers and pumped molecules in and out of the cells, while other proteins assisted in the construction and repair of the membrane. The new plasma membrane was a protein-lipid "sandwich" consisting of a phospholipid interior coated on both sides with thin layers of proteins. It could maintain internal milieus separated from external environments and form a permeability barrier so that it could take up nutrients, ions, gases, water, and other substances, and wastes could be removed.

Fusion of Protocells

Some of the protocells containing proteins would come into contact and fuse together with other protocells enclosing RNA molecules (Figure 2-11). The result would be a larger cell containing both RNA and proteins, enclosed in a plasma membrane. In this sequestered environment, RNA began to replicate, while proteins functioned as enzymes to do the rudimentary work of the protocell, producing energy and signaling changes in the cell's environment. Because of close proximity within a

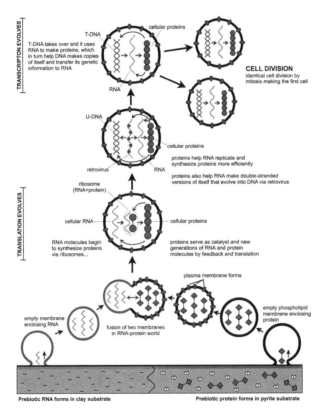

FIGURE 2-11. Outline of proposed pathway for the emergence of the first cell by serial endoprebiosis of RNA and protein molecules. The transitional stages are: (1) plasma membranes; (2) ribosomes and translation; (3) retrovirus and transcriptase enzyme; (4) U-DNA; (5) T-DNA and genetic code; and (6) the first cell capable of reproduction.

membrane, RNA and protein molecules began to interact, cooperate, and catalyze reactions. Such serial endoprebiosis led to the hierarchical emergence of cell components.

Initially, RNAs were capable of directing their own synthesis with exact copies of their own sequence, whereas proteins functioned as catalysts. With time, these cooperating molecules of RNAs and proteins created a hybrid and complex ribosome machine that developed the ability to direct the synthesis of polypeptides. I speculate that the first cooperative molecules to emerge in the protocell were ribosomes, the myriads of production centers, composites of RNA and proteins suspended in the protocell fluid where customized proteins began to form. The ribosome-produced polypeptides that bound back to the same molecular assemblage (i.e., containing the coding of RNAs) would self-reinforce their own formation and lead to production of longer RNAs. The longer the RNA, the longer the polypeptides that would be produced; the longer the proteins and RNA, the more these two molecules would hold together the metabolic web of interactions. Once such a feedback process was established, simpler polynucleotides and polypeptides produced on the mineral substrate became redundant. At this stage, the direction for prebiotic evolution was set to the emergence of DNA, which served as a more reliable repository of genetic information.

Origin of Ribosomes

The origin and evolution of the ribosome is central to our understanding of the RNA/protein world. A ribosome is a complex molecular subsystem, a hybrid of RNA and protein molecules. It can have as many as 80 proteins interacting with multiple RNA molecules. It is a protein-dominated subsystem with an RNA core.[88] Ribosomes consist of two major subunits. The small ribosomal subunit (SSU) reads the mRNA, while the large subunit (LSU) joins amino acids to form a polypeptide chain during protein synthesis. Each subunit is a hybrid, composed of one or more ribosomal RNA (rRNA) and a variety of proteins. The two subunits fit together and work as one to translate mRNA into a polypeptide chain during protein synthesis. The various ribosomes share a core structure, which is quite similar despite large differences in size. Each layer is accreted like the tree rings. The core of the ribosome is universally conserved. The ribosome is an example of endoprebiosis of cooperative molecules, a match of polynucleotide and polypeptide made within a membrane. Ribosomal history is driven by the gradual structural accretion of protein and RNA structure. Translation is a complex and highly coordinated process of protein synthesis that is mediated by a universal ribonucleoprotein (RNP) complex, the ribosome. In modern cells, ribosomes form continuously throughout the life of the cell through self-assembly of proteins and RNAs.

Ribosomes catalyze the peptide bonds that add successive amino acids to a growing polypeptide chain. Both proteins and RNA contributed to the formation of the ancient ribosome.[89] Once ribosomes were formed inside cell membranes, RNA molecules began to guide the synthesis of specific proteins according to their nucleotide sequences in the form of a rudimentary genetic code, while other RNA molecules acted as adaptors, each binding to a specific amino acid. Because RNA molecules could store and replicate genetic information and could produce additional proteins, autocatalytic relationships developed between RNA and protein molecules. RNA began to catalyze reactions leading to the production of more RNA, replacing the primordial RNA synthesized on the mineral substrate. Once the new generations of RNA molecules began to customize specific proteins by translation in accordance with their genetic code, the role of primordial RNA and protein molecules from the volcanic vents became redundant. They would now exist separately as parasites in the form of RNA viruses and prions outside the protocells.

Origin of Retroviruses

The capacity and durability of RNA molecules were severely limited, driving the emergence of a more stable information macromolecule. The protein that ushered in the new, DNA storage format is a special enzyme called reverse transcriptase, capable of transforming the information contained in RNA into a DNA molecule. A retrovirus is an RNA virus with a protein coat, called a capsid, which deploys reverse transcriptase to create a DNA copy of its own genome. During the early stages of cellular evolution, as proteins and RNA molecules were brought together within prebiotic membranes, some proteins accidentally enveloped some RNA molecules to form retroviruses. Retroviruses possess the ability to store information and fulfill one of the major cellular functions (Figure 2-11). They first appeared as reverse catalysts and during the emergence of the first cells would play critical roles in synthesizing DNA from RNA.[90]

Origin of DNA

DNA probably appeared later than RNA in response to the relative instability of RNA molecules, which easily degrade and mutate. DNA is selected over RNA for its expansion of storage capacity and drastic improvement of error-free replication.[91] Once the retrovirus appeared in the protocells, it provided the reverse transcriptase enzyme to convert RNA to DNA. Effective RNA-dependent DNA polymerases would likely emerge only after production of larger proteins became possible. There are just two chemical differences between RNA and DNA molecules: the first is the removal of a single oxygen atom from RNA (ribonucleic acid) to give deoxyribonucleic acid), or DNA. The second difference is the addition of the methyl (CH_3) group on the nucleotide base uracil (U) to give thymine (T). DNA evolved from RNA in two steps.[92] The first was most likely the formation of U-DNA (DNA containing uracil) after the reduction of ribose sugar. Chemicals (reactive free radicals) found in vents could reduce the ribose sugar to deoxyribose. Some modern viruses indeed have a U-DNA genome, possibly reflecting a transition step between the RNA and DNA worlds. The second step would be the conversion of uracil into thymine by methylation to form T-DNA (DNA containing thymine). Again, methyl groups are free radicals of methane gas, available in hydrothermal vents, which could enter through the pore spaces of the membrane. Once DNA molecules came onto the scene, they became the primary genetic molecules, subduing RNA, which became an intermediary between DNA and proteins. RNA is demoted to the role of messenger, carrying information from DNA to the ribosomes (Figure 2-11). DNA began to code for RNA and RNA began to code for protein.

THE BIOLOGICAL STAGE: ORIGIN OF THE REPLICATING CELL

Once assembled, the membrane-enclosed protocells with interacting DNA, RNA, and proteins had to perform two critical functions: metabolic self-maintenance and reproduction. DNA carries the genetic information of a cell; each gene provides a template for building a protein molecule via RNA, whereas proteins perform important metabolic tasks. At some point, the protocells jumped the final hurdle and made a dramatic evolutionary breakthrough: reproduction. In order to reproduce successfully, a cell must be able to grow, duplicate its genetic contents (DNA), and divide into equivalent daughter cells. To be replicated, even the shortest DNA strand needs proteins. The divisions of the first cells would probably anticipate the binary fission of modern bacteria (Figure 2-12). The initial cell divisions may have been accidental bursts or pinchings through pore spaces, as cell size became increasingly large by fusions. Upon reaching a certain size, the cell broke down to form smaller, more stable units. Experiments have shown that membranes can grow and divide like cells.[93] It is likely that early cells divided by binary fission in which a cell pinched into two and

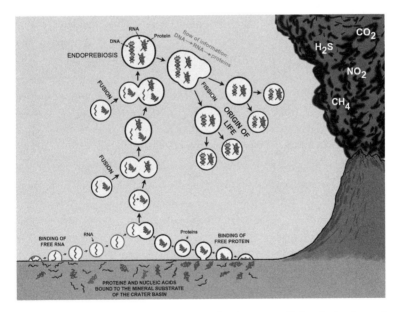

FIGURE 2-12. Proposed endoprebiotic model for the origin of life in hydrothermal vent environments. The cell membranes containing simple RNA and protein molecules established endoprebiotic relationships, where information flows from DNA to RNA to proteins. These primitive cells begin to reproduce to form the first living cells.

divided into two identical daughter cells with identical genes. For division to be successful, each daughter protocell had to include proteins for metabolism and full sets of genes for reproduction. When the genetic material became centralized into a single chain of DNA, and the genetic code was established, reproduction became routine. Over time, the fission processes were perfected: DNA began to regulate the reproduction of not only a single molecule within the cell but also the complex organization of the cell itself. Cells that could provide themselves with energy and components necessary continuously to fuel the growth and function of their autocatalytic webs would become self-perpetuating. Cell division, the ceaseless reproduction of units of living matter, is at the heart of the evolutionary process and continuity of life.

There was a great evolutionary improvement from the first cell to the last common ancestor (LCA). The specific precursor of all subsequent forms of life, the LCA was a sophisticated microorganism characterized by DNA, RNA, ribosomes, multiple enzymes to direct transcription and translation as well as metabolism, membranes with embedded proteins to control ionic and molecular transport, and ATP. The LCA probably had a small genome like those of thermophiles, containing about 1,000 to 1,500 genes, with a minimal organization so that, by taking up nutrients across the plasma membrane from the vent environments, it could reproduce quickly.[94] It had DNA and various types of RNA, including messenger, ribosomal, and transfer RNA to synthesize simple proteins. Eventually, alongside the symbiotic dynamics studied by Margulis, mutations and copying errors also entered into the system, providing more variations for natural selection. In addition, some surviving molecules of the primordial proteins and RNAs became parasites upon the early cells. Viruses and prions evolved concurrently and have maintained these symbiotic relationships since then.

VIRUSES AND PRIONS

If we accept the dual origin of RNA and proteins in the hydrothermal vents of the crater basins, the next question is whether any modern analogs of these two molecules existing outside the living cells may be found. I suggest that the primordial RNA and proteins generated in mineral substrates were precursors to modern RNA viruses and prions. These may be seen as relics of the RNA/protein world. Recent molecular studies suggest that viruses and prions are ancient subcellular particles, which predated the appearance of the first cells.[95] They offer a glimpse of the ancient RNA/protein world. Dyson (2004) considered the possibility that viruses and prions could be early-life particles in support for his dual origin of life, but rejected the idea because he thought these entities are highly derived and appeared much later, after life has flourished. When they invade cells, viruses and prions perform independently of their hosts' replication and metabolism. Are these functional similarities purely a coincidence, or are they evolutionary heirlooms that have survived four billion years? A virus can replicate or a prion can metabolize, but only when invading a living cell. Viruses and prions are the oldest remnants and relics of prebiotic particles that helped to jumpstart the first life forms through trials and errors in innumerable steps. They appeared before the emergence of the first cells, and they are still here.

Unlike actual organisms, viruses and prions are not made up of cells. Virus and prion are subcellular particles, analogous to the electrons and protons in an atom. They are acellular entities that replicate inside the cells of host organisms. A virus or a prion on its own is inert and inanimate. They lie on the edge of life in a twilight zone between life and nonlife, lacking some of the substances needed to live on their own. They have the special symbiotic relationship with living cells called parasitism. Forterre (2005), Villarreal (2005), and Koonin and colleagues (2006) forcefully discussed the role of ancient viruses in the emergence of early life, while Lupi and colleagues (2006) suggested that the prions are relics of an early stage of peptide evolution. In this essay, I have combined both of these views to suggest that ancient viruses and prions are relics of the RNA/protein world. They were precellular components generated within the prebiotic vent environments where the first cells evolved by endoprebiosis. Once life appeared, viruses and prions became redundant, while surviving as parasites of the living cells.

Viral particles are by far the most diverse and abundant biological entities on our planet. Viruses are very ancient microbes, probably originating before LUA, when cells still had RNA rather than DNA genomes.[96] Similarly, a prion is composed of a protein that occurs in a harmless way in living cells. Prions are only functional when associated with a living cell, where they convert cellular protein into new prions. Proteins have a special three-dimensional shape. By folding them into aberrant shapes, the normal prion turns into an infectious prion protein responsible for a number of degenerative brain diseases, such as bovine spongiform encephalopathy ("mad cow disease").[97] When prions invade cells, they act as templates to refold other proteins into their own shapes, making them infectious as well.

The precursors of viruslike and prionlike entities surely predated the appearance of early cells and give some tantalizing clues to the ancient microbial world. It would be exciting to see whether the simplest form of cellular life mimicking LUA could be produced in the laboratory by incorporating viruses and prions into a recipient cell. Using the tools of synthetic biology, it might be possible to create an evolving, replicating primitive cell by enclosing RNA-viruses and prions in cell membranes to simulate critical steps in the primal synthesis of cells as outlined here, including the appearance of ribosomes, translation, DNA, genetic code, and replication. These three components—cell membrane, RNA virus, and prion—may serve as operational functionalities of a minimal cell. And that could provide insight into how life on Earth began. Experiments in creating synthetic cells may settle once and for all the longstanding debates over the origin of life.

CONCLUSION

The role of historical contingency in the origin of life on Earth is an outcome of long chains of unpredictable antecedent stages, one after the other. First, liquid water, one of the prerequisites for the prebiotic synthesis of life, was available on early Earth, delivered by the comets. Water is stable within a certain distance from the Sun, known as the habitable zone. Too close to the Sun, liquid water evaporates to water vapor; too far, water freezes to ice. But Earth, lying between Venus and Mars, lies right where liquid water and life are possible. Earth is a Goldilocks planet—not too hot, and not too cold—an ideal place for gestating and harboring life. Second, the size of our planet is also important, bearing enough gravity to hold down the atmosphere and maintain the hydrosphere. The Moon and Mercury are too small to develop atmospheres. Third, without meteoritic impacts upon the early Earth soon after the crustal formation, life could not have been emerged. The exogenous biomolecules from the comets were the raw materials of life, while endogenous prebiotic synthesis was possible in the unusual protective crater vent environments induced by impacts. Early Earth became a crucible for the synthesis of life, and the underwater hydrothermal vents were the oldest ecosystems.

We have divided the study of the prebiotic evolution of life into four hierarchical stages—cosmic, geological, chemical, and biological. The cosmic stage represents the cometary delivery of biomolecules on Earth. The geological stage witnesses the concentration of biomolecules at the pores and pockets of mineral substrates in hydrothermal crater vents, powered by convective currents, and the appearance of the primitive lipid membranes. The chemical stage represents the selection and concentration of chiral biomolecules such as left-handed amino acids and right-handed sugars and their subsequent polymerization on the mineral substrates as polynucleotides and polypeptides. We suggest that RNA and proteins emerged simultaneously at the chemical stage and were encapsulated to start the RNA/protein world. RNA and protein molecules performed two critical functions within protocells: replication and metabolism, respectively. They began to interact and initiate serial endoprebiosis, leading to hierarchical emergence of cell components including plasma membranes, ribosomes, retroviruses, and finally, DNA. These protocells developed genetic codes that stored, processed, and passed organizational information from DNA to RNA to proteins. The biological stage represents the origin of reproducing protocells. Systemic reproduction is the key breakthrough in the emergence of life. After many trials and errors, natural selection was able to select the cells with the most efficient systems of DNA replication and cellular reproduction machineries. The first cells were probably heat-loving bacteria similar to modern thermophiles that began to replicate so that each daughter cell possessed a true copy of the genetic contents of the parent. Modern viruses and prions probably represent the evolutionary relics of the primordial RNA and protein molecules. They are the oldest cellular components, now unneeded for living processes but surviving independently as parasites since the beginning of life. Thus, we arrive at a fully integrated serial account of the primal foundations of the biosphere.

3.

EXOBIOLOGY AT NASA

Incubator for the Gaia and Serial Endosymbiosis Theories

JAMES STRICH

James Lovelock's Gaia hypothesis underwent its initial gestation in the 1960s, and continued through its adolescence in the 1970s, as a spin-off of NASA's Exobiology Program. This essay explores how the search for life on other planets not only helped to incubate the Gaia hypothesis but also served as the intellectual soil and the sole source of financial support for more than thirty years for the growth of Lynn Margulis's work on serial endosymbiosis theory (SET), which she developed (after 1970) in tandem with her collaboration with Lovelock on Gaia. From the 1960s through the 1980s, the culture of the NASA Exobiology Program was a uniquely favorable environment for both Lovelock and Margulis. It funded much important work that was far too interdisciplinary for more traditional agencies such as the National Science Foundation or National Institutes of Health. Partaking of the broader cultural period in which it appeared, the very interdisciplinarity of their science also brought with it many of the additional, philosophical implications the theories were perceived to have. However, notwithstanding the tenuousness of their early funding and the uncommon range of cultural valences both theories had, much of this work has become central to contemporary biology and the Earth and planetary sciences.

GAIA

After the creation of its Office of Life Sciences in 1960, NASA was moving quickly to recruit the best talent in instrumentation and basic science from all over the world. One man who combined both was research chemist and biologist James Lovelock, who in 1957 had also developed a highly sensitive new device, the electron capture detector (ECD) for gas chromatography. This device allowed for the first time detection of trace organic molecules in the atmosphere down to the parts per trillion range.[1] On May 9, 1961, NASA official Abe Silverstein wrote to Lovelock to invite him to come to the United States and work on development of the gas chromatograph (GC) for the lunar Surveyor spacecraft at NASA's Jet Propulsion Laboratory (JPL) in Pasadena, California.[2] Lovelock eagerly agreed. His first NASA grant, for $30,100, was awarded before year's end and was channeled through the University of Houston, where a tenured professorship for Lovelock at Baylor College of Medicine was arranged, "with a dream salary of $20,000 per annum," as Lovelock expressed it.[3] He was to live in Houston with his family for two and a half years and commute regularly to JPL for much of the next eleven years; he continued to visit JPL periodically as a consultant until just before the launch of the two Viking Mars spacecraft in 1975.[4] Because of ideas he first developed on physical life-detection experiments, in March 1965 Lovelock was also put to work on an early Mars probe design, called Voyager, among other things to develop the GC as a life detection instrument.[5] His description of the discussions between scientists and engineers is highly evocative of the heady sense of mission at JPL during the 1960s, as designing and launching probes to the moon and then the planets became a reality. He recalled some fifteen years later: "As one whose childhood was illuminated by the writings of Jules Verne and Olaf Stapledon I was delighted to have the chance of discussing at first hand the plans for investigating Mars."[6]

As Lovelock describes it, the early meetings at JPL on life-detection strategies for Mars probes had quickly settled into a rut. The strategies all sought to detect Earth-like microorganisms by immersing them in liquid culture broths and then looking for their metabolic by-products.[7] This was true of Wolf Vishniac's Wolf Trap, of Gilbert Levin's Gulliver, and of Vance Oyama's early ideas, all of which eventually became experiments that went to Mars on the Viking mission. Lovelock thought it was far too limiting to make such narrow, "Earthcentric" assumptions about potential Mars organisms. Challenged to come up with a more robust strategy to look for evidence of life, he argued that one ought to look for entropy-reduction phenomena.[8] He suggested the most obvious activity of living things that offsets entropy was that they keep the gas composition of a planetary atmosphere far from chemical equilibrium. For example, if for any length of time a planet's atmosphere contained significant amounts of both methane and oxygen simultaneously, Lovelock argued, this is so far from the equilibrium condition that it is strong presumptive evidence of living processes acting to maintain the disequilibrium. Living things must be constantly replenishing two such reactive gases or their levels would not remain high for long.

By September 1965, geneticist Norman Horowitz had become the new head of the Biology Division at JPL, overseeing much of the planning of life-detection experiments.[9] Congress was not looking favorably at the Voyager mission, and the project was postponed so much by a vote of December 22, 1965, as to effectively kill it.[10] Nonetheless, Lovelock had published a first paper on his atmospheric thinking and was on the verge of a powerful new insight.[11] He realized that the gases living organisms most actively affect, especially carbon dioxide, methane, oxygen and water vapor, are just those gases that most dramatically shape the climate of the planet. He claims to have had a flash of insight one September day at JPL, in which he first wondered if living organisms might actively control the climate of a planet, via feedback mechanisms, to keep the conditions there favorable for their own survival and growth. Immediately blurting out his insight in discussions with Horowitz, Carl Sagan, and Dian Hitchcock, he found them skeptical but sufficiently intrigued to encourage him in his thinking.[12] Indeed, Hitchcock, a philosopher by training, had been collaborating on Lovelock's ideas about physical life-detection for some months already; the two would eventually publish together in Sagan's journal *Icarus*.[13]

Horowitz, according to Lovelock, "was open-minded, and although he disagreed with my views about the Earth and its atmosphere, he thought, as the good scientist he was, that they should be heard." Horowitz arranged for Lovelock to give a paper on his ideas to the American Astronautical Society.[14] He invited Lovelock to the second NASA conference on the origins of life, to be held at Princeton in May 1968, where Lovelock first met Lynn Margulis.[15] Lovelock found the reception of his ideas cool at the NASA meeting, with the exception of the Swedish specialist in chemistry of the oceans, Lars Gunnar Sillén.[16] He recalled that most of the older scientists at the meeting, especially Preston Cloud, were unsympathetic to his concepts.[17] Nonetheless, he worked steadily at the ideas, especially after 1970, when Lynn Margulis began to collaborate with him on the Gaia hypothesis. All the while, he continued as a consultant at JPL, largely on the design of other scientists' instruments. His and Horowitz's concerns notwithstanding, work on the latest versions of Wolf Trap, Gulliver, and Oyama's experiment (now called the "gas exchange" experiment, or GEx) all went ahead on continued NASA funding. So did the development, by Klaus Biemann, Juan Oró, Leslie Orgel, and their team, of a gas chromatograph and mass spectrometer to be sent to Mars to analyze organic compounds present in the regolith. Lovelock came up with the crucial means for hermetically linking the gas chromatograph to a mass spectrometer, when those instruments eventually were sent to Mars on the Viking spacecraft, the next iteration of design after Congress finally definitively cancelled Voyager in the wake of the summer 1967 race riots in many Eastern cities.

Lovelock later called the new field spawned by the Gaia hypothesis "geophysiology." He described its origins thus:

It arose during attempts to design experiments to detect life on other planets, particularly Mars. For the most part these experiments were geocentric and based on the notion of landing an automated biological or biochemical laboratory on the planet. . . . Hitchcock and Lovelock took the opposing view that not only were such experiments likely to fail because of their egocentricity, but also that there was a more certain way of detecting planetary life, whatever its form might be. This alternative approach to life detection came from a systems view of planetary life. In particular, it suggests that if life can be taken to constitute a global entity, its presence would be revealed by a change in the chemical composition of the planet's atmosphere. This change of composition could be compared with the abiological steady state of a lifeless planet. The reasoning behind this idea was that the planetary biota would be obliged to use any mobile medium available to them as a source of essential nutrients and as a sink for the disposal of the waste products of their metabolism. Such activity would render a planet with life as recognizably different from a lifeless one. At that time there was a fairly detailed compositional analysis by infrared astronomy of the Mars and Venus atmospheres, and it revealed both planets to have atmospheres not far from chemical equilibrium. Therefore, they were probably lifeless.[18]

Because of the state of chemical equilibrium in the atmospheres of both Venus and Mars, Lovelock predicted from the first Gaia insight in 1965 that both planets were lifeless. Consequently, he was skeptical about the large expenditures on the Viking biology instruments, above and beyond his earlier skepticism about the conceptual basis of the instruments, now thinking the money could be much better spent on other measurements on Mars.

But now an additional, much deeper insight dawned upon Lovelock. Given the so-called "faint young sun" paradox, the fact that the biota was so actively shaping the chemical environment of the biosphere (including the atmosphere) took on new explanatory power. The sun had been as much as 30 percent cooler at the time when life on Earth first originated. Yet during the entire 3.5 billion years or so since life had appeared, it seemed clear that the Earth's surface temperature could not have varied by nearly as much as 30 percent from present values: Living things could not have survived and proliferated if the Earth had been that much cooler than at present. Either the Earth had been warmer than it should have been at the origin of life, relative to now, or, more likely, living things were regulating the temperature, so that modern temperatures were cooler, relative to how much the sun had warmed, than they would be on a lifeless planet. Since the main means of regulating the Earth's surface temperature known at the time was the so-called greenhouse effect, dependent upon gases given off and consumed by living organisms (CO_2, methane, water vapor, among others), it did not seem impossible that the biota could regulate planetary temperature, decreasing the greenhouse effect slowly over eons, to compensate for the increasing heat of the sun.[19]

Perhaps, Lovelock began to think, the biota acted as a cybernetic system that regulated temperature, pH, oxygen level, and other parameters in just such a way as to maintain conditions on Earth suitable for the survival of life. As mentioned, Lovelock's idea was at first received quite coolly by the scientific community, even at a 1968 NASA-sponsored origin of life (OOL) meeting where interdisciplinary thinking was the norm.[20] (It was so much the norm that the psychiatrist Frank Fremont-Smith and the ethologist Sol Kramer, having been at the 1967 meeting as well, were re-invited.) There was much talk of the origin of life being an epistemological problem as much as a scientific one, invoking Marshall McLuhan's recent slogan that "the medium *is* the message." Kramer described first getting interested in the origin of life problem while studying in a course on cancer, taught by famed psychoanalyst turned natural scientist, Wilhelm Reich.

Though he was not a fan of the Gaia hypothesis, Norman Horowitz agreed with a number of Lovelock's views. Lovelock shared Horowitz's feeling that sterilizing Martian landers was unnecessary, saying: "The concept of contaminating a virginal Mars with Earth-life seemed the stuff of fanatics, not scientists, and the act of sterilization hazarded the delicate and intricate instruments we wanted to send to Mars."[21] (This view was opposed by Carl Sagan, Elliott Leventhal, and Joshua Lederberg.) In a more piquant passage, Lovelock described his view of life detection experiments:

> The engineering and physical sciences of the NASA institutions was often so competent as to achieve an exquisite beauty of its own. By contrast with some very notable exceptions, the quality of the life sciences was primitive and steeped in ignorance. It was almost as if a group of the finest engineers were asked to design an automatic roving vehicle which could cross the Sahara Desert. When they had done this, they were then required to design an automatic fishing rod and line to mount on the vehicle to catch the fish that swam among the sand dunes. These patient engineers were also expected to design their vehicle so as to withstand the temperatures needed to sterilize it for otherwise the dunes might be infected with fish-destroying microorganisms.[22]

But Horowitz and Lovelock also felt that the Wolf Trap, Gulliver, and other designs all shared the basic flaw of assuming that Martian microbes, if they did exist, would do well in a wet environment, since all those designs involved saturating Martian "soil," or regolith—the official geological term—with a liquid broth of nutrients. In Horowitz's way of thinking this produced conditions wildly unlike those of Mars; he thought so still more after July 1965 when the Mariner 4 space probe showed Mars to be a cratered, dry planet.

Mariner's measurements the spacecraft made of the Martian atmosphere found it to be much thinner than previously supposed. The pressure of the air was too low for liquid water to exist on the planet's surface. "CO_2 was its major component, with only a trace of water vapor," recalled Horowitz. "That discovery gave me and my collaborators, George Hobby and Jerry Hubbard, the impetus to design an instrument that would search for life on a dry planet. That instrument was the pyrolytic release experiment. . . . I never applied [to NASA] for funding to develop the experiment, since the funds were provided by JPL."[23] Because of the Mariner 4 results, Horowitz was among those who proposed that Antarctica, specifically the very coldest, driest desert valleys there, was a better analog for Mars than most other sites on Earth; yet even those, he said, were overwhelmingly hospitable places for life compared to the Martian environment.[24] Horowitz and his collaborators Roy Cameron and Jerry Hubbard began to study the microbiology of the driest, most inhospitable parts of Antarctica, to understand whether life could survive there at all.[25] They later claimed to have found some of the only naturally sterile soils on Earth (14 percent of their samples) from these valleys, claiming this made life on Mars still less probable than previously thought and proving that sterilizing spacecraft to be sent to Mars was pointless since conditions there were so much harsher than those sufficient to render some Antarctic soils totally sterile.[26] Cameron and Richard Davies also launched a similar expedition in 1966 to the Atacama Desert of northern Chile.[27] In this regard, Horowitz and Lovelock were largely in agreement: Neither expected Mars to support life.

While he recognized that the interdisciplinary environment of NASA's Exobiology Program had been favorable for a hypothesis as potentially paradigm-shifting as Gaia, Lovelock was also well aware that many in the scientific community looked askance at NASA-funded science and thought it was of lower quality than work funded by NSF, NIH, etc.[28] This created more than a little ambivalence toward his new patron; Lovelock may even have shared this opinion of some Exobiology-funded research. As he expressed it to Margulis in a letter early in 1973, lamenting the difficulty they were having finding a journal that would accept their latest paper on Gaia, "it does seem a pity to have both of our babies fostered by Exobiology, which in my classification is only just one [notch] above psychical research!"[29]

Microbiologist and cell biologist Lynn Margulis collaborated with Lovelock on two major theoretical papers developing the Gaia hypothesis in 1974. Lovelock also presented a paper on the topic to a meeting of the Royal Society of London devoted to extraterrestrial life.[30] The theory received a fairly cool treatment from the scientific community overall, however. More thoughts below, on why this was so. Some years later, in June 1979, a major exobiology meeting convened at NASA Ames Research Center. By the late 1970s new data was pouring in: from Viking, from Carl Woese's recognition of the archaea as distinct from the eubacteria (see Jan Sapp's essay in this volume), from the discovery of thriving biotic communities at deep-sea hydrothermal vents, and so on. Thus John Billingham of Ames saw a need to reconsider the big questions, in origin-of-life research, in what was known of conditions relevant to life on other planets, and in SETI; as a result he arranged the "Conference on Life in the Universe."[31] Since the Viking results had so strikingly borne out Lovelock's prediction that Mars would be lifeless based on its atmospheric chemistry, Lovelock and Margulis and their Gaia hypothesis got a prominent place on the agenda of Billingham's "Life in the Universe" conference.

This was a crucial turning point for the theory. It given a high-profile podium just at the time Lovelock's first book on Gaia was about to appear. And perhaps just as important, Stephen Schneider, a leading atmospheric researcher from the National Center for Atmospheric Research (NCAR) in Boulder, Colorado, was at the meeting as well. Much impressed by the potential power of the Gaia hypothesis, Schneider critically addressed the idea and its promise in a 1984 mass-market book, *The Coevolution of Climate and Life*, and in a television documentary produced in 1985 by the BBC's Horizon and the American NOVA series.[32] In addition, along with Penelope Boston, Schneider organized the first major conference to evaluate the scientific merit of Gaia hypothesis, under the auspices of the American Geophysical Union, in March 1988, after which *Science* described the outcome under the headline "No Longer Willful, Gaia Becomes Respectable."[33] Earlier, in a series of meetings at Ames in 1981–82, convened by Billingham and David Raup, on the evolution of complex and higher organisms, the participants reached the following major conclusion:

Third, and of special interest, is the controversial Gaia hypothesis, which proposes that living things have prevented drastic climatic changes on the Earth throughout most of its history. This view, regarded as highly speculative and tentative by many workers, has yet to be rigorously examined. If it proves to be correct, and if climatic stabilization can be shown to be a likely consequence of the activities of life on other worlds as well, then we may expect that extraterrestrial life is abundant throughout the universe. An effort should be made therefore, to *determine whether the Gaia hypothesis is valid*.[34]

The hypothesis was scrutinized further at a 1981–82 NASA workshop on the Evolution of Complex and Higher Organisms (ECHO); participants who found the hypothesis intriguing said:

Although many of us are skeptical, we agree that the Gaia mechanism approaches one extreme of a spectrum of possibilities (ranging from total control of a planet's environment by its organisms to total lack of control) and that much further study is needed to determine the causes of large-scale environmental stability and change. . . . The Gaia hypothesis in particular could be investigated by seeking to identify evolutionary mechanisms (if any such exist) that are capable of selecting organisms whose activities promote global environmental stability.[35]

Given the potential fruitfulness of the Gaia hypothesis, recognized no later than this time by many in the exobiology community, it is a fascinating phenomenon worthy of study, just how much resistance Gaia generated in the geology, atmospheric science, climatology, and evolutionary biology communities. Charles Darwin had some good rhetorical reasons for clinging so tenaciously to his term "natural selection," despite intense criticism that, to many, it implied an anthropomorphic, voluntaristic "selector" in nature.[36] And in a story with some interesting parallels, James Lovelock's term "Gaia" was attacked from the beginning. The same charges were brought: It is anthropomorphic (no matter how many times he said "I meant it as a metaphor"); you are assigning *agency* to a natural process and therefore secretly slipping a supernatural Creator back in through the back door; and so on. Ironically, this time it was the hardline natural selectionists (W. Ford Doolittle, Richard Dawkins, John Maynard Smith, William Hamilton) who attacked the metaphor for having voluntarist overtones, having themselves worked hard to press the "selfish gene" metaphor to supplement the "natural selection" of their revered forefather Darwin.

From the first, the key technical criticism was how behavior by a microorganism that benefited the biosphere as a whole, but not itself (and might even sometimes be detrimental to its own survival, such as the first release of oxygen by anaerobes) could ever evolve and persist by natural selection. And Lovelock acknowledges that the early versions of the theory, up through his 1979 book *Gaia: A New Look at Life on Earth*, suffered from an inadequate consideration of this question.[37] He developed the "Daisyworld" mathematical model, in collaboration with Andrew Watson of Reading University, to answer these objections.[38] A key intellectual barrier was the idea in geology, evolutionary biology, and environmental science that the environment changes and affects organisms, but that organisms themselves were mostly passive recipients of such selective forces. For most, it required a deep reconceptualization to see living organisms as potent forces, shaping conditions on Earth just as powerfully (or perhaps more so) as they were being shaped by those external conditions.

But in addition, the name "Gaia" drew a great deal of fire for suggesting, via the image of the ancient Greek Earth goddess, everything from vague New Age mysticism to teleology reimported into biology after a 150-year struggle by evolutionary biology to banish it. The first sharp critique in this vein was W. Ford Doolittle's "Is Nature Really Motherly?" published in the *CoEvolution Quarterly* in Spring 1981. Then came Richard Dawkins's *The Blind Watchmaker* (1984). Quite a bit of the criticism is summed up in a piece in *Science* titled "Lynn Margulis: Science's Unruly Earth Mother."[39] The title tells the story: The tone used by critics such as John Maynard Smith and W. Ford Doolittle is rather more harsh and dismissive than is typical for a scholarly scientific exchange. In the ensuing "take no prisoners" firefight, Lovelock has made some modifications in his theory to reflect the valid points his critics have driven home.[40] For example, he acknowledged that his original formulation that living organisms coordinated through Gaia to keep conditions on Earth ideal for their own survival was untenable, and indeed teleological. He later said that Gaia has acted to maintain conditions merely within the bounds of what can be tolerated by life, but need not necessarily continue doing so if human perturbations become too extreme.

Be all that as it may, after the disappointment of Viking, exobiology (and more recently, astrobiology) has fully incorporated Lovelock's insight (though usually without attribution) that life detection strategies need, insofar as possible, to be "non-Earthcentric."[41] After the modifications of the theory as presented in Lovelock's second book in 1988, more workers in the exobiology community found Lovelock's theory acceptable. Harold Morowitz wrote, for instance, that origin-of-life researchers now needed to understand that "in [Lovelock's] sense, life is a property of planets rather than of individual organisms." This view was complementary rather than opposed to the traditional biology view that sought to define life by comparing what all living organisms have in common.[42] Indeed, under the name "Earth system science," the core of the modified Gaia theory is now mainstream science, but, say the critics, "never under the name Gaia."

However, Lovelock tenaciously defends the name "Gaia" and insists that "names are important."[43] Describing one striking episode, he says:

> I stuck with the name Gaia because my Green friends and quite a few scientists regarded a change of name as a betrayal and so do I. I did try the neologism "geophysiology" for scientists and it worked for a while until the snarling dogs realized it was just another name for Gaia. I overheard a distinguished geophysicist at NCAR say to a young scientist, "I will not have you use the word geophysiology—it's just closet Gaia."
>
> Mary Midgley's book *Science and Poetry* . . . deals in full with the name Gaia and why it was rejected by so many scientists. . . . A great deal of the fuss over Gaia is because I work as an independent and only rarely go to meetings of scientists. It is hard to appreciate the work of someone you do not know.[44]

Thus, as with the experience of Carl Woese, even those whose ideas got off the ground in the intense interdisciplinary environment of NASA Exobiology in the 1960s could run into trouble because of plain old disciplinary turf defense, if the main body of the discipline, such as geology or climatology, was still outside of the exobiology context. Lovelock has written at some length on this problem, making it difficult if not impossible for a scientist to operate outside academia as an "independent."[45] He himself barely managed it, even with a long track record of training and research in prestigious British government science establishments prior to transitioning to "independent" status as inventor, and a consultant to NASA and to industry groups.

Lovelock believes that from 1997–2002 or so the climate of reception improved to some extent, but still not enough that many of the neo-Darwinians with whom the vitriolic public conflict occurred would ever openly credit the term "Gaia," even if they accept most of what is now called Earth system science. Lovelock wrote to me in 2002,

> The grandees over here [i.e., in Britain] are ready to admit, even at small meetings, that they were wrong to ridicule Gaia, but apart from Bill Hamilton no one will go public. John Maynard Smith used his powerful influence to have Tim Lenton's article "Gaia and Natural Selection" published by *Nature* as the lead article. Richard Dawkins, at a closed meeting in Oxford of about 25 scientists, said after I had spoken on Gaia and evolution, "Jim has his disciples and I have mine, they both get it wrong." John Lawton, now head of the UK research Council, NERC, had an editorial in *Science* on Earth System science, which generously acknowledged the Gaian contribution. It could be much worse.
>
> I did not discover the name of the NCAR geophysicist who didn't like geophysiology. But I do know that Wally Broecker was for years a strong opponent, even though personally a friend. Robert Dickinson, then at NCAR, joined a meeting in Brazil in the mid 1980s, the meeting included such notables as Paul Crutzen, Robert Harriss, Ann Henderson-Sellers, Ghillean Prance. The book of the meeting had the title *The Geophysiology of Amazonia*. It was published in 1987 by the United Nations University, in Tokyo. . . . An odd thing happened after the meeting, someone at NCAR sent a copy of my paper at the Brazil meeting, to the *Bulletin of the American Meteorological Society*. The first I heard of it was a letter from the journal editor to say that my paper had been accepted and would be published in the next edition. The editor would not tell me who had submitted the paper. I have heard talk of the NCAR underground and this must have been an example of it in operation.[46]

John Lawton's acknowledgment of Lovelock and Gaia is certainly more than many scientists who face such opposition ever see in their own lifetimes. As Lawton put it in 2001:

Physicists have long understood the "Goldilocks effect"—why, in general terms, Earth's natural blanket of atmospheric CO_2 and distance from the sun make the planet "just right" for life, neither too hot (like Venus) nor too cold (like Mars). James Lovelock's penetrating insights that a planet with abundant life will have an atmosphere shifted into extreme thermodynamic disequilibrium, and that Earth is habitable because of complex linkages and feedbacks between the atmosphere, oceans, land and biosphere, were major stepping stones in the emergence of this new science [Earth system science].[47]

Lovelock sees an interesting parallel between the opposition to the new "catastrophism" that broke through during this period and the opposition to Gaia theory. (Kuhn's *Structure of Scientific Revolutions* seems to be widely read among exobiology scientists, especially those who perceive themselves as outsiders.)[48] Both, he claims, were so basically opposed to a powerful Kuhnian paradigm that intense opposition was inevitable:

> The new interest in catastrophism is interesting from a geophysiological view-point. It is easy to model the effect of a sudden loss of species and show how perturbation affects biodiversity and the material environment. The community does not yet seem ready for this approach although they are now moving towards it.
>
> There is far more interest in Gaia in Europe and Japan than in the USA. The old guard seem so much more powerful in America. [This story] resonates strongly with a fine book by Professor Michael Benton of Bristol University, due to be published later this year [2002]. In it he traces the history of Earth Science from the early 19th century. Benton observes the power struggle between the catastrophists and the uniformitarians, with such strong figures as Lyell and Darwin on the side of gradual evolution and shows convincingly how gradualism became the all pervasive dogma of both Earth and Life science. So powerful was this dogma that it persisted, in spite of abundant contrary evidence, until [Luis W.] Alvarez and his colleagues produced almost unequivocal evidence for an impact catastrophe as the cause of the Cretaceous-Tertiary extinction.
>
> During the 150 years from 1830 to 1980, any mention of sudden evolutionary change was treated as if it were heresy and most geologists found it prudent never to speak of catastrophes. It took the hard evi-

dence and the superior rank of the Nobel Laureate Alvarez, to break the ice. Even so he was amazed by the fury and bad manners of those Earth scientists who still continued to attack his research.

So I am indeed naive if I think that the even more heretical theory of Gaia will be recognized by the great Church of Science. Young scientists, who imagine that they have nothing to lose, occasionally break ranks as the article showed but even then only obliquely.[49]

The Alvarez discovery to which Lovelock refers, that an asteroid impact may have been responsible for the extinction of the dinosaurs, came about at least partly, by now perhaps not surprisingly, with help from NASA funding. As I will discuss below, the "saltationist" or antigradualist aspect of Margulis's serial endosymbiosis theory (SET) contributed to a similar reaction against it by neo-Darwinian biologists such as John Maynard Smith and W. Ford Doolittle. And both theories touched upon other cultural hot buttons.

SERIAL ENDOSYMBIOSIS THEORY

Lynn Margulis's serial endosymbiosis theory also developed, beginning well before she met Lovelock, in the exobiology incubator. During her marriage to Carl Sagan, with microbiologist Joshua Lederberg as close family friend and mentor to both young scientists, Margulis was intimately familiar with exobiology research and the very creation of the field, from before she even began her graduate work in protist genetics under Hans Ris at the University of Wisconsin. She met two other crucial mentors, Elso Barghoorn at Harvard University and Philip Morrison at MIT; Barghoorn first invited her to a NASA origin of life conference at Princeton in 1967, just as her first paper on endosymbiosis was going to press in the *Journal for Theoretical Biology*.[50] Thus, her thinking about microbial genetics occurred in an intellectual environment thoroughly steeped in exobiology and origin-of-life thinking.

The chief argument of her 1967 paper was that microscopic cytology, DNA, and other biochemical evidence supported an idea from the 1920s but largely dismissed by cell biologists (see Jan Sapp's chapter for further discussion). This was that mitochondria so closely resembled certain free-living aerobic bacteria and chloroplasts so closely resembled certain free-living cyanobacteria, that the most likely explanation was that these (and perhaps other organelles such as centrioles or kineto-somes) originated as free-living bacteria engulfed serially over time, but not digested, by a mycoplasma-like ancestral cell. A double-layered membrane around each organelle also seemed to support this scenario. The outer membrane was the remnant of the mycoplasma cell mem-brane that formed a phagocytic vacuole when the bacteria were first ingested. Because they were able to set up a symbiotic relationship ben-efiting the host cell, these bacteria became organelles that divided ever afterward in synchrony with the host cell: They went from captured or-ganisms to organelles. This meant that the emergence of the eukaryotic cell—arguably the single greatest breakthrough in evolutionary novelty in the history of life on Earth—was not created by tiny, cumulative mu-tations but by large leaps involving acquisition of an entire microbial ge-nome at once, contradicting a fundamental tenet of neo-Darwinian evolutionary thought. Some of those genes might become incorporated into the DNA of the host cell over time; others would remain as the DNA detectable in mitochondria and chloroplasts today. The idea at first re-ceived quite a cool reception: Philip Abelson was unwilling to publish the paper in *Science*, which he edited, and it was turned down by numerous journals before finally being accepted in the decidedly offbeat *Journal of Theoretical Biology*. Peer review for federal grants was outright hostile, as will be discussed later. Still, Lederberg, Barghoorn, and Morrison encour-aged her to keep at it.[51]

When he first became head of the NASA Exobiology grants program in 1970, Richard S. Young turned up at almost every interdisciplinary exo-biology and origin of life meeting and was always recruiting. Young approached scientists whose work he thought promising and suggested that they apply for Exobiology Program funding at a modest level; "seed money" was what he had in mind. Thus, in 1971 Cyril Ponnampe-ruma recommended Lynn Margulis's work on serial endosymbiosis the-ory to Young after NSF had turned her down, and he encouraged her to apply. At an April 1973 Paris OOL meeting, Young approached Carl Wo-ese and suggested he apply. Young funded both of them immediately, though modestly, and NASA was a critical means of support for both (al-most the sole means for Margulis until it ended in 2002) ever since.

In particular, Young was looking for ideas so interdisciplinary in their breadth that they were having difficulty getting funding from NIH or NSF. The first person with origin of life or exobiology as a major research focus to be elected to the National Academy of Sciences, not until 1973, was Stanley Miller. So the field was still perceived as an odd "borderland" area, not fitting comfortably into biochemistry, geochemistry, microbiology, cell biology, or any other existing disciplinary niche. The large federal science-funding agencies were organized to review proposals pretty much along disciplinary lines. Thus, something far from central to cell biology like Margulis's 1970 proposal for work related to endosymbiosis was likely to be rejected by NSF's Cell Biology Division, often rejected out of hand. As Jan Sapp has shown, by 1970 the rising power of nuclear (chromosomal) inheritance work, especially after Watson and Crick's DNA structure and the consolidation of molecular biology had marginalized study of cytoplasmic inheritance, such as Margulis's study of DNA in mitochondria, chloroplasts, kinetosomes, and other organelles.[52] Margulis recalls:

> I applied for a three-year grant for $36,000. . . . That was exactly when *Origin of Eukaryotic Cells* first edition, Yale University Press, came out. . . . My grant officer calls me up and he says "I'm sorry to tell you we've turned down your proposal . . . you didn't suggest the following controls." He was telling me what was wrong, "you didn't have the following experiment." I said, "look on page seven, that's exactly the experiment we have there so I don't understand." He said, "well, frankly I haven't read the proposal but let me tell you that there are some very important molecular biologists who think your work is shit." He said that on the phone . . . he said, "your work appeals to the small minds in biology." And I said, "well, who are the small minds in biology?" And he said "well, natural historians." And I said "that's quite a compliment." Anyway, he said "don't ever apply to [NSF] Cell Biology again."[53]

Margulis was stymied and quite eager when Young encouraged her to apply to NASA Exobiology. She recalls that, even after the institution of formal review panels, the Exobiology grant application process had a "small town" feel. Young had a fair amount of latitude, if he wanted to encourage a particular investigator, at least with some modest initial funding. In 1971 Margulis received a grant of $15,000.[54] Both she and Carl

Woese say this early seed money was critical to sustaining their research programs, and it gradually increased year by year, as their research proved more fruitful and fulfilled Young's hopes.[55]

The search for and nurturing of interdisciplinary "diamonds in the rough" that had been passed over by NSF and NIH soon became Young's trademark. And the tradition was very much handed down by apprenticeship to his successors. People who first met origin of life workers or first got connected with NASA through these meetings, in addition to Margulis, Lovelock, and Woese, include Elso Barghoorn (1967, 1971), J. W. Schopf (1967, 1970, 1971), Alan Schwartz (1963), Carl Sagan (1963, 1967, 1968, 1971), Leslie Orgel (1967, 1968, 1970, 1971), Jeff Bada (1967, 1971), H. D. ("Dick") Holland (1968), David Buhl (1971), Sol Kramer (1967, 1968), and many others. Barghoorn was a well-established geologist, but Schwartz and Bada were still graduate students when they first attended these meetings, and Schopf had only just finished his PhD. Many others were still quite young scientists (Ponnamperuma, Sagan, Margulis) or were unknown to the few who had dedicated their research primarily to origin of life and/or to exobiology.

Many of the reasons earlier ideas on endosymbiosis had faced opposition were still in play. First, many biologists still considered prokaryotes as primitive and unimportant in evolution, except as agents of disease. René Dubos and others were just beginning to make headway in convincing the world that microbes were crucial *ecological* players. And as Sapp put it: "symbiosis was frequently allied with mutualisms and confronted the emphasis on an incessant struggle for existence between species. Known cases of symbiosis were treated as curiosities, 'special aspects of life,' depicted as 'strange bedfellows.'"[56] In the 1970s (and even today) symbiosis is usually relegated to a single chapter somewhere near the end of biology textbooks, with the implication that it is an unimportant footnote to the grand narrative of nature "red in tooth and claw," evolution propelled entirely by competition. Ludwik Fleck insightfully examines how science uses such "epistemological wastebasket" categories, to conveniently dispose of facts inconvenient for the dominant paradigm.[57] It is surely not accidental that Margulis later became very interested in Fleck's work.[58] Margulis's theory gave a glimpse into an alternate world of "symbiogenesis" (totally different from the existing "thought collective," as Fleck would say), where major evolutionary novelty could result from acquisition of whole genomes at once via the "cooperative" arrangement of symbiosis.[59]

For those who followed Darwin in imagining a "tree of life," the discovery in the 1970s of horizontal gene transfer among microbes was unsettling and made such trees much more difficult to construct. Endosymbiosis seemed to hold out the possibility that branches and whole trunks could "randomly" graft onto one another—rendering the entire "tree constructing" enterprise of dubious value. Stephen Jay Gould had tried to calm "tree makers" by suggesting that an evolutionary "bush" would work better. Margulis seemed to be showing a glimpse into an evolutionary process that was more like a hopelessly tangled jungle.[60] In those thickets, evolutionary novelty could be acquired in huge saltations, not merely via tiny, gradually accumulating steps by mutation. More than that, symbiosis could be construed as a *cooperative* process rivaling competition in driving evolution forward by acting as a powerful generator of novelty for selection to work upon.

It should be noted that many biologists might simply have felt there was insufficient evidence in hand in 1970. Until more proof of, for instance, kinetosome DNA was forthcoming, skepticism was an appropriate response (though the visceral quality of the skepticism is rather telling). When A. Allsopp of Manchester University dismissed Margulis's argument that the entire mitotic spindle apparatus and the flagella of eukaryotes were of prokaryotic origin, he saw the similarities between the two as the very thing one would expect if eukaryotes had evolved from prokaryotes in traditional Darwinian fashion. Thus, "Occam's razor" would opt for that hypothesis as less complicated than endosymbiosis. Similarly, Rudolph Raff and Henry Mahler opposed the endosymbiotic account of the origin of mitochondria, arguing that Margulis's theory looked strong if one privileged cytological evidence but considerably more awkward if one concentrated on comparative molecular studies. The protoeukaryote by this evidence was already likely to respire aerobically, they argued. So it would not gain any large competitive advantage by incorporating a free-living aerobic bacterium. Furthermore, they pointed out that mitochondria did not have a full complement of genes for independent existence and their genes were rather tightly coordinated with nuclear genes; for example, most of their proteins were coded for by DNA within the nucleus.[61]

Not only were her skeptics rather more visceral in the intensity of their criticism than one usually sees in scientific debate; when Margulis became a public supporter of Lovelock's Gaia hypothesis in 1973, this seemed to confirm the critics in their harsh judgment of her interest in ideas that "can't possibly be true."[62] The reaction to both Gaia and SET show many parallels that suggest wider cultural themes were involved. Recall that when Ford Doolittle first attacked Gaia theory, the title of his paper "Is Nature Really Motherly?" and its publication in the *CoEvolution Quarterly* suggests that neo-Darwinian biologists saw Margulis's involvement in both ideas as not coincidence. Edited by Stewart Brand of *Whole Earth Catalog* fame, *CoEvolution Quarterly* had championed Lovelock's Gaia hypothesis. Brand had placed a favorable review very visibly in the first few pages of the October 1980 *Next Whole Earth Catalog*, as well as a favorably slanted interview, "Lynn Margulis: Unlike Most Microbiologists," in the Spring 1980 *CoEvolution Quarterly*.[63] While affiliated with "sustainability" and "appropriate technology" ideas, the *Catalog* and the *Quarterly* avoided, shall we say, "politically correct, party line" environmentalism and feminism. Brand celebrated how "Lynn's rampant non-feminist style in male-dominated science reminds me somewhat of Margaret Mead—she never slights someone for being male, only for being stupid. And, like Lovelock, she's a fine collaborator."[64] Brand was in particular taken with Lovelock's somewhat curmudgeonly remarks about environmentalists and his cybernetic-engineering point of view. Lovelock decried all the hoopla about chlorofluorocarbon propellants from spray cans (it was, after all, he who had first discovered the accumulating CFCs in the atmosphere), saying:

> I am not unconcerned about CO_2 or acid rain or even the ozone war. I just think that the environmentalists have their priorities misplaced. They agonize over a still unproven hypothesis, that the current production of chlorofluorocarbons might slightly deplete ozone, when the tropical forests are being ripped off at a rate which could eliminate them in two decades. This to me is truly straining at the gnat while swallowing the camel.[65]

Predating the nuclear winter debate by three years, Brand took up Lovelock's claim that Gaia would probably survive a nuclear exchange between the United States and the Soviet Union, though humankind might not.

Nonetheless, to many, both Gaia and SET seemed to portray a nature in which "cooperative" processes vied for a role of equal importance in nature to competition. This impression was surely reinforced when historian Carolyn Merchant's 1980 ecofeminist, anticapitalist book *The Death of Nature* in a new 1990 edition celebrated Gaia as a new paradigm that might help heal the environmental crisis she laid at the feet of the capitalist, mechanistic Scientific Revolution.[66] The cultural moment in terms of the feminist movement, the ecology movement, the post-Vietnam sense that scientists had contributed to an unjust war gave both Gaia and SET a set of cultural valences Lovelock and Margulis never intended, but which may well have conditioned the response to these theories. In many ways, like Vernadsky before him, Lovelock's vision of Gaia was a vision of a Whole Earth. Though at first an engineer's view, a cybernetics view, among his readers Lovelock's idea touched more—and very different— chords than that. Coming at the flowering of the environmental movement, it resonated strongly with those groups (and not only with the clique involved in the *Whole Earth Catalog*). And since, unlike Vernadsky, Lovelock's chosen metaphor reimagined the Whole Earth in a female image, perhaps it is not so surprising that Gaia became an appealing topic for the feminist movement just gaining prominence, particularly for those who would become "ecofeminists." As Fleck observed, once words become "battle cries," they take on whole new layers of meaning for the communities that take them up. To Carolyn Merchant, Gaian science looked like just the kind of less mechanistic, reductionistic science—reimagining nature as feminine—she had prescribed in *The Death of Nature* as needed to get us out of our modern environmental crisis.

Although, like Gaia, SET was initially greeted with a fair amount of hostility, it is now more firmly accepted even than Lovelock's insights. By 1980–81, new molecular tools for DNA sequencing had shown that the DNA within chloroplasts was practically identical to that of a particular cyanobacterium; similarly, the DNA of mitochondria was found to be extremely similar in sequence to that of a family of α-proteobacteria. Thus, Margulis quickly went from being scorned to attaining membership in the National Academy of Sciences in 1982. By 1986, *New Scientist* was celebrating serial endosymbiosis theory as the triumphant new paradigm, one of the most important contributions to life science of the twentieth century.[67] Historian of science and biophysicist Evelyn Fox Keller wrote a profile of Margulis and her discoveries in this issue of *New Scientist*, "One Woman and Her Theory," in which she argued (quoting a reviewer of Margulis) that championing SET was "a job for a woman."[68] The reviewer had written, "It has to be a young scientist and a woman who dared to challenge the scientific establishment by writing such a book"—that is, such a book as Margulis's 1970 *The Origin of Eukaryotic Cells*.[69] Keller had achieved fame by making a similar argument in a 1982 biography of geneticist Barbara McClintock's discovery of gene transposition, just a year before McClintock was awarded the Nobel Prize. When this notion was parodied in *Science* in a profile on Margulis and her work five years later, the title used (as I mentioned earlier) was "Science's Unruly Earth Mother." While the idea of a "gendered science" enjoyed an academic heyday for a decade or so, Margulis herself strongly resisted such an idea, at least as a description of her own work:

> They say it's a kind of science only a woman could do—that 'Earth Mother' crap—and that's a back door way of putting it down, of saying it's not really top quality science. I don't get my ideas because I'm a woman, but I have to work harder than most to produce the evidence in just as hard-boiled and mechanistic a way as any other scientist, or they won't believe what I say.[70]

Whatever else Margulis believed, she believed there was only one kind of good science, and it had no gender. Yet, the gender valence of SET and of Gaia has resonated with many students of mine over the past fifteen years, reminding us once again that the cultural reception of scientific ideas can often be more complex (and even the opposite) than what the scientist intended.

BACK TO GAIA AND STILL LARGER CULTURAL ISSUES

What of Merchant's claim that Gaian science was the kind that could help change the attitudes that had led to the environmental crisis? Though at first pooh-poohing alarmist environmentalists, in later years, after becoming alarmed over global climate change, Lovelock himself wrote that much scientific resistance to Gaia might be rooted, consciously or unconsciously, in the fact that Gaia contradicted another Darwinian view: that humans, like other creatures, are mere passive responders to environmental change imposed upon them from without. The Gaian model by contrast puts living organisms in the driver's seat, as joint creators of the environment in which they live. Lovelock had begun to think along these lines almost from the first exchange with Ford Doolittle in the spring of 1981.[71] Life changes the environment; "the options then open to life as a whole are constrained by its past activities."

By 2001, Lovelock further developed this idea, arguing that the Gaian view thereby implies greater ethical responsibility—at least for sentient creatures capable of understanding their role—for the problems our activities create that might destabilize Gaia. Avoiding accepting greater responsibility was a significant motivation for skepticism, Lovelock opined. He added,

> We now know enough about living organisms and the Earth System to see that we cannot explain them by reductionist science alone. The deepest error of modern biology is the entrenched belief that organisms interact only with other organisms and merely adapt to their material environment. This is as wrong as believing that the people of a village interact with their neighbors but merely adapt to the material conditions of their cottages. In real life, both organisms and people change their environment as well as adapting to it. What matters are the consequences: if the change is for the better then those who made it will prosper; if it is for the worse then the changers risk extinction. . . . Early in the history of civilization, we realized that overreaching self-worship turns self-esteem into narcissism. It has taken almost until now to recognize that the exclusive love for our tribe or nation turns patriotism into xenophobic nationalism. We are just glimpsing the possibility that the worship of humankind can also become a bleak philosophy, which excludes all other living things, our partners in life upon Earth.[72]

Nor were these merely Lovelock's own idiosyncratic views about the theory's implications. For moral philosopher Mary Midgley, "the idea of Gaia . . . is a useful idea, a cure for distortions that spoil our current worldview. Its most obvious use is, of course, in suggesting practical solutions to environmental problems."[73] But she argued that Gaia could help us out of much deeper intellectual and cultural tangles, most of all our atomistic, reductionistic view of nature using machine imagery—the legacy of the Scientific Revolution of the seventeenth century—and the parallel idea that humans are completely apart from (and charged to have dominion over) nature—over all other life on Earth. The environmental crisis forced modern science to accept the importance of at least one less reductionistic science, ecology, "which always refers outwards from particulars to larger wholes."[74] Thus, the Gaian view of life on Earth as a unified whole that includes humanity is, she says "not just a useful but actually a scientifically necessary one."[75] Like Lovelock and, more still, Margulis, Midgley criticizes the neo-Darwinian rhetoric of "selfishness, spite, exploitation, manipulation, investment, insurance, and war-games" because it reinforces social-atomist, every-man-for-himself ideology that incentivizes continued depredation of Gaian life-support systems, but also because it gives an incomplete picture of how the living world works, leaving out, for example, symbiogenesis. Instead, "by using a different imagery and a different basic pattern, Gaian thinking tends to correct this outdated bias. It does not reject the central scientific message of neo-Darwinism about the importance of natural selection. It simply points out that it is not the whole story."[76]

Midgley noted that the extreme reductivism in biology had led orthodox scientists to reject Gaia outright in the early 1970s. She believed, far more than Lovelock (as quoted earlier), that when he shifted from the name "Gaia" to describing his idea as "geophysiology," this helped quite a lot in making the idea more acceptable, more than the steadily growing proof of individual Gaian mechanisms. As she put it:

> The main difficulty . . . was never about the scientific details. It concerned the imagery, the vision of a wider whole, in some sense a living whole, of which we are a part. It became clear how much this imagery mattered to the scientists when Lovelock introduced a slightly different image, namely the medical model of the earth as a sick planet needing our care and attention—needing, in fact, a science of *geophysiology* to study its health and sickness.
>
> This way of talking greatly reassured the scientific public. It has allowed detailed discussion of the implications to go on without alarm and outrage. This was reasonable because the metaphor of illness is indeed a valuable one, leaving room for the sense of urgency that we so badly need. . . .

Yet it is interesting to ask just what is the difference between this image and the original one of Gaia the earth-goddess. Why is one of them acceptable while the other still causes alarm? After all, the medical model too accepts that the earth should be treated as if it were a living whole, which was what was originally supposed to be objectionable.[77]

For philosopher Jacob Needleman, Gaia was a new vindication of thoughts he had since the 1970s, when he worked on a team under physicist David Langmuir on an English translation of Vladimir Vernadsky's classic 1926 book *The Biosphere*.[78] Vernadsky was a geochemist who saw "life as a cosmic phenomenon which was to be understood in terms by the same universal laws that applied to such constants as gravity and the speed of light."[79] The laws the NASA Exobiology program sought to explicate. Although Lovelock was unaware of Vernadsky's work until well after publishing his Gaia theory, Needleman had a profound impression from reading Vernadsky, that the Earth is alive, a unity, and moreover, that life is the Earth's *response* to an "infinite number of radiations" that were "pouring on to the Earth from 'every part of celestial space.'"[80] Much as Aldo Leopold in the 1920s linked such ideas with the work of Russian philosopher and mathematician P. D. Ouspensky, Needleman also sees the problem of man's relationship to the Earth through the lens of the teachings of Ouspensky and his teacher G. I. Gurdjieff.[81] He sees the symbiotic "microcosmos" of the eukaryotic cell and its parallel with symbiotic Gaia as a reflection of the ancient dictum "As above, so below."[82] Needleman felt that the Gurdjieff-Ouspensky system of ideas was validated and given environmental relevance when he read Lovelock and Margulis's work on Gaia: "If Earth is a living being, as it surely is, then like everything that lives it is either growing or dying. But perhaps in ways that we do not understand, in order to grow, the Earth needs our uniquely conscious human energy."[83] Or, rendered more plainly, it is humanity's low level of planetary consciousness—including about the impact of our activities upon the stability of Earth's ecosystems—that is pushing the biosphere to the verge of collapse. So, Needleman proposes, we must increase our consciousness in order to avert that tragedy. The cultural valences of these ideas are indeed powerful and widespread.

The NASA Exobiology Program of the 1960s–1980s created a unique intellectual environment. Despite predictable discomfort from many scientists, ideas of breathtaking scientific sweep including both the Gaia hypothesis and Margulis's serial endosymbiosis theory could be nurtured there and given a chance to prove their merit. By wedding the science of searching for life on other planets with research on the origin and early evolution of life on Earth, exobiology facilitated conversations among specialists from a wide variety of disciplines, most of which Lovelock and to a lesser extent Margulis necessarily drew upon in first formulating the Gaia hypothesis and SET. Ideas of such breadth and interdisciplinarity, perhaps not surprisingly, provoked multivalent responses from the broader culture as well, not least because they were embedded in NASA's scientific search for life on other worlds. That context also highlighted implications ranging from the nascent environmental movement, to feminist theory, and even to humanity's place in the cosmos. Many scientists remain skeptical about pursuing theories across the boundary of science into cultural implications. Thus, despite the abiding importance of the sciences of Gaia and symbiosis, the range of general interests they generated as well as the occasionally enthusiastic and unruly receptions they received also contributed to initial scientific resistance to these ideas.

4.

ON SYMBIOSIS, MICROBES, KINGDOMS, AND DOMAINS

JAN SAPP

Over the past two decades, a dispute quietly persisted over the highest ranks in the natural order of living things. As advocated by Lynn Margulis, should the great diversity of life on Earth be organized into five Kingdoms: Plants, Animals, Fungi, Protists, and Monera (bacteria)? Or, as proposed by Carl Woese, should one recognize taxa above the rank of kingdoms, comprising three domains: Archaea, Bacteria, and Eucarya? No other twentieth-century biologists were more important for emphasizing the evolutionary diversity of the microbial world than Margulis and Woese. However, Margulis focused on the diversity of protists—that is, eukaryotic microbes—and the origin of eukaryotic cell organelles; Woese focused on the diversity of bacteria in the broad sense, and on the origin of the genetic code. Margulis did not support the molecular revolution in phylogenetics; Woese led it. Margulis upheld the distinction between Prokaryote and Eukaryote as the fundamental dichotomy of life created through symbiosis, Woese denied that prokaryotes represented a coherent evolutionary grouping, and argued for a deeper trifurcation based on differences in molecular genetic systems. Their differences, I suggest, reflect the specific core problems they investigated, more than any ultimate disagreement about the role of symbiosis in evolution. And at a larger scale, their combined efforts to rewrite the foundations of evolutionary thought are crucial because they focus attention on deep evolution, the major transitions of life from precellular entities to bacteria, and from prokaryote to eukaryote, and in so doing, highlight new modes of evolutionary change through symbiosis and the inheritance of acquired genes and genomes.

To root this discourse and dispute over biological kingdoms, I first offer a brief overview of the place of microbes in evolutionary biology in the context of the general principles of post-Darwinian classification. I then situate the conflict over kingdoms and domains in the context of evidences for the fundamental role of symbiosis in evolutionary change on the one hand and the molecular revolution in phylogenetics on the other.

LIFE IN TWO KINGDOMS

From the nineteenth century throughout most of the twentieth century, the world of biology was essentially one of two kingdoms: Plants and Animals. A problem remained: where to place microbes in this dichotomy? Linnaeus had put them in one species of *Vermes*, which he drolly called *Chaos infusorium*. The microbial classification remained in chaos for generations. Post-Linnaean biologists recognized that there were organisms that did not fit the two-kingdom world. Jellyfish—polyps, for example—seemed to possess the essential characteristics of both, and were classified accordingly as phytozoa, that is, as plant-animals. In his *Philosophie zoologique*, Jean-Baptiste Lamarck argued that life should be arranged in terms of evolutionary relationships and processes.[1] There were two general modes of evolution for Lamarck. One was an evolutionary tendency to progress from simplicity to ever-increasing complexity of organization. That complexity could be seen in what he called "the essential system of organs" of the higher taxa of animals, but it was less apparent in plants, and not at all among the infusorians. But the progression from the simple infusorians to "invertebrates" and vertebrates was clear enough, and, Lamarck argued, it was driven by an unknown though not metaphysical "force of life." The second mode of evolutionary change for Lamarck resulted from adaptations of organisms to their immediate surroundings. These branches on the phylogenetic tree did not reflect increasing complexity, but resulted from the inheritance of characteristics acquired by interaction with changing environments.

Fifty years later, in *On the Origin of Species*, Darwin argued similarly to Lamarck in two principle ways. First, he suggested that classification should be based on evolutionary relationships, that is, in terms of genealogies in accordance with divergence from a common ancestor, and that the order of things so arranged would resemble a branching tree. "The natural system is founded on descent with modification; that the characters which naturalists consider as showing true affinity between any two or more species, are those which have been inherited from a common parent, and, in so far, all true classification is genealogical; that community of descent is the hidden bond which naturalists have been unconsciously seeking, and not some unknown plan of creation, or the enunciation of general propositions, and the mere putting together and separating objects more or less alike."[2]

Second, like Lamarck, Darwin distinguished between adaptive characteristics tied to environmental contingencies and "essential" organismal characteristics far removed from the vicissitudes of life. Those highly preserved essential characteristics were most useful for establishing lines of descent, not mistaking "adaptive or analogous" characteristics of no use for classification from "homologous characters" based on common ancestry. The similar characteristics of a whale and a fish are *analogous*, but they are of no importance for a taxonomic classification, any more than are the wing of an insect and that of a bird. In contrast, the hand of a man and the fin of a porpoise are *homologous*; they are of taxonomic significance. "The generative organs being those which are most remotely related to the habits and food of an animal, I have always regarded as affording very clear indication of its true affinities. We are least likely in the modifications of these organs to mistake a merely adaptive for an essential character. . . . It may even be given as a general rule that the less any part of the organization is concerned with special habits, the more important it becomes for classification." "Embryological characters," he added, "are the most valuable of all."[3]

Before the revolution in molecular phylogenetics, beginning in the 1960s and '70s, the, relationships between plants and animals were based on comparative embryology and comparative morphology. But microbes lacked complex morphologies. One could not distinguish between ancient well-preserved "essential characters" and those that were more recent adaptations in the microbial world. Their classification would remain outside the evolutionary fold of post-Darwinian biology throughout the nineteenth and most of the twentieth century.

HINTS OF HIDDEN KINGDOMS

Darwin said little of the microbial world in *Origin*. In the concluding sentence he hinted at the possibility of multiple microbial origins when he famously wrote, "There is grandeur in this view of life, with its several powers, having been originally breathed into a few forms or into one."[4] Indeed, whether life as we know it originated once or several times would remain unknown for over 100 years. Darwin's world was of two kingdoms, but he seemed also to hint at more when he wrote to his friend T. H. Huxley on September 26, 1857:

> But as we have no written pedigrees, you will, perhaps, say this will not help much; but I think it ultimately will, for it will clear away an immense amount of rubbish about the value of characters & will make the difference between analogy & homology, clear. The time will come I believe, though I shall not live to see it, when we shall have very fairly true genealogical trees of each great kingdom of nature.[5]

As the microbial world became better known with improvements in microscopy in the nineteenth century, discourse arose over a third kingdom of life. Paleontologist Richard Owen spoke of the "protozoa" (first animals) as a kingdom unto themselves, as organisms lacking the sophisticated anatomy of animals.[6] Still, others objected that the term protozoa was zoocentric—after all, that grouping surely included the precursors of higher plants, too. John Hogg renamed the group the "Primigenium," "the first creatures of the Creation," members of the kingdom he called Protoctista.[7]

Ernst Haeckel took the argument a step further in 1866 when he proposed a wholly new kingdom, the Protista.[8] It would comprise protozoa (one-celled animals), protophyta (one-celled plants), and a "neutral" phylum neither plant nor animal. Haeckel's kingdom also included another previously unconsidered group of creatures: the Monera comprised bacteria and cyanophycae (blue-green algae), which he said were not true cells. Monera lacked chromatin and were perhaps more like crystals; they were to bridge the gap between the living and the non-living. But none of the third-kingdom proposals were well accepted in the nineteenth century. Critics argued that the forms in that third grouping were far too diverse to be included in one kingdom. It was unlikely that they were monophyletic, that is, that they had one common ancestor that was also a member of the group. It seemed more likely that the grouping was polyphyletic. Monera as defined by Haeckel was also rejected by many of his contemporaries. Whether or not bacteria and blue-green algae were "true cells," defined as possessing chromosomal genes and dividing by mitosis, would be disputed for most of the twentieth century.[9]

Microbiology was primarily an applied science. Pathologists conceived of microbes as disease-causing entities, as "germs," as Joseph Lister first called them in 1874.[10] Since that time, the word "bacteria" (from the Greek meaning little rod or staff) was frequently employed for the smallest of germs, "all those minute, rounded, ellipsoid, rod-shaped, thread-like or spiral forms."[11] Otherwise, the classical animal-plant dichotomy persisted. Microbes were divided according to the highest institutional organization of biologists themselves within universities. Botanists studied those one-celled creatures that possessed green bodies, zoologists those that lacked them. Bacteria were the domain of botanists; they were conceived of as plants, and referred to as the fission fungi: Schizomycetes. One can still hear physicians today refer to our bacterial symbionts as "gut flora."

In any event, bacterial classification remained outside the general evolutionary paradigm based on comparative morphology and embryology. It was impossible to know which characters were ancient and highly conserved over the course of evolution and which were more recent adaptations, what characters were homologous and what were analogous. Instead, bacteria were named according to as many characteristics as possible: disease-causing properties, usefulness for industry, and biochemical and culturing properties. This ad-hoc procedure was called "statistical taxonomy." As bacteriologist Robert Breed lamented, "A review of the literature will show that the most popular term that has been used to describe systematic bacteriology is 'chaos'; and this irrespective of the period of history under consideration."[12]

The "evolutionary synthesis" of the 1930s and '40s, based on a merger of Darwinian natural selection and Mendelian genetics, focused on the evolutionary processes and principles of classification among plants and animals. Bacteria were not included. The neo-Darwinian two-step of random gene mutation and the selection of organisms possessing favorable variations did not seem to (fully) embrace bacterial evolution. Inheritable variation among them did not seem to arise randomly and then get selected. Rather, neo-Lamarckian principles seemed evident, as heritable variations in the bacterial world seemed to arise in direct response to the environment—as environmentally induced adaptive changes. Bacteria were typically thought to belong to "a Lamarckian rather than a Darwinian world, or at least a world in which both Lamarckian and Darwinian evolutionary mechanics are operative."[13] Indeed, it was still uncertain if bacteria had genes. The minimalist conception of the bacteria Julian Huxley presented in *The Modern Synthesis* was not unlike Haeckel's Monera concept seventy-four years earlier:

> They have no genes in the sense of accurately quantized portions of hereditary substances. . . . The entire organism appears to function as soma and germplasm, and evolution must be a matter of alteration in the reaction-system as a whole. . . . .We must, in fact, expect that the processes of variation and evolution in bacteria are quite different from the corresponding processes in multicellular organisms. But their secret has not yet been unraveled.[14]

A NEW DICHOTOMY

Bacteria came to the fore of biology after World War II, when geneticists were able to show that they possessed genes and how they could be domesticated for molecular genetics. Still, bacterial geneticists were generally no more interested in the natural history of bacteria than, say, classical geneticists were in the natural history of the guinea pig. Microbiologists at mid–twentieth century knew no more about bacterial relationships than they did in Pasteur's day. Proposals for kingdoms for the bacteria were made outside the mainstream of biology. In 1938, Herbert Copeland, an instructor at Sacramento Junior College, proposed four kingdoms when he distinguished the kingdom Monera from Protists, not on Haeckel's grounds that Monera lacked chromatin, but on the grounds that bacteria lacked a membrane-bound nucleus and did not divide by mitosis, but by simple fission.[15] Copeland's four-kingdom proposal was treated with circumspection by the few who considered it.[16] Protists seemed to be too diverse a group for one kingdom, and Copeland's Monera were defined negatively. When defined by things they lacked it was a grouping like "invertebrates," an unwieldy category which embraced many unrelated organisms, from worms to insects. Even when distinguished as "Monera," it was not certain that bacteria and "blue-green algae" could be grouped together. It was difficult to distinguish small bacteria-like rickettsia from larger viruses, and it was uncertain if blue-green algae reproduced by simple fission or divided by mitosis. Such problems were resolved with the deployment of the electron microscope. In a 1957 paper titled "The Concept of the Virus," André Lwoff articulated the difference between a virus and cell.[17] Viruses did not divide by mitosis, had either DNA or RNA but never both, and were practically devoid of enzymes; they did not reproduce by division like a cell; and they reproduced only within a susceptible cell.

A definition of the bacteria on the same basis followed suit when Roger Stanier and C. B. van Neil wrote "The Concept of the Bacterium" in 1962. "The abiding scandal of bacteriology," they said, "is the absence of a clear definition of a bacterium."[18] Using terms proposed by Lwoff's former mentor Édouard Chatton, they distinguished "eucaryotic" cells (from the Greek: true nucleus) from the "procaryotic" cells.[19] Eukaryotic cells, divided by mitosis, had a membrane-bound nucleus, a cytoskeleton, an intricate system of internal membranes, mitochondria that perform respiration, and in the case of plants, chloroplasts as well. Prokaryotic cells divided by fission and lacked eukaryotic structures. Just as there were no transitional forms between virus and bacterium, they could observe no transitional forms between prokaryote and eukaryote. In the second edition of their seminal text *The Microbial World*, Stanier, Michael Doudoroff, and Edward Adelberg declared that, "In fact, this basic divergence in cellular structure, *which separates the bacteria and blue-green algae from all other cellular organisms*, probably represents the greatest single evolutionary discontinuity to be found in the present-day living world."[20] Stanier and colleagues emphasized that a natural, genealogical classification was impossible. Life, as we know it, could have evolved once, twice, or indeed many times. No one could know. Nonetheless, they still posited that all bacteria had a common origin: "All these organisms share the distinctive structural properties associated with the prokaryotic cell . . . and we can therefore safely infer a common origin for the whole group in the remote evolutionary past."[21]

FIVE KINGDOMS

The clear distinction on cytological grounds between eukaryotes and prokaryotes immediately led to proposals of superkingdoms of Prokaryota and Eukaryote.[22] That suggestion was generally overlooked for decades. But a kingdom for Monera or Prokaryotae was proposed immediately and well accepted. Based on the prokaryote-eukaryote distinction, plant ecologist Robert Whittaker added the kingdom Monera to his earlier four-kingdom scheme, in which, like others before him, he distinguished a kingdom for fungi as well as for protists.[23] Whittaker's scheme was not meant to be genealogical. It represented the direction of evolution, and was "based on three levels of organization—the prokaryotic (kingdom Monera), eukaryotic unicellular (kingdom Protista) and eukaryotic multicellular and multinucleate."[24] He knew that the Protista were a highly diverse grouping, most likely with multiple origins; he suspected the same to be true for plants and animals. The highly diverse fungi were distinguished as a kingdom because, unlike ecological producers and consumers, they were decomposers. But Whittaker was confident that the bacteria had a common ancestry, that they were monophyletic. Lynn Margulis followed Whittaker and further articulated the five-kingdom scheme.[25] Recognition of microbial kingdoms was crucial in heralding the great diversity of life in the microbial world and in pushing evolutionary theory beyond the classical investigation of the origin of plant and animal species over the past 450 million years to the origins of those eukaryotic cells that comprised them, and conditioned their existence. Margulis distinguished animals as organisms that developed from a blastula, plants as developed from an embryo, and fungi from spores. Monera lacked eukaryotic structural properties and did not divide by mitosis. She changed the name Protists to kingdom Protoctista to recognize that not all protists were unicellular, as the name had come to denote. Over the next four decades, Margulis integrated these five kingdoms into her discussions of symbiosis, and into her pedagogical imagery of the hand of Gaia.[26]

Fundamental for Margulis was the distinction between prokaryote and eukaryote. It was at the basis of her symbiotic theory for the origin of eukaryotic cells. She breathed new life into the idea that symbiosis played a fundamental role in evolutionary change when she proposed, in her now famed paper "On the Origin of Mitosing Cells," that the presumed greatest discontinuity in nature, that between a prokaryote and a eukaryote, occurred not gradually but by leaps, by a series of symbioses.[27] As Margulis was always the first to admit, however, that symbiosis played a fundamental part in evolution was already an old idea. It originated in Europe in the late nineteenth century when botanists studied various forms of symbiosis. The evolutionary force of symbiotic relationships was supported by lichens being dual organisms of fungi and algae; by mycorrhizal fungi in the roots of forest trees, nitrogen-fixing bacteria in the root nodules of legumes, and evidence that "animal chlorophyll" in translucent "green animals" such as hydra and sea anemones was actually symbiotic algae.

Let us continue with this short history of symbiosis in biological thought. That chloroplasts might have originated as symbionts was proposed German botanist Andreas Schimper in 1883 when the word "chloroplasts" was coined. In Russia in the 1890s Andrei Famintsin conducted experiments to extract and culture zoochlorella and zooxanthella from animals with the hope of learning to culture chloroplasts. Ernst Haeckel also postulated that chloroplasts originated from a symbiosis with Chromacae (cyanobacteria).[28] Russian botanist Constantin Merezhkowsky also championed the idea.[29] Merezhkowsky coined the word "symbiogenesis" in 1909, to mean the origin of organisms by symbiosis. He also proposed that that nucleus and cytoplasm originated as a symbiosis of two different kinds of microbes.[30] In a Woods Hole lecture of 1893, Japanese zoologist Shōsaburō Watasé suggested that nucleus and cytoplasm of might have once been symbionts, and perhaps centrioles, too, if they could be shown to be self-reproducing.[31] The symbiotic origin of cilia or flagella was proposed by Boris Kozo-Polyansky in his 1924 book *Symbiogenesis: A New Principle of Evolution.*[32]

In 1918 in his book *Les symbiotes*, Paul Portier created an elaborate theory of symbiosis in heredity and development based on the idea that mitochondria originated eons ago as bacterial symbionts.[33] Although they have become adapted to intracellular life and could not now be cultured, new bacteria entering cells rejuvenated them. And Portier reasoned that bacteria in the root nodules of legumes could be cultured: they were protomitochondria. In *Symbionticism and the Origin of Species*, Ivan Wallin at the University of Colorado developed a theory of mitochondria as acquired symbionts.[34] He conjectured that mitochondria were continuously acquired into a heterogeneous population, were the source of new genes, and that mitochondrial differentiation resulted in many other cell organelles including Golgi bodies, cilia and flagella, as well as chloroplasts. Wallin also claimed to have definitive proof that he had cultured them outside the cell.[35] Symbiosis had also been proposed for the evolution of bacteria themselves. In 1917, Félix d'Herelle reported on "an invisible microbe" that he named "bacteriophage" (bacteria eater) that decimated a colony of the dysentery bacillus. Two years later, he noticed that not all bacteria were destroyed by bacteriophages. Mixed cultures of phage and bacteria could be subcultured indefinitely, and there were transformations in the morphology and physiological properties of the infected bacteria. D'Herelle saw the analogy with lichen symbiosis. He referred to the mixed cultures as "microlichens." "Symbiosis," he said, "is in large measure responsible for evolution."[36]

Despite these views, biologists generally considered symbiosis to be a rare, exceptional phenomenon of little importance to evolutionary processes. One can identify a number of reasons why hereditary symbiosis remained on the edge of evolutionary biology. The disease-causing aspects of many microbial infections and the success of the germ theory of disease overshadowed the evidence for the beneficial outcome of other microbial "infections." "It is a rather startling proposal that bacteria," Wallin commented, "the organisms which are popularly associated with disease, may represent the fundamental causative factor in the origin of species."[37] Hereditary symbiosis, the inheritance of acquired genomes, conflicted with the one genome–one organism doctrine of biology reinforced by the Mendelian-chromosome theory of heredity and nucleocentric genetics. Classical geneticists defined heredity in terms of the sexual transmission of (nuclear) genes between individuals of a species. Excluded from this concept was the inheritance of acquired genes and genomes, or "infectious heredity."[38] The basic tenets of the neo-Darwinian evolutionary synthesis based on gradual transformations resulting from gene mutation and recombination within species effectively denied a prominent role for hereditary symbiosis. Studies of symbiosis were also at odds with the biological emphasis on conflict and competition in nature. The phenomena of symbiosis were treated as curiosities, "special aspects of life," of little significance for general biology.[39] And symbiotic theories for the origins of cell organelles were far too speculative for empirically oriented biologists. As E. B. Wilson commented prophetically in *The Cell in Development and Heredity*: "To many no doubt, such speculations may appear too fantastic for mention in polite society; nevertheless it is within the range of possibility that they may someday call for some serious consideration."[40] All of these issues were reignited in the 1960s and '70s when concepts of symbiosis in cell origins emerged in light of new molecular evidence.

ENTERING THE REALM OF POLITE SOCIETY

The turning point for theories of the symbiotic origin of organelles occurred in the early1960s when mitochondria and chloroplasts were shown to possess their own DNA and their own protein-synthesis apparatus. This new information shot to the center of cell biology. Virtually all of those biologists who showed, in the 1960s, that mitochondrial and chloroplasts possessed their own DNAs, also suggested that they might have evolved as bacterial symbionts. When, in 1963, Sylvan and Margit Nass at the University of Stockholm reported evidence of mitochondrial DNA, they commented that a "great deal of modern biochemical and ultrastructural evidence that may be interpreted to suggest a phylogenetic relationship between blue-green algae and chloroplasts and bacteria and mitochondria."[41] The international Society for Cell Biology hosted a symposium on the issue in 1966.[42] But this rather saltational scientific change was not simply a case of new evidence for old ideas, "long neglected" or "premature." Symbiosis theories of the 1960s and '70s differed fundamentally from some of those that preceded them[43]: mitochondria were not the source of new genes; they no longer represented a diverse community of symbionts that could be cultured independently; they did not give rise to other organelles such as chloroplasts and centrioles; nor were they the primary agents of cellular differentiation during ontogenetic development. As the new studies clearly showed, they were organelles of respiration, rendering the eukaryotic cell aerobic.

But no one developed symbiosis theory and amassed so much data from so many specialties to bear on it than did Margulis.[44] Her interest in symbiosis originated as a Master's student at the University of Wisconsin in Hans Ris's advanced cytology class.[45] In 1962 when Ris and his colleague Walter Plaut considered the electron microscopic evidence for the DNA-containing area of the chloroplast of the unicellular alga *Chlamydomonas*, they commented: "With the demonstration of ultrastructural similarity of a cell organelle and free living organisms, endosymbiosis must again be considered seriously as a possible evolutionary step in the origin of complex cell systems."[46]

Margulis went further to argue for an independent symbiotic origin of the centrioles (from which mitotic spindles were then thought to emanate) and the related organelles kinetosomes (motility organelles at the base of eukaryotic cilia).[47] There was suggestive though contested evidence that centrioles/kinetosomes may also possess DNA, as mitochondria and chloroplasts had just been shown to do.[48] Margulis correlated that evidence with evidence of spirochetes attached to the protist *Mixotricha paradoxa* in the hindgut of termites, as newly reported by Cleveland and Grimstone.[49] Like the "animal chlorophyll" of old, those spirochetes had actually been mistaken to be organelles of the protist before it was shown that they were in fact independent organisms. The functional similarity of spirochetes and cilia was a most remarkable case of convergent evolution, and Cleveland and Grimstone suggested that this particular similarity between organism and organelle could only have evolved in special sheltered environments such as the hindgut of the termite. But for Margulis those spirochetes, once mistaken for cilia, held fundamental phylogenetic meaning.[50] She thus suggested that centrioles, kinetosomes, and related structures emerged when spirochetes attacked a primitive amoeboid cell. If so, mitosis would have evolved from symbiosis: If this were the case, eukaryotes would be wholly chimeric. Serial symbiosis would be the mechanism responsible for what Stanier, Douderoff, and Adelberg said was the greatest single discontinuity in the living world—that between prokaryote and eukaryote.

Margulis's 1970 book *The Origin of Eukaryotic Cells* heralded the revolution to come in the next decade regarding mitochondria and chloroplast origins. In that decade, the idea that organelles were symbionts was met with various degrees of skepticism and acceptance. Some critics protested against all of it, arguing that because the symbiotic origin of organelles was saltational—that is, not grounded in Darwinian gradualism—it was comparable to a belief in Special Creation.[51] Others found it easier to accept that chloroplasts arose as symbionts, but more difficult to believe the same for mitochondria. Margulis's most daring proposal, that centrioles and related structures (kinetosomes and eukaryotic flagella) arose as symbionts, was looked upon the least favorably of all. It met with fierce resistance, for a number of reasons: (1) unlike mitochondria and chloroplasts, the evidence for DNA in centrioles/kinetosomes was shaky (today, it is accepted that there is no DNA within centrioles), (2) centrioles lack a double membrane possessed by mitochondria and chloroplasts, (3) centrioles lack a genetic translation apparatus for synthesizing proteins, which mitochondria and chloroplasts possess, (4) they do not divide by fission, and in some cases arose de novo.[52] Undeterred, Margulis would maintain throughout her career as central to her writings that centrioles/kinetosomes and related structures originated from spirochetes.[53] And she would elaborate this claim into conjectures regarding the origins of the cognitive capacities of the senses and the nervous system.

STILL METASCIENCE

When all was said and done in the 1960s and early '70s, there was still no definitive proof for any theories of organellar origins, whether they arose endogenously or exogenously. At first Margulis thought it might be possible to culture mitochondria and chloroplasts independently outside the cell.[54] But geneticists showed that these organelles were so well integrated into the cell as a whole that their genomes possessed only vestiges of what might have been their symbiotic origins. The only way to prove that organelles were derived from free-living microbes was to show their genealogical relationships—that is, to demonstrate their direct lines of descent. But that seemed impossible. The main method for bacterial classification of the 1960s and '70s was that of numerical taxonomy, which like "statistical taxonomy" before it, was a nonphylogenetic ordering that did not rely on distinguishing homologies from analogies.[55] Instead, by a "'majority rule' strategy," it grouped organisms together based on as many characteristics as possible regardless of whether those characteristics had a common origin.

However, without a genealogical ordering of bacteria, the whole debate over organellar origins was sterile. Paraphrasing Wittgenstein, one commentator remarked: "Whereof one cannot know, thereof one should not speak."[56] Roger Stanier, one of the main architects of the prokaryote-eukaryote dichotomy, spoke for many cell biologists in 1970 when he asserted that cell origins were beyond the realm of science: "Evolutionary speculation constitutes a kind of metascience, which has the same fascination for some biologists that metaphysical speculation possessed for some medieval scholastics. It can be considered a relatively harmless habit, like eating peanuts, unless it assumes the form of an obsession; then it becomes a vice."[57] In 1975, Margulis agreed that such "historical theories" could "never be directly tested." Evolutionary biologists, she said, were in the same logical predicament as historians, and "can only present arguments based on the assumption that of all the plausible historical sequences one is more likely to be a correct description of the past events than another."[58] But in fact, all of these views were belied with the emergence of the field of molecular evolution, which closed the debate over the symbiotic origin of chloroplasts and mitochondria in the early 1980s.

MOLECULAR EVOLUTION

Genealogical trees of great size and scope, as Darwin had argued, required comparisons of highly preserved ancient features far removed from the everyday life of the organisms. In bacteria, those highly conserved characteristics could be found, but only at the molecular level, in the most conserved parts of the genome. And since the mid-1960s, those searching for a bacterial phylogenetic classification had argued that genes coding for components of the translation mechanism were the likely candidates for a universal phylogeny.[59] After all, being universal, the genetic code would have evolved at the dawn of life, together with certain components of the mechanisms for translating nucleic acids into the amino acids of proteins. Once that system was optimized, there would be selection pressure against any major changes to it. The genes for the translation system would interact with so many others such that piecemeal change would be difficult.

The argument for a highly conserved core through which organismal genealogies could be discovered was also crucial because of ever-increasing evidence of the prevalence of horizontal gene transfer between bacterial taxa.[60] Many bacterial genes can be transferred between unrelated bacteria though viruses and through conjugation between species. Therefore, similarity in a gene may not be a measure of genealogical relationship. If organism type A and organism type B carry the same gene, it may not be because both belong to the same taxonomic group because of common descent, but rather because one, or both, of them acquired that gene. For bacteria, gene histories were not necessarily organismal histories. In principle, horizontal gene transfer could scramble the phylogenetic record.

The importance of lateral gene transfer became apparent in the early 1960s with evidence that multiple antibiotic resistance was due not to bacterial mutation but to the inheritance of acquired genes coding for that resistance. That horizontal gene transfer was indeed a principal means of bacterial adaptation and evolutionary innovation became especially clear with the rise of genomics in the late 1990s. Because many genes can be transferred between taxa, much of bacterial evolution resembles a web more than a tree. In principle, horizontal gene transfer can erase the phylogenetic record because gene histories would not be organismal histories. Indeed, if all bacterial genes were transmitted between taxa, a taxonomy based on genealogy would be impossible. The argument for a highly integrated genetic core of informational genes concerned with DNA replication and translation, not readily transferred between species, was thus fundamental for establishing a molecular evolutionary taxonomy of bacteria.[61]

The concept of a highly conserved genetic core was central to Carl Woese's research program based on comparisons of ribosomal RNA sequences.[62] Woese's core research interest was in the evolution of the genetic code and the origin of molecular translation. For instance, in the system for translating information in nucleic acids to amino acids, why does CCC encode the amino acid proline? Francis Crick suggested that the genetic code was a meaningless "frozen accident," and once codon assignments were established they would be difficult to change.[63] But Woese had a different thought: the code and its translation apparatus would have evolved in steps. With the aim of constructing a universal genealogical tree, and perhaps of tracing life back to a time when the translation apparatus first evolved, he turned to ribosomal RNA as the "ultimate molecular chronometer."[64] Ribosomes—the cellular organelles housing the translation apparatuses—comprise two subunits, a smaller one slightly cupped inside a larger one; both contain RNA and protein. The smaller subunit of ribosomes contains about 1,450 nucleotides. Focusing primarily on comparisons of the RNA of the small subunit (SSU RNA) of about 150 nucleotides, Woese's research program was an extraordinary success. Comparisons of SSU rRNA sequences would provide a phylogenetic taxonomy of bacteria, and allow the construction of a universal tree of life, hitherto deemed impossible. According to the SSU rRNA phylogenies, most of the bacterial taxa above the level of genus would be reclassified.

Woese had read Margulis's 1970 book *The Origin of Eukaryotic Cells*. He favored the symbiotic origin of organelles and he set out to prove it with molecular methods. Indeed, the methods Woese developed were crucial for closing the general debate over the origin of mitochondria and chloroplasts.[65] Based on rRNA comparisons, mitochondria and chloroplasts were of alpha-proteobacterial and cyanobacterial origin respectively.[66] The basic question of the bacterial ancestry of plastids and mitochondria was rigorously resolved. However, since no genetic material could be definitively located within them, that centrioles/kinetosomes originated from spirochete ancestors could not be tested by those methods. Be that as it may, for Margulis the origin of organelles through symbiosis was no exception, it was a far-reaching phenomenon, a principal means of evolutionary innovation with the rise of eukaryotes from microbes to macrobes, a view she championed beginning in the late 1980s.[67] While recognizing the importance of symbiosis in eukaryotic evolution, Woese also applied symbiosis to evolution at the dawn of life at what he called the progenote phase of evolution, when early protocells were in the throes of developing the translation apparatus.[68] Given an error-prone translation system, he reasoned, symbiotic fusions would speed up the evolutionary process during that era. However, there would be no lineages as such during the progenote phase—before the fundamental lines of descent emerged.

In Woese's program, horizontal gene transfer between "species" and symbiosis were central features of the evolutionary process beginning with the evolution of the progenote, in the throes of developing the cellular translation machinery. As he wrote to Emile Zuckerkandl, editor of the *Journal of Molecular Evolution*, when explaining his progenote concept in March 1977:

If we are obliged to start with typical bacteria . . . then you have to re-move the wall and develop phagocytosis. This makes endosymbiotic interactions: (1) late on the scene (i.e. evolving after the grand plan of the bacteria has begun to unfold) and (2) a special occurrence, as the capacity evolved (in a rather unlikely way) in only one group of bac-teria. In other words, endosymbiosis seems *ad hoc*; it needs to be excused. . . . What we now take as endosymbiosis is only the tip of the iceberg. Therefore, what we now define as eucaryotic is only a re-stricted segment of the real class thereof. Endosymbioses have been a major force in evolution for over three billion years.[69]

Herein lies a major difference between Woese and Margulis. Whereas Margulis saw symbiosis as emerging after the origin of life and after the initial radiation of the prokaryotes, Woese saw symbiosis as an evolu-tionary force active from the dawn of life. Certainly other bacteriologists had also considered the relationship between some viruses and hosts as symbiotic.[70] But Margulis drew the line with eukaryotes. In her view it was improper to refer to the beneficial effects of viral infections on pro-karyotes and eukaryotes as symbiogenic. For her, symbiogenesis was a phenomenon that occurred exclusively in the evolution of eukaryotes, and indeed it was a feature distinguishing prokaryotic from eukaryotic evolution.

CONCEPTS IN CONFLICT

Woese's research program would clash head on with Margulis's views when he confronted the classical prokaryote-eukaryote dichotomy and the five-kingdom model. Although now canonical, the prokaryote con-cept carried with it two untested assumptions: that all prokaryotes had a common prokaryotic origin, and that prokaryotes preceded eukaryotes and gave rise to them. Based on differences in their molecular genetic or-ganization, Woese and his collaborators challenged both assumptions. From SSU ribosomal RNA comparisons, Woese and his collaborators grouped together a phenotypically diverse group of little-studied organ-isms: methanogens, which live in coalmine refuse, swamps, and the guts of rumens; extreme halophiles, which live in brines five times as salty as the ocean; and thermophiles, which live in geothermal environments that would cook other organisms. Woese and Fox named the group formed from these organisms the "archaebacteria," and reconceptualized them as an ancient lineage of organisms from the dawn of life, an ur-kingdom as phylogenetically distinct from the typical bacteria as typical bacteria were from eukaryotes.[71] Though at first subjected to skepticism and criticism, the archaebacteria group was shown to possess many common characteristics: their walls lacked peptidoglycan, a defining feature of "prokaryotes"; the lipids in their cell membranes were unique; and so too were their transfer rRNAs, transcription enzymes, introns, and viruses. In 1990, in a formal taxonomic proposal to replace the classical bipartite division of life into prokaryotes and eukaryotes, Woese, Otto Kandler, and Mark Wheelis introduced three "domains" of life as a rank above kingdoms: the Archaea, the Bacteria, and the Eucarya.[72] The three-domain model heralded the great predictive powers of the SSU rRNA methods, the great diversity in the bacterial world, and the triumph of molecular phylogenetics.

In proposing three Domains above the level of kingdoms, Woese and colleagues confronted the five kingdom model head on. The Protista was far too diverse to be in one kingdom: the SSUrRNA data from other labs indicated that they were more diverse than plants and animals combined. The kingdom Monera, they said, was faulty for two reasons. First, it failed to recognize that the difference between Monera and the other four kingdoms was far more significant than the differences among those four. Second, it failed to recognize that the Monera itself was an ill-defined grouping defined negatively "by their lack of characteristics that define the eukaryotic cell."[73] Although Archaea and the typical bacteria were cytologically similar in the prokaryotic form, they said that at the molecular level, the Archaea resembled the "eubacteria" no more (probably less) than they do the eukaryotes. "*Prokaryotae* (and its synonym *Monera*) cannot be a phylogenetically valid taxon."[74]

Neither the five-kingdom scheme nor any other scheme of the 1970s or 80s had considered taxa above the level of kingdom. Discussions over the kingdoms, superkingdoms, or domains and whether "prokaryote" held any biological meaning heated up over the next fifteen years.[75] Superkingdoms Prokaryotae and Eukaryotae, proposed two decades earlier,[76] were reborn this time to counter the three-domain proposal. Those who maintained the value of the prokaryote-eukaryote distinction as a dichotomous depiction of cell structure and gene organization typically placed the archaebacteria as a kingdom within the *Prokaryotae*.[77] Classical evolutionist Ernst Mayr recast the prokaryotes and eukaryotes as "domains" or "empires."[78] Many evolutionists, including Margulis, shared his views on the three-domain proposal. In his many books, Mayr had given little attention to the microbial world, considering the question of other kingdoms matters of taste and convenience until the arrival of the three-domain proposal.[79]

By 1998, Mayr recognized that Woese's research program had revealed an unknown world equivalent to a new continent.[80] Still, his views about classification differed from those of Woese in two regards. First, he insisted that biological classification need not be based on genealogy. Instead, he argued that classification should be based on as many characters as are available, and that even defined negatively, the prokaryote was fine because the absence of characters itself was valid. "The classification supported by me is based on the traditional principles of classification, which biology shares with all fields in which items are classified, as are books in a library or goods in a warehouse."[81] Nothing separated Woese and Mayr more than the importance of phylogeny. Woese's aim was to understand life's course, the primordial genomes: knowledge of the manner in which the genetic apparatus evolved could only be deductions of phylogeny. Genealogy was primary, and phenotypic differences could at best corroborate taxonomy based on phylogenetic, that is, molecular characteristics. Because it was a polyphyletic grouping overriding the differences between the Archaea and the Eubacteria, the concept of the prokaryote was anathema to what Woese considered "a natural classification." Woese, Kandler, and Wheelis countered Mayr in stating, "A global classification should reflect both principal dimensions of the evolutionary process: genealogical relationship and quality and extent of divergence within a group. The ultimate purpose of a global classification is not simply information storage and retrieval."[82]

Second, for Mayr evolution was "an affair of phenotypes," not of genes. The kinds of molecular and biochemical differences between bacteria and archaebacteria did not compare to the great phenotypic differences between "palms, oaks, and orchids; mice, bats, and whales; and hummingbirds, chickens, and ostriches."[83] Woese articulated the differences between himself and Mayr well in reply.

Mayr's biology is the biology of visual experience, of direct observation. Mine cannot be directly seen or touched; it is the biology of molecules, of genes and their inferred histories. Evolution for Dr. Mayr is an "affair of phenotypes." For me, evolution is primarily the evolutionary process, not its outcomes. The science of biology is very different from these two perspectives and its future even more so.[84]

Margulis agreed with Mayr on the principles of classification, and she further denied that molecular methods were the primary basis for revealing evolutionary relationships. In her classic book *Symbiosis in Cell Evolution*, she brought the new molecular evidence together with other kinds of morphological data for the symbiotic origin of mitochondria and chloroplasts, as well as that of centrioles, while depicting many facets of the evolutionary drama.[85] She did not see nucleic acid sequences and amino acid sequences as being any more important that other characters that might be used for a genealogical classification. As she and Ricardo Guerrero wrote: "The meaningfulness of any phylogenetic tree as a guide to evolutionary history depends critically on what, and how many characteristics were used to construct it."[86] They then listed "appearance, anatomical organization, development, mode of nutrition, metabolic pathways, gas emissions, pigments." But to molecular evolutionists, Margulis's position was difficult to understand. It made no distinction between analogous and homologous characteristics. It was essentially the same as the numerical taxonomy of bacteriologists in the 1960s and '70s, who had abandoned a classification based on evolutionary principles of homology. For molecular evolutionists, molecular data were not merely one type among other kinds of phylogenetic information. Rather, only nucleic acids and proteins held the genealogical record in the microbial world.

Margulis and Mayr's views in regard to methods of classification were similar, but not so their views in regard to Kingdoms and Domains. Mayr was willing to assign the archaebacteria to a rank above kingdom, a "subdomain" of the Prokaryotae; he also thought the protists should be a subdomain on the basis of their diversity.[87] Margulis held fast to five kingdoms. In 1977, Woese sent her a reprint announcing the methanogens as archaebacteria, and informing her of his results indicating that homophiles and thermophiles were also members of the group. She replied, "The archaebacteria question needs a discussion."[88] At the time that the three-domain proposal was formally made in 1990, she adamantly opposed it. However, Margulis's response changed as she grappled with the upwelling of interest in early evolution and the origin of kingdoms. She recognized superkingdoms in the second edition of *Symbiosis in Cell Evolution* of 1993. Therein she sometimes referred to the kingdom "Prokaryotae or Monera" and sometimes to the superkingdom "Prokaryota" along with Eukarya.[89] She placed the Archaebacteria and Eubacteria as subkingdoms in the kingdom Monera. She argued that Woese's three domain system was based mainly on SSUrRNA of nuclear origin, but those methods would not readily apply to classifying the eukarya because of the symbiotic origin of their organelles:

> Primarily on the basis of nucleocytoplasmic sequence data, Woese and colleagues place all nucleated organisms together in the Eukarya domain. Because all Eukarya have more than a single type of 16S ribosomal RNA as a consequence of their polygenomic ancestry, the three domain system proposed by these molecular biologists privileges one (the nucleocytoplasmic) over the others (e.g. plastid or mitochondrial 16S ribosomal RNA). Thus, the scheme which has been so useful for systematizing the bacteria—leading to the recognition of two subkingdoms (Archaebacteria and Eubacteria)—is invalid as a taxonomic tool to organize eukaryotes, all of which have more than a single genetic system because of their derivation from symbionts.[90]

A new kingdom was proposed on the grounds of symbiosis: kingdom Chromista was suggested for those protists whose chloroplasts were derived, not from cyanobacteria, but secondarily from eukaryotic algae.[91] Still, Margulis held onto the five-kingdom model.

Perhaps more than any other biologist, Margulis was reluctant to grant the archaebacteria kingdom status, and she never did. She did change the title of the fourth edition of her *Five Kingdoms* to *Kingdoms and Domains* in 2009.[92] But in so doing she changed the meaning of "Domain" from a category above kingdom to a category below. Thus, she maintained five kingdoms within two superkingdoms distinguished on the basis of their nonmolecular cellular organization and on the basis of their origin by symbiogenesis: Superkingdom Eukarya, Superkingdom Prokarya (origins not by symbiogenesis), Subkingdoms (Domains) Archaea and Eubacteria.

CONCLUSION

In light of Margulis's indifference to molecular phylogenetics, it is perhaps ironic that those methods came to verify one of her main theories in regard to the evolutionary process: organelle genesis aside, that symbiogenesis may be a major driving force of macroevolutionary change among multicellular eukaryotes.[93] In addition to evidence that horizontal gene transfer is pervasive in the bacterial world, there is growing evidence of widespread transfer of genes and genomes between microbes and eukaryotic organisms. Nothing illustrates that better than the diverse bacteria of the genus *Wolbachia*, inherited maternally through the egg of about 25 percent of insect species as well as nematodes. Genome fragments and whole genomes have been transferred to hosts.[94] As Margulis and others have long maintained, the saltational tempo and mode of evolutionary change resulting from hereditary symbiosis conflicts with the neo-Darwinian tenets of gradual change resulting from gene mutation and recombination within species.[95]

Symbiosis extends beyond the inheritance of acquired genes and genomes to embrace an organismal unity of multispecies coexistence. Indeed, as long been argued, the meaning of symbiosis lay in widening the concept of the organism beyond the limitations of the genetic unity to embrace all those intimate multispecies associations whose persistence from generation to generation is not ensured by inheritance through sex cells, but which nevertheless comprise the functional whole.[96] The organismal unity by which we recognize the persistence of form within plant and animal species is one of interlocking functions involving microorganisms. Only in this way can we understand, for example, the mycorrhizal fungi in the roots of 95 percent of plant species, the nitrogen-fixing bacteria in the root nodules of legumes, the diverse populations of bacteria and protists in the digestive system of xylophagous insects, and the complex microbial populations in the digestive systems of cellulose-digesting ruminants. Symbiosis thus belies the classical concept of the organism comprised of cells whose cooperation results from a common genome.[97] The symbiotic organism represents a functional unity or functional field of diverse species genomes.[98] The classical "one-genome–one organism" concept is thus superseded by a conception of the multispecies, multigenomic holobiont or symbiome.[99]

5.

THE WORLD EGG AND THE OUROBOROS

Two Models for Theoretical Biology

SUSAN SQUIER

> The existence of the quantum discontinuity means that the past
> is never left behind, never finished once and for all, and the future
> is not what will come to be in an unfolding of the present
> moment; rather, the past and the future are enfolded participants
> in matter's iterative becoming.
>
> —KAREN BARAD

In 1967, twenty people gathered at the Rockefeller Foundation Research
Center on the shores of Lake Como, Italy. They had come there for ten
days at the invitation of C. H. Waddington, the embryologist and developmental biologist, to participate in the second of four meetings known
as the Serbelloni Symposia, sponsored by the International Union of
Biological Sciences between 1966 and 1969. Their goal was to formulate,
"at least in broad outline, the structure of a discipline of General Theoretical Biology."[1] Waddington was also at work during these years on another
project: a book on modern painting and modern science based on the Greg-
ynog Lectures he had presented in 1964 at the University of Wales, published as *Behind Appearance: A Study of the Relations between Painting and the
Natural Sciences in This Century*.[2] By temperament and practice a profoundly

interdisciplinary thinker who "simply loved thinking about biological principles and analyzing their implications for the living world," he was also
a maverick, whose interdisciplinary impulses did not always sit well with
his contemporaries.[3] The collected proceedings of the Serbelloni Symposia reflected this maverick status, and were dismissed by one biologist as
"a bizarre miscellany of articles about whether there can be general theories in biology, and some specific speculations about development and
evolution."[4] Similarly, the reviews of *Behind Appearance* also damn with
faint praise, ranging from Martin Kemp's "suggestive juxtaposition of modernist works with visual images in twentieth-century science" which still
lacks "a more substantial historical foundation than he was able to provide" and Wikipedia's "has wonderful pictures but is still worth reading,"
to Brian K. Hall's "enormously ambitious analysis of the relations between
painting and the natural sciences in the twentieth century."[5]

Despite their lukewarm reception, however, these projects merit another look. Although I can only sketch out here the far more extensive
analyses that still beckon, a comparative reading reveals another possible past for modern biology, a past that (to borrow from Karen Barad)
"hasn't happened yet."[6] In what follows, I argue that we can glimpse the
far future of a broader, systems-level science in the liminal space between
Waddington's two endeavors. While the Serbelloni Symposia reveal how
modern biology shifted its ruling metaphor from the visual to the verbal, giving precedent to the genecentric model that would dominate biological research for decades, Waddington's crossover study of science and
painting provides tantalizing glimpses of the alternative vision of development and evolution explored by contributors to this volume, and particularly in the visionary work of Lynn Margulis.

THE SERBELLONI SYMPOSIA

Interdisciplinarity was the hallmark of the Serbelloni Symposia as a whole.
Throughout the four years, Waddington chose guests strictly on the
principle that they "would have something interesting to say."[7] Given that
Waddington, a wide-ranging intellect whose publications spanned "paleontology, population genetics, developmental genetics, biochemical
embryology and theoretical biology," had convened the meeting, this was
certainly an opportunity for the participants to think broadly.[8] Wadding-

ton himself had a research trajectory so wide-ranging that one contemporary biologist believes it would be impossible today. "No modern funding agency would allow any individual to undergo so many changes of interest and direction."[9] Charged with arriving at a unified biological theory, a "General Theoretical Biology" that would achieve an "intellectual and academic stature comparable to Theoretical Physics," the participants devoted the first Symposium to determining that unlike physics, this new field "would not seek for universal and eternal laws."[10] Rather, they agreed that living systems are particular examples of some kind of "organized complexity," and they saw their task as asking "what kind of complexity?" and "what are the principles of its organization?"[11]

By the Second Serbelloni Symposium in 1967, the participants included the theoretical physicist David Bohm, a molecular biologist, three neuroscientists, a geneticist, seven physicists, a theoretical chemist, a chemical engineer, a systems analyst, a philosopher, an automata theorist, and a secretary.[12] Rather than try to corral them into producing one coherent theory of biology, Waddington simply invited them to suggest some of the questions such a new field might pose.[13] However, despite Waddington's characteristic embrace of a wide range of views, a tension, perhaps disciplinary or perhaps simply the chafing of the orthodox against the maverick, is evident in the published proceedings of this Second Serbelloni Symposium. In his preface to the volume, subtitled "Sketching Theoretical Science," Waddington explains he had decided to include not only the talks of the participants, but also "such non- (but I hope not sub-) standard items as . . . my autobiographical Note."[14] He hoped thereby to "bring together a number of elements which, we thought, would be useful in the eventual construction of a coherent and comprehensive Theory of Biology."[15] Nonstandard indeed. In ten brief gemlike pages, Waddington's contribution, "The Practical Consequences of Metaphysical Beliefs on a Biologist's Work: An Autobiographical Note," both illuminates the forces that propelled the scientific project at the center of the Serbelloni Symposia and reveals some of the pressures on Waddington that may have narrowed its outcome.[16]

Waddington's talk followed that of theoretical physicist David Bohm, whose presentation some of the participants found disturbing, owing to its evocation of metaphysics. "Order may well be a fundamental notion that underlies both physics and biology," Bohm argued, providing a "common language and concepted [*sic*] structure for the formulation of both."[17] Indeed, he went on, "order may well be the basic factor which unites mind and matter, living and non-living things, etc."[18] Looking back in an article written after the conference, Bohm seems unaware of the discord these remarks produced. As he remembered it, the most important aspect of the exchange of views at Bellagio was "a common realization that metaphysics is fundamental to every branch of science."[19]

However, Waddington was aware of the tensions Bohm ignored. In his opening salvo he even embraced the drama, siding with Bohm against "several of the more 'hard-headed' characters at the Second Symposium [who] expressed from time to time, at cocktails or after dinner, a suspicion that metaphysical considerations . . . have ultimately no real impact on the directions in which science advances."[20] Dissenting from their view of metaphysics as merely "part of the froth churned up while the theoretical physicists flounder and thrash about trying to find a firm footing in the deep and dangerous waters of quantum theory, sub-nuclear particles, and the like," Waddington proclaimed: "a scientist's metaphysical beliefs are not mere epiphenomena, but have a definite and ascertainable influence on the work he produces." He put his own experience on the line as evidence: "I am quite sure that many of the two hundred or so experimental papers I produced have been definitely affected by consciously held metaphysical beliefs, both in the types of problems I set myself and the manner in which I tried to solve them."

Perhaps Waddington's simultaneous involvement in writing *Behind Appearance* while coordinating the Second Serbelloni Symposium accounts both for the volume's art-inflected title ("Sketching Theoretical Biology") and for the remarkable jumble of artistic references in his defense of metaphysics. He seems still to be searching for an apt metaphor to describe how philosophy shapes his approach to science. Metaphysics is "something more than a set of decorative flourishes on the proscenium arch, giving on to the stage in which the real action takes place. [It is] a sort of poetry," he asserts, recalling David Bohm's observation, and as poetry, "metaphysics can be absorbed through communication-channels other than extended rational exposition." However, the visual and spatial aspects begin to dominate his text as he warms to recounting his own "metaphysical-experimentalist's autobiography," detailing the "two (or perhaps three) notions which infiltrated into my thinking at a very early stage, without much benefit of academic dignity, and which have remained there ever since." He acquired these beliefs in his schoolboy days, he explains, when a tutorial in chemistry designed to prepare him for the university entrance examination branched into more esoteric studies. His tutor, E. J. Holmyard, was obsessed with the way that Greek philosophy and technology were transmitted across space and time, from the Alexandrian Gnostics and the Arabic Alchemists to fourteenth century Europe. Holmyard gave him lessons in Arabic and exposed him to "a large number of very odd late Hellenistic documents."[21] From that rich mixture, "two ideas stuck":

→ *The world egg* "Things" are essentially eggs—pregnant with God-knows-what. You look at them and they appear simple enough, with a bland definite shape, rather impenetrable. You glance away for a bit and when you look back what you find is that they have turned into a fluffy yellow chick, actively running about and all set to get imprinted on you if you will give it half a chance. Unsettling, even perhaps a bit sinister. But one strand of Gnostic thought asserted that *everything* is like that.

→ The Ouroboros, the snake eating its tail. This famous symbol . . . expressed the whole gist of feedback control almost two millennia before Norbert Wiener started "creating" ' about the subject at MIT and invented the term "cybernetics."[22]

The published proceedings include a schoolboy drawing he made of the Ouroboros which offers a glimpse of what he considers the third metaphysical belief: "I reproduce it because you will see that inscribed within the ouroboros is a third subsidiary notion . . . 'the one, the all,' a phrase which implies (in a cybernetic context, be it remembered) that any one entity incorporates into itself in some sense all the other entities in the universe."[23] The image is dominated by three globes: on the left, the snake swallowing its tail, a vital image of circularity that Waddington explicitly connects to the cybernetic notion of feedback; on the right, two circles referencing the ancient alchemical art of abiotic creation, as base metals are transmuted into gold. The image gestures to the World Egg while never actually picturing it, referring instead to the circular processes that characterize both organic and inorganic materiality.

Let us pause in our analysis of Waddington's talk to consider the project on which they were all engaged during their stay at Lake Como:

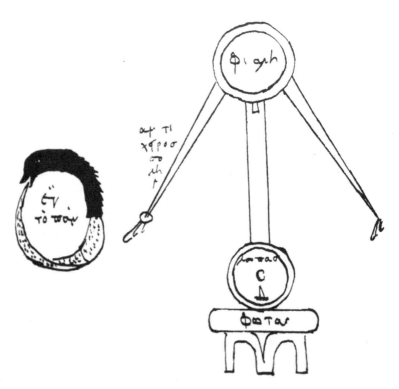

FIGURE 5-1. The World Egg and the Ouroboros were metaphysical beliefs that shaped Waddington's experimental papers. From C. H. Waddington, *Towards a Theoretical Biology 2: Sketches* (1969).

brainstorming the properties of a new field of General Theoretical Biology. There was an urgent need for this intervention, Waddington would explain in *Nature* the following year: "Theoretical physics is a well-recognized discipline, and there are departments and professorships devoted to the subject in many universities. . . . It is widely accepted that theories of the nature of the physical universe have profound consequences for problems of general philosophy. In contrast to this situation, theoretical biology can hardly be said to exist as an academic discipline."[24]

It is tempting to explain this desire for a discipline with the precision and academic status of theoretical physics as simply a case of P.E. I borrow the acronym from Lynn Margulis, who recalled a time when Richard Lewontin, speaking at an economics class at the University of Massachusetts, engaged in "a kind of Neo-Darwinian jockeying," detailing the "Fisher-Haldane mechanisms" that produced evolutionary changes: "mutation, emigration, immigration, and the like."[25] When she asked him why "none of the consequences of the details of his analysis had been shown empirically," and why his elaborate mathematical model was "devoid of chemistry and biology," she recalls that he responded enigmatically, "P.E." What could he mean, she thought. "Population explosion? Punctuated equilibrium? Physical education?" " 'No,' he replied, " 'P.E. is 'physics envy' . . . a syndrome in which scientists in other disciplines yearn for the mathematically explicit models of physics."[26]

Waddington's yearning for a general theoretical biology may indeed be an example of the phenomenon of physics envy Margulis mocked. Yet at play with him also was surely a P.E. of a different kind—philosophy empathy. Returning to Waddington's autobiographical talk at Serbelloni enables us to parse this need more deeply. Responding to his own rhetorical question, "What did I actually do as a practicing biologist, and how was this influenced by this metaphysical background?" he goes on to recount a career that moved from investigations of the ammonite as a paleontologist, to experiments on *Drosophila* evolution, until his final focus on experimental embryology, most notably in the series of experiments on chick embryos that made his name and led him to the formulation of the field of epigenetics, his attempt to meld the insights of embryology and genetics.[27]

Yet "before these highly poetic metaphysics had any practical influence on my scientific work," he explains, they were overlain by "a large body of much more explicitly rationalized thinking," the writings of philosopher Alfred North Whitehead, and other dialectical and biological materialists.[28] A "philosopher of systems in the process of becoming," Whitehead inspired Waddington, an embryologist-in-the-process-of-becoming-a-developmental-biologist, particularly in his attention to "system, process, and the creative advance into novelty."[29] It was Whitehead who first suggested "new lines of thought" that offered alternatives to the Newtonian "conceptions of billiard-ball atoms existing at durationless instants in an otherwise empty three-dimensional space" dominating science at the time.[30] As he began to investigate an interactionist ontology, he explains, he found particularly important Whitehead's notion that "the raw materials from which we start to do science . . . 'are occasions of experience,'" each "essentially infinite and undenumerable," a "unity" comprising "an experiencing subject and an experienced object" that can be differentiated only heuristically and artificially and that has "a duration in time." He suggested that there was in Whitehead's metaphysics an anticipation of quantum theory: "the idea that a particle must also be thought of as a wave function extending through the whole of space time." And he embraced Whitehead's central notions of "concrescence"—"the coming together of the constituent factors in an event" and "prehension"—"the way . . . an event here and now incorporates into itself some reference to everything else in the universe . . . in accordance with its own 'subjective' feeling." Most of all, he learned from Whitehead the "replacement of 'things' by processes which have an individual character which depends on the 'concrescence' into a unity of very many relations with other processes."[31]

In fact, as Waddington explains to the symposium participants, Whitehead had influenced his scientific work as early as the late 1930s, when he "began developing the Whiteheadian notion that the process of becoming (say) a nerve cell should be regarded as the result of the activities of large numbers of genes, which interact together to form a unified 'concrescence.'"[32] In a series of landmark experiments with cultured chick tissues in the 1930s and 1940s, Waddington was able to explore how embryonic cells are induced to develop into a specific type of organ or tissue.[33] Significantly, these studies were inherently spatial in nature; Waddington was still working and thinking as an embryologist, and they relied on the embryologist's technique, transplantation.[34]

Those studies revealed that embryonic development was guided not merely by the DNA in the nucleus of the cell but also by the interactions between the cell nucleus and its surroundings, both the cytoplasm and the extracellular environment. As Hannah Landecker has documented, their impact on biology was transformative; they demonstrated the plasticity of living matter and the possibility that biological time and space could be operationalized—that is, detached from context, manipulated and engineered to a specific end.[35] They also led Waddington to propose a new area of study: epigenetics, the causal analysis of development. He felt that biologists needed a theory to explain the relationship between the genotype and the phenotype, and unlike both embryology and gene-

FIGURE 5-2. John Piper's enigmatic image of the epigenetic landscape. Although it was intended to represent the water as flowing away from us, toward the sea at the high horizon, the strange configuration of the valleys actually makes it seem that the water is flowing toward the viewer. Frontispiece of C. H. Waddington, *Organisers and Genes* (1940).

tics, epigenetics could account for this relationship both at the cellular level and at the level of whole organisms, and even of populations. To convey the broader theoretical and pragmatic capability of this new field, Waddington created "a visual depiction of a set of developmental choices that is faced by a cell in the embryo."[36] He called this the *epigenetic landscape*, a schematic image of the way these relations could shape and channel development over time. He introduced this concept in his 1940 book *Organisers and Genes*, using as the frontispiece an illustration of the epigenetic landscape drawn by his friend, artist John Piper, whose work he would later include in the modern painters discussed in *Behind Appearance*.[37]

At first glance, this image seems a naturalistic landscape: Water flows between banks, towards what the puffy clouds at the top of the image suggest may be an ocean at the high distant horizon. Yet, Waddington's caption marks the curious spatial doubleness of this image. What seems to be an image of a river flowing toward the sea also appears to be a counterfactual (Escher-like) image of the same river *flowing up into the mountains*, a detail of importance to which we will return.

The concepts of epigenetics and the epigenetic landscape were philosophically framed and visually represented in ways that anticipate the findings of contemporary biology, as I will explain in what follows. However, their scope and reception were curtailed by the context in which he was developing them, as a return to Waddington's Serbelloni talk will indicate. The central metaphysical notions that influenced the experimental work giving rise to those concepts, as he describes them to the audience, seem to trigger in him a dynamic between timidity and convention, receptiveness and revolt. "Since I am an unaggressive character, and was living in an aggressively anti-metaphysical period, I chose not to expound publicly these philosophical views," he explains. "Instead, I tried to put the Whiteheadian outlook to actual use in particular experimental situations."[38] His responses to the World Egg and the Ouroboros enable us not only to glimpse Waddington's growing conception of biological development as an interactive and recursive set of systems, but also to understand his reasons for resisting that vision at Serbelloni, splitting off experimental science from philosophy, while linking them (as we will see) in *Behind Appearance*.

Although Waddington asserted in his formulation of epigenetics that model experimental subjects (like the chick embryos) should be understood not as autonomous entities but as entangled processes, always changing in response to their environments, his response to the World Egg belies that understanding. Instead, he finds it "unsettling, even perhaps a bit sinister" when it hatches a "fluffy yellow chick, actively running about and all set to get imprinted on you if you give it half a chance." Not only is the chick's receptivity frightening, but there also is no sense that the experimenter/viewer is similarly receptive to influence and affiliation. Note that here Waddington omits the reflexive turn, the acknowledgement of self-reference that would entail exploring the impact of scientific work on the experimenter and the broader human community as well as on the experimental subject. This omission excludes affect, attention, and prehension, the subtle valences attracting one being to another; it also removes certain phenomena from consideration, limits his concept of epigenetics, and (for a time, at least) narrows the applications of epigenetic theory.

Yet if Waddington resists the two-way potential of the chick's imprinting in the World Egg image, precisely that recursivity is central to the other image of central importance to his experimental work, the Ouroboros. His response to that ancient alchemical symbol had a valence that extends beyond its narrow implementation in the several decades after Serbelloni. Waddington notes that this image has inscribed within it the phrase "'the one, the all,' a phrase which implies (in a cybernetic context, be it remembered) that any one entity incorporates into itself in some sense all other entities in the universe." "This famous symbol . . . expressed the whole gist of feedback control almost two millennia before Norbert Wiener . . . invented the term 'cybernetics,'" he goes on to observe.[39] There is a lineage to the Ouroboros image that links it not only to first-order cybernetics, with its stress on information over materiality, but also to the neocybernetic stress on environments and embodiment that would follow.[40] The Macy Conferences were still of recent memory when the Serbelloni Symposia were occurring, and they had spawned a debate between Wiener's focus on control and communication and a more broadly focused attention on interactive behavior that would persist into the 1970s and beyond.[41] As taken up by molecular biology in the years to come, reading of the Ouroboros prefigures the stress on algorithms that enable the prediction of the parameters of individual development, offering the promise of certainty and control. Yet, as Clarke and Hansen have observed, in the work of neocyberneticians it would in contrast generate a model of emergence. The epigenetic landscape, too, contained anticipations of contemporary molecular embryology. Indeed, only four years after the publication of the final Serbelloni volume,

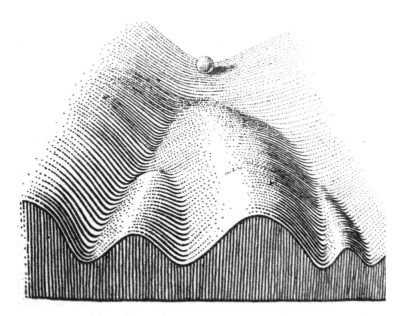

FIGURE 5-3. "Part of an Epigenetic Landscape." Waddington's caption explains: "The path followed by the ball, as it rolls down towards the spectator, corresponds to the developmental history of a particular part of the egg. There is first an alternative, towards the right or the left. Along the former path, a second alternative is offered; along the path to the left, the main channel continues leftwards, but there is an alternative path which, however, can only be reached over a threshold." From C. H. Waddington, *The Strategy of the Genes* (1957).

collaborating on an edited volume, *Evolution and Consciousness*, Waddington would acknowledge "the inadequacy of purely informational theories to deal with developing and evolving systems."[42]

Yet, in the decade of the Serbelloni Symposia, Waddington shifted theoretical models. Although the epigenetic landscape arguably allowed a broad and flexible representation of evolutionary change, allowing for ambiguity and play at both temporal and spatial scales in the passage from genotype to phenotype, by 1962, his desire to bring his embryological interests into conversation with genetics had led him to a model capable of more certainty: the operon, with its ability to predict developmental outcomes with specificity based on newly identifiable chemical triggers for cell differentiation.[43] Just as the Second Serbelloni Symposium records Waddington's discomfort at the World Egg, so the conclusion to the final volume of the symposium papers also seems to signal his decision to abandon the visual imagery of development it embodies. Summarizing four years of conversations, Waddington does not even pretend to claim a consensus has been achieved. Instead, observing that they were working throughout to formulate a theory that would provide "a greater comprehension of the general character of the processes that go on in living as contrasted with non-living systems," he cavils that their goal *should have been* not a "General Theory of Biology" but a more focused "Theory of General Biology."[44]

The most challenging problem of that more focused theory, he continues, should have been the interaction of the two processes most requiring explanation: "complexity-out-of-simplicity (self-assembly), and simplicity-out-of-complexity (self-organization)." Initially he affirms the catastrophe theory of René Thom as one of the most fully explanatory approaches to complexity. As he describes it, this is a spatial and topographic intervention: "a general theory about discontinuities, which may divide a multivariate phase-space into regions . . . bounded against each other by the catastrophe surfaces." His spatial commitment continues as he describes the "the focal point" of the problem as "the confrontation of microstates and macrostates." But then, surprisingly, despite the perspectival metaphor of 'focal point,' he concludes that the best model for exploring self-assembly and self-organization would be not a picture plane or an image but "the analogy of language." "The 'structures mediating global simplicity' which we have to search for in the theory of general biology are, then, perhaps profitably to be compared with languages," he argues, because "the primary biological disjunction between genotype and phenotype" is best understood "as the analogue of symbol-symbolized."[45]

In the "Epilogue," Waddington did make one last-ditch attempt to recast language as neither deterministic nor dispositive but rather interactive. Posing the rhetorical question: "What . . . is this thing, a language, you refer to? . . . What is a language essentially?" he described human language as something that evolves, based on the continuing relationship between being and environment. In other words, language is neither soliloquy nor pure representation, but inducement and interaction, attributes no longer unsettling but now worth advocating for, because of their force and necessity.

> The fundamental form of "generative grammar" on this view is not: Noun Phrase—Verb Phrase. It is: *You* → *Do*. It must be the Second Person: if it were the First, *I* do, there is no point in saying anything; if it is the Third, *He* do, it becomes again a mere description, a statement. And that would conflict with the essential necessity for the '*Do*'—if that is omitted, no effect is produced, and natural selection can take no interest.[46]

Waddington's epilogue to the Serbelloni Symposia narrows their chosen model of science down to a language modeled on the predictive value of mathematics and the control of first-order cybernetics. "Perhaps I should have foretold," Waddington quips, "when I wrote the introductory essay to the first volume, and stressed that biology is concerned with algorithm and programme, that this would be the company in which we should eventually find ourselves."[47] The prediction and control-based "language-metalanguage analogy" that Waddington saw as having the greatest potential to illuminate developmental processes would, of course, translate relatively easily into the gene-centric view of molecular biology in the years to come.

BEHIND APPEARANCE

As formulated in the conclusion of the Serbelloni Symposia and then put into practice during the several decades of ascendency of molecular biology that followed, theoretical biology would avoid the metaphysical and symbolic resources Waddington invoked in his "autobiographical Note" at the Second Symposium. Instead, it became an approach to biology that was mathematical, mechanistic, and instrumental. Even now, theoretical biology usually refers to biology employing "classical mathematical-analytical models, often explicitly inspired by theoretical physics," statistically based models, and "intensive computer modeling."[48]

Yet we will find the persistence of another perspective if we turn to the second project Waddington was occupied with during the same period, *Behind Appearance*. There, as in the edited volume that would follow it, *Evolution and Consciousness,* we can see traces of the "systems counterculture" that as Bruce Clarke points out would soon motivate people as diverse as Heinz von Foerster and Lynn Margulis to "move beyond mainstream doctrines and institutions . . . to detoxify the notion of 'system' . . . and to redeploy it in the pursuit of holistic ideals and ecological values."[49] This massive, three-part, lavishly illustrated, oversized volume characteristic of the 1960s era efflorescence of coffee-table art books makes no promise to offer "new theories about the nature of aesthetic experience" or to provide justifications for any hierarchy of painterly prominence.[50] Rather, it argues that the new vision of the world introduced by quantum physics and illuminated by the writings of Alfred North Whitehead has shaped both scientific knowledge and human experience, leading to a radically different mode of understanding and representing individuals and society.[51] Focusing on the counterintuitive findings of quantum physics and the "retreat from likeness" characteristic of modern painting, Waddington finds a commonality: they are both "revolts against old-fashioned common sense," reflecting "changes in the world view that have been occurring in the last fifty years [and which] are amongst the most far-reaching in the whole history of human thought."[52] Unlike the visual models Pigliucci identified with molecular biology, the set of visual images in painting and scientific practice that Waddington discusses in *Behind Appearance* are heuristic and performative, and so reveal a set of possibilities for visual representation as a model for development broader than simply representational or deterministic uses.

In order to clarify its role in Waddington's developing vision, we can ask several questions about this endeavor, particularly in relation to the simultaneous Serbelloni effort to "formulate some skeletons of concepts around which Theoretical Biology can grow."[53] What is the model of science that Waddington juxtaposes to modern painting in this ambitious volume? What do we make of the fact that Waddington focuses on the relation between science and painting in *Behind Appearance*, when his epilogue to the final volume of Serbelloni Symposium proceedings instead affirms the power of language to explain, by analogy and as an aspect of the new potency of algorithms and computer code, "the general underlying nature of living systems"?[54] What does it mean for our interpretation of this volume that the structural principles of *Behind Appearance* are figures from ancient mythology, and how does this perspective inflect the vision of the world the volume describes within scientific and artistic practice? And finally, how stable is Waddington's assertion that scientific paradigm shifts drive cultural transformations rather than vice versa, considered in light of the comparison of these works?

What is the model of science in *Behind Appearance*? Waddington sets out his model of science in the wide-ranging first chapter, "The Image of Our Surroundings." He offers a chronology of advances in scientific knowledge proceeding from the *First Science* (Euclidean and Pythagorean science), to the *Second Science* (Renaissance experimental science), and finally to M. C. Goodall's *Third Science* of the present day, the worldview he finds represented by both the painters and the scientists under consideration in his study.[55] His description of the characteristics of the Third Science might well have been moved over, whole cloth, from the scientific doctrines shared at the Serbelloni Symposia: "the conception of biological evolution by natural selection," the "esoteric notions of quantum mechanics and relativity," and finally "the study of such general properties as information or organization . . . not entities in any usual sense but . . . characteristic of systems." Yet in contrast to the tensions between classical and theoretical physicists and between metaphysics and science that Waddington identified at Serbelloni, *Behind Appearance* emphasizes that both scientists and artists have come to acknowledge the extent to which they are implicated—culturally, epistemologically, positionally, perspectively, and ontologically—in their research. For scientists, he explains, this has meant drastically revising the notion that "science is completely 'objective,'" while for artists it has meant deepening their involvement in their work to the point that they think of themselves not as "delineating a scene, but of performing an activity in co-operation with their media and tools."[56]

The chapter "The Scientists" opens with a remarkable adaptation of the epigenetic landscape as a metaphor for human life on Earth: "Man in the world is like a caterpillar weaving its cocoon. The cocoon is made of threads extruded by the caterpillar itself and is woven to a shape in which the caterpillar fits comfortably. But it also has to be fitted to the thorny twigs—the external world—which supports it."[57]

Now, taking us "behind appearance" in the context of developmental biology, the solid pegs and taut guywires of this alternative image of the epigenetic landscape remind us that the very landscape of development, scaled from the size of a caterpillar to that of the entire biotic realm, is one complex intertwined system, as is the human web of meanings spun

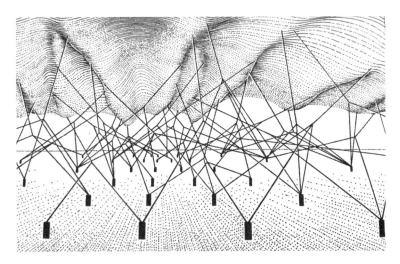

FIGURE 5-4. A behind-the-scenes view of the epigenetic landscape. It shares the attention to the fibrous connections supporting and constraining development central to Waddington's caterpillar metaphor. However, the epigenetic landscape image shows those wires forming the landscape itself, and thus shaping the developmental pathways activated in a caterpillar in the process of metamorphosis. From C. H. Waddington, *The Strategy of the Genes* (1957).

from art and culture to science and technology. Waddington's image suggestively frames what he sees as the central scientific issue of the day—the debate over whether proper science should concern itself with measurable facts or struggle to achieve intuitive understandings—in terms of whether words or images are the foundation of scientific thought. In contrast to Watson and Crick's identifying the structure of the DNA molecule ("seeing a word that will fit into a crossword puzzle") he offers Einstein: "The words of the language, as they are written or spoken, do not seem to play any role in my mechanism of thought. The psychical entities which seem to serve as elements in thought are certain signs and more or less clear images which can be 'voluntarily' reproduced and combined."[58]

Dispatching Wittgenstein's strategy of exploring facts through the medium of language, which he pronounces "not very rewarding territory to anybody, and . . . far from the two interests with which we are concerned," Waddington instead explores scientific facts through a survey of the images produced by the Third Science. From "Electron density contours

The World Egg and the Ouroboros 143

for part of the myoglobin crystal" and "Weather map of the Northern Hemisphere calculated by computer from theoretical equations supplied to it," to "Computer simulation of turbulent flow" and "Two spectrographs of bird song," these images embrace biotic and abiotic systems scaling from the very small to the very large. These examples serve, he argues, as "a new 'landscape of thought,' a new climate of form." Not only do artists now pay attention to such scientific images, they do so because they have realized that "hard scientific analysis—from which conclusions could be drawn . . . led to images of the same general kind." Instead of a unified theory, whether of art or of painting, whether scientists or artists, all humans share a process that sounds very much like Whitehead's concrescence and prehension: the search for "a unit of thought, feeling, and action, to which he can attach himself."[59]

Why focus on the relations between science and painting rather than science and language? As he explains in "The Image of Our Surroundings," Waddington has chosen to focus on painting rather than writing because while both modes of human endeavor share the same goal—to produce syntheses of broad scope—scientific language is so technical and specific that it risks putting writers at a disadvantage. "One cannot discuss in words the implication of physical theories of indeterminacy, or of biological theories of genetic determination, without the question being asked whether the writer can understand in words what those theories state."[60] Note that the point here is not simply the inability of a novelist, wielding a nonscientific vocabulary, to grasp the technical vocabulary of science, but rather the inherent impossibility of expressing in *language* what scientific research reveals.

Yet, as he goes on to argue, comprehensive representation, whether in language or image, is not the ultimate goal. Just as science requires "the exercise of the faculties of insightful perception of natural phenomena and of the imaginative creation of new concepts," neither of which, he argues, is an inherently verbal process, so, in his view, is painting closer to performance than to pronouncement. If painting does comment "on the world," it does so "not by logical or even visual analysis of it, but by a process of 'showing' similar to that which Wittgenstein claimed was the only way of exposing to view the most profound truths of philosophy."[61] Note how different this position is to his comment in the epilogue to the final Serbelloni Symposium, which falls in with the vogue for information theory so prevalent during the period of the first cybernetics: "To a biologist . . . language is a set of symbols, organized by some sort of generative grammar, which makes possible the conveyance of (more or less) precise commands for action to produce effects on the surroundings of the emitting and the recipient entities."[62]

What might be the shaping effect of mythology on the vision of the world in *Behind Appearance*? Mythic titles designate the three sections comprising the bulk of the volume: "Part One: The Binocular Cyclops," "Part Two: The Hybrid Argus," and "Part Three: Sucklings of Diana of the Ephesians." Waddington progresses from identifying the monstrously one-sided visions of the universe provided by science and painting from the time of Cubism on, to arguing that Science, like painting, is "no sort of a Cyclops— monocular or binocular—it is more of an Argus, with a hundred eyes."[63] But there is a deeper role to the mythology in Waddington's view of science in this volume, which we can see even in its introductory chapter, "The Image of Our Surroundings": it expresses a still dormant aspect of his understanding of development as involving a powerful maternal presence that shapes and nurtures all living beings. In hindsight, the Gaian undertones here are unmistakable.

We can sense this presence as the first chapter rounds to its conclusion. The rhetoric recalls and exceeds the idiosyncratic vision of Waddington's autobiographical Serbelloni presentation as it participates in the contemporary fascination with the theories of Melanie Klein, D. W. Winnicott, and other object-relations theorists. Interesting indeed to speculate whether Winnicott, innovator of the "squiggle technique" of free association, might lie behind not only Waddington's interest in sketching and painting, but also this deeply primary-process-laced vision of the world as a "good-enough" nurturing mother.

> Psycho-analysts argue that the basic impulse for such strivings is the desire to find our way back to the good breast from which we were perforce weaned. Perhaps the main conclusion this book will come to is that, at least in the fields of science and painting, there is no one good breast to be discovered or rediscovered. The world has too much to offer for us to take in . . . our terrestrial nourishments from any single source. We should be worshippers of the many-breasted Diana of the Ephesians—Diana Polymastigos.[64]

This reference to the ancient image of nature, the Mother of All Being, marks a decisive difference from the masculinist gender dynamics of the Serbelloni Symposia.[65] Waddington's Foreword to *Behind Appearance* suggests what may lie behind that different perspective. Although this work, like the Serbelloni Symposia, also emerged "from a whole host of conversations with my friends," Waddington expresses his greatest debt to one person as being particularly important to his work on art. That is Ruth Sager, "whose genuineness of response, coupled with a challenging capacity to expose the weaknesses of any argument or line of talk which didn't quite make the grade, did more than anything else to force me to find time to put this all down on paper."[66]

Sager was a plant geneticist and starting in 1966 a professor in the biology department at Hunter College (CUNY). Her work on cytoplasmic inheritance and chloroplast genetics would earn her election to the National Academy of Sciences in 1977, and in 1988, the Academy's Gilbert Morgan Smith medal.[67] More significantly still, that research would eventually lend support to Lynn Margulis's theory of the endosymbiotic origin of organelles: "If an organelle originated as a free-living cell, it is possible that naturally occurring counterparts can still be found among extant organisms."[68] Waddington's call out to Sager reveals the extent to which he exceeded the scope of his Serbelloni vision in this project, taking more seriously the notion of environmental influence, expanding the concept of development beyond the realm of genetic determinism to include the more ambiguous possibilities of cytoplasmic inheritance, and even opening himself up to the exploration of a systems-based approach at all scales.

Waddington develops this image and these concepts at greater length in "The Profits of Plurality," in "Part Three: Sucklings of Diana of the Ephesians." Leaving aside the distinctively sixties-era politics here, which advocate a shift from a philosophically unified system of hierarchical culture to "an Egalitarian Democracy of ideas and activities," the most notable aspect of this chapter is the model of scientific thinking it adapts, one that accommodates precisely the reflexivity he seems to have shied away from at Serbelloni.[69] With the distinction between observer and observed breaking down, he reiterates, scientists of this Third Science understand the ways they are part of, indeed implicated in, their experiments in ways never envisioned during the height of Second Science experimentalism.

Rather than being unsettled by the activity and potentially symbiotic engagements of the World Egg hatchling, with its Gnostic overtones, *Behind Appearance* describes this revelation as part of the "profits" associated with the new pluralistic worldview. As opposed to the stultifying perspective of Newtonian physics, which preserves a kind of "vacuous actuality," in its allegiance to an image of "unequivocal impenetrable atoms, each simply located at a precisely defined position in an unambiguous framework of space and time," this insight from biology draws together the arts, the sciences, and other human endeavors in "some type of 'organicist' view, which emphasizes the unity and reciprocal interaction between man and nature, and between the various aspects of human existence."[70] The vision fulfills the point he makes somewhat whimsically in the Preface to *Behind Appearance*: "Our picture of human nature must be in quite other dimensions if we consider that the basic structure even of the physical world is such that *everything is really everywhere, though in some places more than others.*"[71]

Waddington argues in the concluding chapter that language offers an insufficient entry point to the understanding of the universe shared by modern painting and modern physics. We must move beyond conceptual categories, including the literature/science divide, to realize the "essential artificiality" of disciplinary fragmentation. He reminds us approvingly of Aldous Huxley's critique of the narrowing effects of "book-learning" in contrast to performance. For Huxley, he points out, "training in the sciences is largely on the symbolic level; training in the liberal arts is wholly and all the time on that level. When courses in the humanities are used as the only antidote to too much science and technology, excessive specialization in one kind of symbolic education is being tempered by excessive specialization in another kind of symbolic education."[72] Just as no scientist can fully grasp "the full load of meaning" contained in a work of Jackson Pollack while looking at the work "through scientific eyes," the reverse is also true: "We have been led, by a consideration of one apparent discontinuity in human experience, that between painting and natural science, to recognize that there is continuity between them after all, and that this continuity extends out into wider fields."[73]

A SYSTEMS-BASED EPIGENETIC LANDSCAPE

How stable is Waddington's assertion that scientific paradigm shifts drive cultural transformations rather than vice versa? Far from it. I have been arguing that Waddington's interdisciplinary projects of the 1960s revealed his own conflicting allegiance to two different models for understanding the living world: the model of language ultimately affirmed in the Epilogue to the last Serbelloni Symposium and the model of painting with its spatial logic explored in *Behind Appearance*. Gilbert maintains that during this same decade he was also slowly abandoning the evocative image of the epigenetic landscape central to his attempt to synthesize embryology, genetics, and evolutionary theory.[74] Yet to view the epigenetic landscape as having been abandoned because "the channels and spheres had no physical reality" is to frame it in linguistic terms, as a metaphor inadequate because of the ambiguity of nonscientific language, and to disregard the Ouroborotic insight that what counts as reality is co-constituted by the observer.[75] Such a view underestimates the extent to which Waddington was able to keep multiple perspectives alive by moving in multiple disciplinary worlds. As we have seen, even as the appreciation of contingent and interconnected developmental systems occupied the "nonstandard" margins of theoretical biology at Serbelloni, Waddington was extending and elaborating it in *Behind Appearance*.

Waddington would return to the debate over appropriate models for biology in a volume he coedited with astrophysicist and forecaster Erich Jantsch, *Evolution and Consciousness: Human Systems in Transition*. There, in his "Concluding Remarks," he apologizes for coming late to the volume, which Jantsch had "assembled in typescript before I had got rid of some other commitments sufficiently to give much attention to it."[76] And in recompense he promises an assessment of "whether the kind of systems we are interested in can be adequately discussed in terms of information, and if not, what other terms would be more appropriate to the phenomenon of self transcendence." Returning to "a series of discussions on theoretical biology which I organized several years ago at the Villa Serbelloni in Italy," he recalls Howard Pattee's notion, "put most aphoristically . . . that we regard certain biological molecules as a message; that is to say, we consider them as conveying instructions of a kind comparable to the instructions which can be given in a natural language." Now, although he still feels "this is an extremely important manner of regarding living things and their evolution," he acknowledges that he feels "not at all clear" about "its biological basis and its consequences." What is the difference, he asks, between those instructions that amplify fluctuation and instability and those that tamp it down? Temporizing on the question of whether information serves as an adequate model, he returns to the philosophical perspective on life familiar from his Serbelloni talk, concluding: "We can perhaps leave it to the professional philosophers to decide what characteristics these sets must necessarily possess." And he joins his fellow contributors to *Evolution and Consciousness* in calling for "a conscious, devoted, but critical attempt to create new guiding images of man and the future."[77]

Recanting in *Evolution and Consciousness* his earlier affirmation of a language/metalanguage model, Waddington advocates instead the adoption of multiple viewpoints, or what Jantsch called the "symbiotization of heterogeneity."[78] More significantly still, in that later work he abandons the attempt to create a synthesis of embryology and genetics so evident at Serbelloni and earlier. Instead, he defiantly—and presciently—grounds his theoretical position in his work as an embryologist and then extends the insights won from that position to encompass a wider world. As an embryologist, he has "extensive experimental and theoretical acquaintance [with] . . . the analysis of embryonic development." Based on this professional knowledge, he maintains that "embryos, like ecosystems, are multifactorial."[79] By implication, we require a model that can incorporate the tensions and constraints on development embodied by both an embryo and an ecosystem.

What should that model be? Waddington's Serbelloni Symposium contributions to the project of Theoretical Biology understandably reflected the deterministic and eukaryotic focus of mid- to late-twentieth-century biology. In their metaphysical forms as the World Egg and the Ouroboros, the nucleated cell and the cybernetic principle linking "the one" to "the all" were central to Waddington's biological research. Yet inspired by the work of Ruth Sager on cytoplasmic inheritance and chloroplast genetics, in *Behind Appearance* Waddington also reached beyond those concepts to explore what he called "a new 'landscape of thought,' a new climate of form." He observed that what biology studied as systemic behavior, rather than being called "developmental" (a word frequently used by physicists), should really be called "epigenetic." And he emphasized the predictive power of epigenetics: "There is no reason in principle why epigenetic processes should not be completely deterministic, or why chance should play any role in them. In fact, it is doubtful if it does play any important part in most biological epigenetic systems, such as those presented by developing eggs."[80]

The qualifications in this statement—"there is no reason in principle" and "it is doubtful"—suggest that once again Waddington gave himself an opening to think otherwise, to return to the full possibilities of the epigenetic landscape as contained in its very first image, the painting of his friend John Piper and its curious caption, "as the river flows away into the mountains."[81] As his exploration of modern art in *Behind Appearance* may have revealed, rivers *can* reverse themselves, and run uphill. Development is neither as unidirectional nor as deterministic as that ball drawn down the channels by gravity might initially have suggested. In contemporary epigenetics and molecular embryology, the epigenetic landscape model continues to be relevant, both as an epistemological space where scientific hypotheses can be laid out and worked through visually and as a performance space where the nonlinear relations of development can be modeled.[82] Contemporary revisions of the epigenetic landscape incorporate new discoveries in stem cell science. To take one crucial example, applying techniques used in molecular and developmental studies of slime molds (*Dictysotelium discoideum*) and bacteria (*E. coli*), molecular embryologist M. Monk detailed the mechanisms of deprogramming and reprogramming that returned cells to their totipotent state, the nature of genomic imprinting, and the processes of transgenerational epigenetic inheritance.[83] Contemporary science weaves links between Waddington's vision of the World Egg and the Gaian insights of Lynn Margulis celebrated in this volume.

6.

THE PLANETARY IMAGINARY

Gaian Ecologies from Dune *to* Neuromancer

BRUCE CLARKE

The space-borne photographs of Earth that first arrive in the mid-1960s prompt an epochal reimagining of our planet as a whole system.[1] Astronautic, computational, and communicational technologies combine to mediate and publicize unprecedented images of the planet, spurring further speculations regarding unseen, previously unimagined processes of the Earth system. As with the *Whole Earth Catalog*, which begins in 1968 with a space-borne Earth image on its outside front cover, recognizing the Earth as a whole system also calls forth a multitude of creative technologies to send that message into social communication, individual interrogation, and cultural application. The suite of newly mediated images of the Earth seen from space form a new planetary imaginary, one that renovates intuitions of the actual Earth in its complex operations or that inspires new fictive worlds to refract those processes for us.[2]

Nonetheless, even space-borne photographs have no way of showing, for instance, the Earth's carbon cycle in its total systemic operation as a planetary phenomenon. One must still visualize that for oneself, study or construct a diagram or a narrative for it.[3] If one aims to hew to an evidentiary replica of the carbon cycle's dynamics, ecosystem ecology or Earth system science provide the data to do so, but in any event, that construction will also participate in the planetary imaginary of the moment. As with a work of art, such as a narrative fiction—say, a science-fiction novel—the planetary imaginary renders the mundane more complex. It communicates otherwise unknown objects. Such constructions may actualize something potential but until then nonexistent, such as a previously unrecognized Earth process. They may concoct a picture of something long past—say, the wild proliferation of microbial life in the Archean eon—or forecast something not yet extant—say, the coming evolution and passing establishment of new ecologies. Or they may offer sketches of things whose full reality cannot be contained in a single view or grasped at any single scale—for instance, the biosphere altogether reconceived as an interlocking complex of self-regulating geobiological systems.

The planetary images arising in the 1960s render the state and dynamics of the actual Earth and its observation in detailed descriptions that possess both technoscientific heft and speculative or heuristic potential. These newly self-referential views and visions of the Earth continue to frame our efforts to rethink the human in relation to the biosphere.[4] This essay explores the planetary imaginary of our era through Gregory Bateson's ecological writings and the Gaia discourses of James Lovelock, Lynn Margulis, and Dorion Sagan. The planetary imaginary of any era is also an abiding creative resource that constitutes itself whenever an artist invents and communicates fictive images of living worlds—perhaps, also, of the cosmos that contains those imagined planets, or of the ecologies they sustain—and bodies these forth in some workable medium. The critical reading of such images may decipher them as conveying and perhaps clarifying discernible Earthly connections or plausible worldly value commitments. Staying with the culture of the 1960s and its close aftermaths, I will discuss the novels *Dune* and *Neuromancer* in this light.

The *planetary* imaginary differs from the *global* imaginary. Barring outright fantasy, a planet will be conceived as a substantial cosmological or ecological phenomenon. If it possesses life, a planet potentially bears processes capable of constituting its own observers. In contrast, a globe is literally a hollow sphere, the model and representation of a planet's surface. The map is not the territory.[5] And a globe's surface typically presents demarcations corresponding more to parochial human interests than to natural features. For instance, near the beginning of Frank Herbert's classic science-fiction novel *Dune*, the villain of the piece, Baron Vladimir Harkonnen, is introduced first behind and then next to a massive globe.

It was a relief globe of a world, partly in shadows, spinning under the impetus of a fat hand that glittered with rings. . . . The fat hand descended onto the globe, stopped the spinning. . . . It was the kind of globe made for wealthy collectors or planetary governors of the Empire. . . . Latitude and longitude lines were laid in with hair-fine platinum wire. The polar caps were insets of finest cloudmilk diamonds.

The fat hand moved, tracing details on the surface. "I invite you to observe," the basso voice rumbled. . . . "Nowhere do you see blue of lakes or rivers or seas. And these lovely polar caps—so small. Could anyone mistake this place? Arrakis! Truly unique. A superb setting for a unique victory."[6]

This oblique view of the planet Dune presents it, so to speak, in luxurious effigy. Expressing a feudal will to exploit Arrakis for political and economic gain and personal glory, the Baron can be said to voice a global imaginary, one that responds not to planetary matters or processes and their interrelations but to maps, grids, and lines of control.[7] In contrast, the planetary imaginary corresponds to a necessarily partial observation of whole and interconnected territories. The former is political and economic, all too human; the latter is ecological and geocentric, more than human. And whereas a globe can be spun so as to have no dark side, no shadow of the unknowable, at least within its limited notion of the known world, an actual planet such as Earth in the full intricacy of its biosphere cannot really be spun about and controlled. It will surpass our technological as well as our epistemological grasp. As a speculative faculty, then, the planetary imaginary is a significant mode of human self-observation that undercuts or goes beyond its own capacity to observe, and through which, for that very reason, the theoretical construction of a planetary body for artistic, discursive, or scientific ends decenters the human in relation to its worldly situation.

Moreover, just as the novel *Dune* depicts both global conflicts and planetary coordinations, the modern era of world history is one not just of globalization, but also of planetization. Considered as an epistemological stance, the globalizing process operates through an instrumental humanism driving commodity extraction amidst political and economic consolidations, negotiating corporate and cultural differences and the mobility of their interrelations relative to a humanity splintered into feudal or ideological tribes. In contrast, I would describe planetization as the effort to readjust the view of the human toward its common relation to the Earth as an ultimately incalculable and uncontrollable system. A philosophy adequate to planetization is still under construction, but some promising outlines for it are being set by meditations upon ecology and systems theory, for instance, by Edgar Morin, Michel Serres, and Peter Sloterdijk.[8] For another instance, through his own writings and direction of symposia sponsored by the Lindisfarne Association since the 1970s, cultural historian William Irwin Thompson has been leading important conversations about the formation of a "planetary culture." Connecting thinkers in science, art, and spiritual traditions, these meetings and their subsequent publications have brokered significant personal and professional connections among a number of broad-minded natural and social scientists, from Gregory Bateson, Heinz von Foerster, Henri Atlan, and John Todd to Humberto Maturana, Francisco Varela, Lynn Margulis, James Lovelock, and Susan Oyama.[9]

As planetization emerges in the work of thinkers such as these, various strands of systems theory intertwine with varieties of ecological thought to theorize the embeddedness yet noncentrality of the human within a wider natural or cosmological scheme. For instance, Gregory Bateson's *Steps to an Ecology of Mind* has been a seminal manifesto for such a systemic viewpoint. It concedes its status as metaphysics: "The cybernetic epistemology which I have offered you would suggest a new approach. The individual mind is immanent but not only in the body. It is immanent also in pathways and messages outside the body; and there is a larger Mind of which the individual mind is only a subsystem. This larger Mind is comparable to God and is perhaps what some people mean by 'God,' but it is still immanent in the total interconnected social system and planetary ecology."[10] Bateson develops a conceptual shift, starting from ecology as a natural-scientific metadiscipline on a par with cybernetics and specifically focused on the interrelations of life and environment, and moving to ecology as a mobile figure for any situation of systemic complexity and interdependent self-maintenance.

Dune offers a concurrent and comparable example of this philosophical turn. One of the first great American ecological novels, serialized starting in 1963, published entire in 1965, thus arriving in the vanguard of 1960s counterculture, Herbert's fiction presents mind expansion and alternative communities in a context of planetary environmental concerns. Both *Steps to an Ecology of Mind* and *Dune* participate in, as well as further promote and refine, a larger body of ecological discourse that comes into specific conversation with the cybernetics that first emerges in the later 1940s and gathers intellectual and cultural momentum coming into the 1960s.[11] As both are being written, the wider cultural reception of cybernetic discourses—captured in phrases and concepts such as whole systems, self-organizing systems, informatics, computation, communication, artificial intelligence, noise, entropy, and synergy—is reaching a critical mass. In their own ways, both *Dune* and *Steps* mark two highpoints of this particular cultural crest. Concurrently, thinkers such as those associated with Lindisfarne are crossing over from mainstream scientific agendas to the systems counterculture, or in Bateson's phrase, "the new epistemology which comes out of systems theory and ecology."[12]

The discourses of cybernetics and ecology combine to form ecosystem ecology, another key conceptual incubator for the planetary imaginary that first crests in the 1960s. Historian Joel B. Hagen notes that the "ecosystem is an intuitively appealing concept for most ecologists, even for those critical of the way ecosystem ecology has developed as a specialty. It is the only ecological concept that explicitly combines biotic and abiotic factors and places them on roughly equal footing."[13] In *Symbiotic Planet*, Lynn Margulis gives an economical reformulation of this incisive ecosystemic conception of the relation between abiotic and biotic elements, physical flow and biological cycle: "Sunlight moves through life, empowering cyclic work."[14] Ecosystem ecology's physicalist orientation to "abiotic factors" in relation to living systems is also depicted in *Dune*, through an appendix to the narrative proper, "The Ecology of Dune," which contributes the backstory of the first imperial planetologist of Arrakis, Pardot Kynes: "To Pardot Kynes, the planet was merely an expression of energy, a machine being driven by the sun."[15] In both scientific and fictive instances, the planetary imaginary frames a cosmology that connects solar energy to terrestrial life through the material cycles of its planetary environment.

In its global dimension, *Dune* depicts the machinations of royal families wrangling over the control and exploitation of melange—the precious spice as indispensable to this storyworld as coffee, tobacco, and alkaloids of coca are to ours, and to be mined only from the sands of Arrakis.[16] The novel's planetary imaginary emerges alongside these tribal elements. Its ecological themes are immediately tangled up with both. Just as the Harkonnens vie to control the spice trade, the will of the ecological mentors of the indigenous Fremen is focused on control of the planet's ecology. These crosscurrents dovetail upon the protagonist Paul Atreides, who nonetheless struggles to achieve and sustain a planetary rather than a global attitude. He does so, it may be said, with the aid of Dune's unique pharmacopeia, the mind-altering spice that blows in the planetary wind, and the Water of Life drawn from the Fremen symbiosis with the spice-producing sandworms. That is, while the planet *Dune*'s global commodification is a scene of multiple contestations, Paul's visionary career registers its planetary imaginary as the increment of racial and ecological mystery that is playing itself out in his person.

However, one of the most important scenes of self-realization in the story is not psychedelic at all, but rather, a sober moment registering connections among the mundane details of Fremen social ecology. At one point, Paul wins a duel of honor and with that, his defeated Fremen challenger's wife, Harah, for a consort. Harah then shows the newcomer Paul through the sietch or cavern commune and environs of their underground society. As they pass along the corridors of the sietch, Harah has occasion to describe Fremen lifeways. These have already incorporated techniques imported to the naturalized inhabitants of Arrakis by the imperial planetologists Pardot and Liet Kynes. First the father and then the son have crossed over to the side of the Fremen and guided their society. Bringing first-world scientific and technological knowledge to their adopted native culture, they have also infused the Fremen with an ecological prophecy, the vision of a desert world rendered verdant by proper and deliberate environmental management. As Paul and Harah walk along, she informs him about the dew collectors: "Each bush, each weed . . . is planted most tenderly in its own little pit. The pits are filled with smooth ovals of chromoplastic. . . . It cools with extreme rapidity. The surface condenses moisture out of the air. That moisture trickles down to keep our plants alive."[17] When they pass a classroom, Harah explains how, now that both of them are dead, the Kynes's ecological teachings are being kept alive by these lessons: " 'Tree,' the children chanted. 'Tree, grass, dune, wind, mountain, hill, fire, lightning, rock, rocks, dust, sand, heat, shelter, heat, full, winter, cold, empty, erosion, summer, cavern, day, tension, moon, night, caprock, sandtide, slope, planting, binder.' " Eventually they reach Harah's living space, and as Paul hesitates at her threshold, he has an encompassing thought: "It came to him that he was surrounded by a way of life that could only be understood by postulating an ecology of ideas and values."[18]

The story has already established that the young nobleman has received ecological training on a par with the Kynes's science. So the implication is that, in this thought "postulating an ecology of ideas and values," Paul has added together his comprehension of the ecological significance of the dew catchers with the way that "sandtide, slope, planting, binder" concern the ongoing Fremen efforts to control the wayward tides of the dunes. Both are part of a wider plan to shape their total environment. What Paul seems to realize is that the natural environment and the engineering techniques that would cultivate and control it, together with its inhabitants and the pedagogical program to inculcate them with an ecological sensibility that comprehends the Kynes's environmental prophecy, are all mutual elements within an encompassing social and spiritual system.[19]

With the phrase "an ecology of ideas and values," and in this phase of the novel, *Dune* has already climbed up a number of Bateson's "steps to an ecology of mind." Its fore-echo in *Dune* suggests that the ecological sensibility Bateson hopes to cultivate in his later writings is itself a discursive variant of a planetary imaginary such as that Herbert exercises in a mode of fictional narrative. In the front matter of *Steps to an Ecology of Mind*, Bateson states: "Broadly, I have been concerned with four sorts of subject matter: anthropology, psychiatry, biological evolution and genetics, and the new epistemology which comes out of systems theory and ecology."[20] Ecology is treated here as a variety of the discourse of cybernetic systems theory broadly construed. As such, it can partake of the strategic generality of cybernetic formulations of information and communication. In particular, it can pass beyond strictly geobiological and evolutionary reference and provide the natural-scientific foundation and systemic occasion for the immanentist metaphysics of his ecology of mind. The later papers of *Steps*, especially those clustered at the end, all written between 1967 and 1971, are largely and explicitly engaged in the project of redescribing and repositioning human culture and communication as part of a mental or informational ecology that is planetary in scope.

Bateson proposes "a new way of thinking about ideas and about those aggregates of ideas which I call 'minds.' This way of thinking I call the 'ecology of mind,' or the ecology of ideas." His conceptual approach is not just new, he states further, but necessary, for it makes possible an otherwise unobtainable way of understanding the manifold complexities of worldly relations extending from natural evolutionary forms to human cultural behaviors: "Such matters as the bilateral symmetry of an animal, the patterned arrangement of leaves in a plant, the escalation of an armaments race, the processes of courtship, the nature of play, the grammar of a sentence, the mystery of biological evolution, and the contemporary crises in man's relationship to his environment, can only be understood in terms of such an ecology of ideas as I propose."[21] It is a remarkable circumstance and striking resonance that, as composed in 1971 and first published in 1972, Bateson's ecophilosophical declaration in this passage repeats nearly verbatim the passage we just looked at from *Dune*.

So, on one hand, for Bateson in 1971, the comprehensive interrelatedness of planetary matters could "only be understood in terms of such an ecology of ideas as I propose." And, on the other, for Herbert, who composes this piece of text for his hero Paul sometime before 1965, the current, renovated culture of the Fremen "could only be understood by postulating an ecology of ideas and values," period. Now, Bateson's every use of the term *ecology* documented in *Steps* occurs in an article postdating *Dune's* publication.[22] I would explain this mutual resonance and wide overlap in literary and philosophical approaches as concurrent responses to the same cultural paradigm that by the 1960s names itself, simply, *ecology*. Both authors receive the American inflection on the twentieth-century shift in the discipline of ecology from a descriptive form of natural history to a theoretical systems science. Positioned outside the science of ecology proper, both are free to move its systemic descriptions of humanity's wider interrelations with its worldly environments toward their cultural implications. Or again, the planetary imaginary of both authors belongs to the same *ecology of ecology*. For instance, both draw from the particular line of ecosystem ecology. "The Ecology of Dune" makes this disciplinary identification explicit through one of Pardot Kynes's exhortations: "'The thing the ecologically illiterate don't realize about an ecosystem,' Kynes said, 'is that it's a system. A system! A system maintains a certain fluid stability that can be destroyed by a misstep in just one niche.'"[23]

In such a passage, the planetary imaginary takes the form of a pedagogical exhortation to maintain an encompassing ecological vision. In light of these fictionalized problematics of scientific literacy, the final paper in

Steps, "Ecology and Flexibility in Urban Civilization," contains a closing section, "The Transmission of Theory," that offers another remarkable echo of *Dune*. In this striking meditation, Bateson notes that a "first question in all application of theory to human problems concerns the education of those who are to carry out the plans."[24] We saw the narrative of *Dune* depict a comparable concern when Paul, being guided by Harah, was first made cognizant of the Fremen's ecological planning, and then encountered the chanting children, a scene from the pedagogical regime designed to imprint on its future executors Pardot and Liet Kynes's plan for eco-engineering the environment of Dune. Bateson connects a related issue of ecological citizenship to the insight that the larger system constituted by the human social organism *plus* its worldly environment is traversed by mental and informational as well as physical and geobiological circuits. Regarding this systemic description, he asks:

> Is it important that the right things be done for the right reasons? Is it necessary that those who revise and carry out plans should understand the ecological insights which guided the planners? . . . The question is not only ethical in the conventional sense, it is also an ecological question. The means by which one man influences another are a part of the ecology of ideas in their relationship, and part of the larger ecological system within which that relationship exists.[25]

By such reasoning, *Dune* would also be a systemic element within the ecology of ideas that yields Bateson's *Steps*, presenting a fictionalized ecophilosophy that resonates with the sort of holistic ecosystemic ensemble to which Bateson gives expression. In "The Ecology of Dune," we read further:

> "There's an internally recognized beauty of motion and balance on any man-healthy planet," Kynes said. "You see in this beauty a dynamic stabilizing effect essential to all life. Its aim is simple: to maintain and produce coordinated patterns of greater and greater diversity. Life improves the closed system's capacity to sustain life. Life—all life—is in the service of Life. Necessary nutrients are made available to life by life in greater and greater richness as the diversity of life increases. The entire landscape comes alive, filled with relationships and relationships within relationships."[26]

But whereas Kynes's fictive discourse may be taken to represent ecology proper and so to be contained within a geobiological frame, Bateson's ecology of ideas extends the concept beyond its home reference just to the natural world. His reinterpretation of ecology as a systemic epistemology allows for variety regarding, one, the level at which one draws the boundaries of the system one intends to observe against, two, the environment constituted by that distinction. In this way, a planetary system composed from the cooperation of life *and* its ecological environments may be enlarged, in cosmological perspective, to encompass immanent Mind as well: "I now localize something which I am calling 'Mind' immanent in the large biological system—the ecosystem. Or, if I draw the system boundaries at a different level, then mind is immanent in the total evolutionary structure."[27]

This is the planetary imaginary doing philosophy, turning upon ecology to make the conceptual shift by which "the large biological system— the ecosystem" may be taken to include as well those metabiotic matters—the mediations of consciousness and communication—for which life per se is the precondition if not precisely the operation. But granting this interdependent nesting of systemic conditions, Bateson presses its implications for cultural evolution to some profound conclusions. For one, ecological mind is conditioned by the body of ecology about which it reasons: "the problem of how to transmit our ecological reasoning to those whom we wish to influence in what seems to us to be an ecologically 'good' direction is itself an ecological problem. We are not outside the ecology for which we plan—we are always and inevitably a part of it."[28] Paralleling the second-order cybernetics concurrently in embryo at Heinz von Foerster's Biological Computer Laboratory, this is a classic self-referential insight placing the observer within the system constituted by their observation.[29] Bateson does not minimize the vertiginous quality of this realization: "Herein lies the charm and the terror of ecology—that the ideas of this science are irreversibly becoming a part of our own ecosocial system."[30] However, having realized that we are inescapably bound to the Earth's evolving ecological conditions, as we negotiate our own trepidation, it is bracing to note how Bateson concludes with a sublime ethical edict not to get hung up on hasty activisms to the point of losing touch with the primary vision of planetary participation: "If this estimate is correct, then the ecological ideas implicit in our plans are more important than the plans themselves."[31] Whereas cultural strategies need to adapt to fluctuating social circumstances, the higher wisdom to be won from the hard science is what will continue to count.

The same could be said about Gaia theory: the best planetary ideas are more important than the cultural programs that come and go in their name. Hagen recounts Gaia's congruence with ecosystem ecology despite its emergence outside of ecology proper: "The idea that the biosphere is a homeostatic or cybernetic system of living and nonliving components was a central feature of the ecosystem concept. . . . Although Lovelock rarely used the term ecosystem, many of his ideas meshed perfectly with [a] broad systems approach to ecology."[32] Lovelock had formulated the Gaia hypothesis before teaming up with Margulis in 1971.[33] But their collaborative description of Gaia, joining atmospheric chemistry to evolutionary microbiology, environment to life, placed it on the most fundamental biological basis of all, the planetary grip of the microbes. Unlike this or that eukaryotic species, the unspecific bacteria are everywhere all at once, perfuse the biosphere from top to bottom. The microbial microcosm is the most planetary component of any life on the planet.[34] In the phylogeny that follows from Margulis's serial endosymbiosis theory, symbiogenesis, or permanent mergers among the microbes, determine the basic framework of evolutionary relationships.[35] Indeed, as she writes in one of the most poetic statements in *Symbiotic Planet*: "Symbiogenesis was the moon that pulled the tide of life from its oceanic depths to dry land and up into the air."[36] By such Gaian interactions, the domains of life have unfolded into five kingdoms interrelated by their origins in microbial symbiogeneses. And when placed into planetary view, this phylogeny yields not a branching tree but a reticulated web, woven from strands composed by the prokaryotes, of interactive life coevolving with its abiotic and postbiotic environments. "Planetary physiology—Gaia . . . is symbiosis seen from space."[37]

Gaia theory in its current scientific understanding couples Margulis's description of symbiogenetic dynamics to the evolution and maintenance of the biogeochemical cycles arising out of and returning to a modulated planetary environment. But this redescription of the Gaian system means that evolutionary biology proper, considered as a discipline tightly bound to life per se, is overly restrictive to a subsystem always already contingent upon a larger metasystem. The coevolution of life altogether with the Earth as a whole, networked by the planetary mesh of the microbes, comes together in Gaia's own evolution, of which each succeeding viable variation is a self-similar image. As near as I can see, mainstream science has yet to wrap its head fully around this reticulated conceptual sphere of Gaian coevolution. For instance, providing a quick fix on mainstream disciplinary coordinates, my weekly e-alert from Nature.com continues to separate "Biological sciences" from "Earth & Environmental sciences," and Gaia still falls between the cracks. In contrast, Gaia theory as Margulis formulates it in *Symbiotic Planet* is thoroughly ecosystemic in its interpenetration of life and its planetary and cosmic environments: "The sum of planetary life . . . is not an organism directly selected among many. It is an emergent property of interaction among organisms, the spherical planet on which they reside, and an energy source, the sun."[38]

While the associations of the microbes have certainly driven the major evolutionary developments and continue as always to prop up the rest of the biosphere, then, life itself is not the whole story. When Gaia began, and for over a billion years afterwards, a strictly bacterial biosphere coevolved with its previously nonbiotic yet increasingly postbiotic milieu. Three billion years later, the forms of living organisms continue to coevolve along with and in constitutive relation to both the abiotic and the biogenic factors of their environments, what Tyler Volk has felicitously named the "gaian matrixes" of air, ocean, and soil—the geobiological media of the biosphere's compounded processes.[39] Gaia evolves along with and in constitutive relation to both its biotic and its metabiotic productions. Lovelock has recently restated this extended Gaian perspective: "When Darwin came upon the concept of evolution by natural selection he was almost wholly unaware that much of the environment, especially the atmosphere, was a direct product of living organisms. Had he been aware I think he would have realized that organisms and their environment form a coupled system . . . *what evolved was this system, the one that we call Gaia.* Organisms and their environment do not evolve separately."[40] This theoretical view goes beyond the constructions of the earlier Gaia hypothesis. Lovelock's comprehensive rethinking of Gaia as an evolving metasystem, first codified in *The Ages of Gaia* of 1988, pivots on a prescient passage from Alfred Lotka's 1925 text *Elements of Physical Biology.* On October 22, 1986, Lovelock writes Margulis about his discovery of this Gaian precursor: "I have been reading Alfred Lotka's beautiful book 'Physical Biology' . . . A photocopy of page 16 is enclosed. He said it all."[41]

Indeed, Lotka strongly anticipates the complex holism or ecosystemic totality of the Earth system as now seen by Gaia theory: "The several organisms that make up the earth's living population, together with their environment, constitute one system, which receives a daily supply of available energy from the sun." To this sentence Lotka appends a footnote contra reductionism. Lovelock and Margulis will go on to reclaim the planetary implications of his systemic insights, envisioned well before their own planetary time, from the parenthetical status to which Lotka consigns them in his text:

> This fact deserves emphasis. It is customary to discuss the "evolution of a species of organisms." . . . We should constantly take in view the evolution, as a whole, of the system [organism plus environment]. It may appear at first sight as if this should prove a more complicated problem than the consideration of the evolution of a part only of the system. But . . . the physical laws governing evolution in all probability take on a simpler form when referred to the system as a whole than to any portion thereof. It is not so much the organism or the species that evolves, but the entire system, species and environment. The two are inseparable.[42]

THE PLANETARY IMAGINARY OF *NEUROMANCER*

Forms of life from bacterial colonies to beavers, termites to technological humanity, are also systemically bound to their *built* environments. And once Gaia locks its planetary processes into place, all Earthly environments are "built up" due to their henceforth incessant cycling together with living processes. Taking a Gaian evolutionary view, Sagan and Margulis draw an enlarged concept of technics into the planetary ecology: "The machinate world that appears so new and unprecedented, so quintessentially and exclusively *H. sapiens*', is really not that at all. . . . These new human-fostered technologies are in a direct line with the old."[43] By "old" technologies they mean the ancient propensity of living systems to evolve by incorporating and repurposing abiotic components or even organic excrement (such as the exuded calcium out of which vertebrates fashioned their bones and teeth) in their environments: "All arose from precedents—prehuman precedents—in an evolutionary and ecological context."[44] And in "Gaia and Philosophy" they argue that one of the pre-eminent prehuman precedents of human technologies is Gaia itself, understood cybernetically as a natural technology for planetary homeostasis: "On the one hand, Gaia was an early and crucial development in the history of life's evolution. Without the Gaian environmental modulating system, life probably would not have persisted. . . . On the other hand, the full scientific exploration of Gaian control mechanisms is probably the surest single road leading to the successful implementation of self-supporting living habitats in space."[45]

These Gaian reflections of Margulis and Sagan can help us draw out the planetary imaginary of a novel that is not often consulted for its ecological witness, William Gibson's *Neuromancer*.[46] Gibson famously names the virtual world within his near-future storyworld *cyberspace*, placing it alongside his noir vision of a disnatured natural world in which biological bodies are in submission to technological manipulations and digital prosthetics. Gibson's brilliant way of playing cyberspace straight, coupling it to a material Earth and rendering it available for digital construction, exploration, and exploitation—all this sets the paradigm for the cyberpunk manner in subsequent fiction and cinema. When, after a season of forced exile, Gibson's protagonist Case jacks into cyberspace, he sees

> in the bloodlit dark behind his eyes, silver phosphenes boiling in from the edge of space, hypnagogic images jerking past like film compiled from random frames. Symbols, figures, faces, a blurred, fragmented mandala of visual information . . . flowered for him, fluid neon origami trick, the unfolding of his distanceless home, his country, transparent 3D chessboard extending to infinity.[47]

This is the environment that readers of *Neuromancer* tend to fixate on, an exclusively optical expanse, "bright lattices of logic" seemingly detached from planetary contingencies. But on occasion the text throws cyberspace into relief by referring back to its flip side, its embodied preconditions. Following the passage just cited, the next sentence reads: "And somewhere he was laughing, in a white-painted loft, distant fingers caressing the deck, tears of release streaking his face."[48] Between the digital virtuality of cyberspace and the actual place where Case is bodily present, we have the two ostensible ecologies of *Neuromancer*. One is the virtual environment we have just surveyed, a domain for the mind with a variable and malleable ecology of digital entities. The other is the degraded environment of the outer storyworld.

What green metaphors come forward are flipped on their heads, as here, in the single appearance of the word "ecology" in the text: "The dubious niche Case had carved for himself in the criminal ecology of Night City had been cut out with lies, scooped out a night at a time with betrayal," and here, with "the mall crowds swaying like windblown grass, a field of flesh shot through with sudden eddies of need and gratification." In ironic homage to Paolo Soleri's once hopeful projects, cityscapes are "dominated by the vast cubes of corporate arcologies."[49] "Arcology" is Soleri's own contraction of "architecture" and "ecology," coined to name his earnest efforts since the 1960s to retool urban design. The text deploys the word to gain its tone but travesties its sense. In a similar fashion, Buckminster Fuller's retro-futuristic architectural signifier appears in "towers and ragged Fuller domes," as when Case remembers from his adolescence "the rose glow of the dawn geodesics."[50] We are to think that the cities absorbed into the Sprawl have largely been encased in geodesic domes against a noxious outer world. *Neuromancer*'s occasional open vistas, redolent of "blasted industrial moonscape . . . broken slag and the rusting shells of refineries" reprise the entropic horizons of J. G. Ballard and Philip K. Dick stories in the 1950s and '60s. Mentioned once in passing are a "pandemic" rendering horses and presumably other large animals extinct, and "the rubble rings that fringe the radioactive core of old Bonn," placing the Earthbound setting of the novel in the aftermath of a reactor meltdown or limited nuclear exchange.[51]

However, beyond the initial opposition of these inner and outer storyworlds—one beckoning, the other broken—*Neuromancer* does travel to something like another planet altogether, when the storyworld shifts from the Earthbound settings of the earlier action to the high-orbital space colonies where the main and climactic events will be staged. Gibson brilliantly imagines these orbital settings and provides some significant accounting for their operation. But residing somewhat in the glare and shadow of the wider celebration and exaltation of cyberspace, *Neuromancer*'s high-orbital environments tend to slip by without special notice. In fact, they have a remarkable, specific, but largely unrecognized provenance, one that tethers *Neuromancer* back to a fascinating chapter of American technoculture, as well as to the milieu of the earliest general publication of the Gaia hypothesis.[52] Following Margulis and Sagan, we will see how the planetary imaginary in Gibson returns in contemplating the artificial Gaian systems of his high-orbital space colonies.

Part III of the novel begins with the following incantation:

Archipelago.

The islands. Torus, spindle, cluster. Human DNA spreading out from gravity's steep well like an oilslick.

Call up a graphics display that grossly simplifies the exchange of data in the L-5 archipelago. One segment clicks in as red solid, a massive rectangle dominating your screen.

Freeside.[53]

The L of "L-5," it turns out, is for Joseph-Louis Lagrange, the French mathematician and astronomer, and L5 is one of several Lagrangian libration points, prime locations in space where the relative gravities of the two-body system of Earth and Moon balance such that an object placed in orbit there, near the line of the Moon's own path around the Earth, will not decay but be stable indefinitely. All the satellites, shuttles, and space labs we are familiar with, the ones that fall back to Earth on occasion, are in low orbit. An object at L5 is in high orbit.

The Fall 1975 issue of *CoEvolution Quarterly* (*CQ*) devotes its first thirty pages to the serious proposals for space colonies then being put forward by Princeton physicist Gerard O'Neill. Now that the Apollo space program has wound down, a "high frontier" is ready to go beyond JFK's "new frontier." O'Neill has been at this project for several years and has already received modest support from NASA. *CQ* provides a long transcript of his recent testimony before a congressional subcommittee seeking long-term NASA commitment to his blueprints for space colonies to be positioned at L5. In its Spring 1976 number, *CQ* publishes eighty pages detailing the controversy that erupts among its readers over its positive presentation of O'Neill's vision as a potential form of countercultural commune in the sky. Stewart Brand—the creator and editor of the *Whole Earth Catalog*, of which *CQ* is a major spin-off—works the space-colony debate for all it's worth, pitting postpsychedelic space-oriented technophiles such as himself against Whole Earth–identifying environmentalists and green technophobes. Brand publishes the responses of a vast roster of supporters and detractors, including novelists Ken Kesey and Wendell Berry, the poets Gary Snyder and Richard Brautigan, astronaut Rusty Schweickart, cultural observer William Irwin Thompson, scientists Lynn Margulis, Paul and Anne Ehrlich, and Carl Sagan, and also some notables with cameo roles in *Neuromancer*, Buckminster Fuller and Paolo Soleri. Then in 1977, Brand gathers, republishes, and expands these materials as the freestanding paperback volume *Space Colonies*.[54] If one looks carefully through these pages with *Neuromancer* in mind, the conclusion in inescapable. High orbit, L5, torus, spindle, cluster—it's all here, served up on a hip countercultural platter from which Gibson has liberally helped himself.[55]

For instance, take the narrator's statement, "Human DNA spreading out from gravity's steep well like an oilslick." The spilling of human DNA into a cosmic oilslick is a great image whose originality and perversity, as far as I know, is all Gibson's. The equally striking phrasing "gravity's deep well" is Gibson's as well, but the idea is O'Neill's. In his interview with Brand under the title, "Is the surface of a planet really the right place for an expanding technological civilization?" O'Neill warms to his answer, which is—no, it is not. Here is a curious sort of planetary imaginary in reverse. Why, O'Neill asks, just when we have emerged from Earth's gravity hole, should we want to drop down another one?

> The classic science fiction idea of colonization is always you go off and you find another planetary surface, like the moon or Mars. . . . They're the wrong distance from the sun. . . . The sort of analogy I like to use nowadays is to say that, "Here we are at the bottom of a hole which is 4,000 miles deep. We're a little bit like an animal who lives down at the bottom of a hole. And one day he climbs up to the top of the hole, and he gets out, and here's all the green grass and the flowers and the sunshine coming down. And he goes around and it's all very lovely, and then he finds another hole, and he crawls down to the bottom of that hole." And if we go off and try to get serious about colonizing other planetary surfaces, we're really doing just that.[56]

In other words, gravity has been holding us down upon the surface of Earth, but we can now escape it by going into space. Free space is not only there for free, but it is also free of gravity, which we can turn to our advantage in an artificial high-orbit environment. In his article "The High

Frontier" that opens the *Space Colonies* volume, O'Neill declares that "the L5 Earth-Moon Lagrange libration point . . . could be a far more attractive environment for living than most of the world's population now experiences."[57] But, he cautions, we lose that advantage if we colonize the surface of the Moon or Mars and so capitulate to another gravity hole. It is striking how O'Neill denigrates life on Earth as occupying the bottom of a dark hole in contrast with life in a high-orbit space colony as a virginal pastoral Eden, an idealized Earth with "the green grass and the flowers and the sunshine coming down." I take it that O'Neill is not aware of the ironies that attend this turn of phrase in his argument by analogy, but that would simply underscore the larger problems with his scheme, which are abundantly documented in the critical commentaries Brand also publishes alongside O'Neill. It is simply that the knowledge needed to engineer materially closed ecologies making the proposed space colonies permanently habitable does not yet exist, not then, not even now. Their viability rests on the successful manufacture of massive self-recycling ecosystems fit for large populations. While, as Margulis and Sagan speculate in 1984, "the full scientific exploration of Gaian control mechanisms is probably the surest single road leading to the successful implementation of self-supporting living habitats in space," in 1975 O'Neill simply takes their availability for granted. This issue, while also largely scanted in *Neuromancer*, does eventually peer out from its later pages.

Let us blast off for a moment to enjoy the glorious fantasy O'Neill sets forth for his space colonies with the graphic assistance of NASA's artists, since these images cycle directly into the high-orbit storyworld of *Neuromancer*. The outside back cover of the Fall 1975 *CQ* has an image captioned "O'Neill Space Colony, Model III (6.2 miles long—1.24 miles diameter) at the L-5 Lagrangian Point" (Plate II). Here is the "Archipelago," its "islands" made up of a "cluster" of such mega-constructions. Here is *Neuromancer*'s "blunt white spindle, flanged and studded with grids and radiators, docks, domes."[58] Spindle is Gibson's substitution for what O'Neill terms a cylinder, the primary "living habitat." Beneath the full image on *CQ*'s outside back cover is a schematic view naming its components, the "twin cylinder to compensate for gyroscopic action" to produce gravity, "agricultural cylinders," "zero-gravity industries," and "mirrors to bring natural sunlight into habitat." Another NASA illustration (Plate III) shows what the interior could look like. In *Neuromancer*, Case gets a briefing on Freeside's terrain: "Casinos here . . . Hotels, strata-title property, big shops along here. . . . Blue areas are lakes. . . . Big cigar. Narrows at the ends. . . . Mountain effect, as it narrows. Ground seems to get higher, more rocky, but it's an easy climb. Higher you climb, the lower the gravity."[59] Brand had already explained in the first *CQ* space-colony number how the artificial gravity produced by the centrifugal rotation of the cylinder around its long axis fades to nothing at the very center, where the tip is located. NASA's artists appear to have humored O'Neill's notion of "the green grass and the flowers and the sunshine coming down" by envisioning a moist and verdant landscape basking in reflected natural sunlight and happily reproducing a recognizable terrestrial ecosystem (Plate IV). The artist renders an orbital San Francisco complete with bays, harbors, sailboats, rolling hills, seagulls, and a Golden Gate Bridge, all gently bent by the inside of the circular shell of the cylinder within which this promised land has been planted.

Of course, Gibson turns all this into cyberpunk by converting O'Neill's earnest and wholesome San Francisco into a debauched orbital resort: "Freeside is many things, not all of them evident to the tourists who shuttle up and down the well. Freeside is brothel and banking nexus, pleasure dome and free port, border town and spa. Freeside is Las Vegas and the hanging gardens of Babylon, an orbital Geneva."[60] Given the debased status of the Earth from which Freeside is an escape, it is no surprise that it, too, is no ecological showcase. Like the mind traveling in cyberspace at a virtual distance from the immobile body from which it emerges moment by moment, Freeside is largely although not entirely imagined in detachment from its basis in some ecosystemic coupling of living and nonliving components. Similarly, with some exceptions, science fiction in the 20th century is just as oblivious to the fabulous complexity, planetary in origin, of any particular viable ecosystem, natural or artificial. But this circumstance is just a special case of a more pervasive and common humanistic presumption about our dominion over the Earth, about its owing *us* a livelihood, of the convenient ecological naiveté and environmental impunity to which many in our culture are still trying to retain their individual property rights. Only in a sort of afterthought, which we will examine in a moment, does *Neuromancer* bring the very ecological possibility of a space colony such as Freeside forward as an issue. That it does so at all is further evidence, to my mind, that Gibson has in fact attended thoughtfully not just to O'Neill's designs, which were available elsewhere, but also to the full airing of the issues surrounding them provided by the *CQ* space-colony volumes.

Gibson hides these ecological considerations in plain sight by placing them in relation to Zion, the high-orbital Rastafarian shantytown within the same archipelago as the luxurious space resort of Freeside: "Zion had been founded by five workers who'd refused to return, who'd turned their backs on the well and started building. They'd suffered calcium loss and heart shrinkage before rotational gravity was established in the colony's central torus." The torus is in fact one of the three forms O'Neill envisions for space-colony living habitats—cylinder, torus, and sphere. And Zion, in contrast to the stratified society of Freeside, is clearly a people's colony. With a mural showing "a painted jungle of rainbow foliage," it houses an egalitarian population of space Jamaicans speaking patois, passing spliffs, and providing Case's crew with off-grid transportation and bodyguard services: "Case gradually became aware of the music that pulsed constantly . . . dub, a sensuous mosaic cooked from vast libraries of digitalized pop; it was worship . . . and a sense of community. . . . Zion smelled of cooked vegetables, humanity, and ganga."[61] Zion's role in the ecology of *Neuromancer* is not just to provide a nature preserve—a vestige of green space in the midst of largely abiotic built environments—but also to suggest an alternative artificial environment that could actually insure its own viability for the long term.[62]

The ecological specifics come late in the novel. They are brief, and would seem to be afterthoughts that the narrator sends through Case's mind while he is otherwise occupied with his current caper, the Straylight run. And yet, their coming at the start of a new chapter subtly emphasizes them. All at once, the narration explicitly addresses the ecologies of three different habitats distributed throughout Freeside and Zion. At the moment, Case and his crew are at the tip of Freeside's spindle with a "Steep climb out of gravity," penetrating the Villa Straylight. In the midst of more pressing matters, it occurs to him that

> The Villa Straylight was a parasitic structure. . . . Straylight bled air and water out of Freeside, and had no ecosystem of its own. . . . Freeside's ecosystem was limited, not closed. Zion was a closed system, capable of cycling for years without the introduction of external materials. Free-

side produced its own air and water, but relied on constant shipments of food, on the regular augmentation of soil nutrients. The Villa Straylight produced nothing at all.[63]

The likely sense of these ecological details for most readers would be the metaphor suggested between the Villa Straylight's materially parasitic relation to Freeside and the parasitic as well as self-consuming decadence of the Tessier-Ashpool clan immured there inside their private hive. But taking it at face value, this passage literally and cogently addresses the ecological status of its high-orbital environments, and it does this in terms that are fully developed in the space-colony debates published in *CQ*.

These debates range widely.[64] For instance, as collected in *Space Colonies*, a number of commentators there worry the vexing sociological issues that would be involved in the constitution of a specific human population for a particular space colony. Gibson deflects that problem by setting forth Freeside as a tourist destination and Zion as an ethnically unified subculture. In any event, such demographic problems do not seem insurmountable. Granting that engineering solutions enable high-orbit space colonies to be built in the first place, the more challenging issue confronting them is precisely the ecology to be initiated and established there and then shared and maintained by their full ensemble of living inhabitants as a closed ecosystematic whole. The most learned commentators on this score are near unanimous in their conclusion that the current state of research and technique is not adequate to the ecological problems space colonies would confront.

Paul and Anne Ehrlich send *CQ* a statement to this effect. Paul Ehrlich is famous for authoring *The Population Bomb*, a much-debated popular text of 1968 warning of coming resource depletion on an overpopulated Earth. Those touting the space colony program are of course talking up their utility for siphoning off excess human population, or even for providing a last refuge for a remnant of humanity after it destroys the bulk of itself along with the viability of the Earth. The Ehrlichs acknowledge these arguments: "The prospect of colonizing space presented by Gerard O'Neill and his associates has had wide appeal especially to young people who see it opening a new horizon for humanity. The possible advantages of the venture are many and not to be taken lightly." Nonetheless, they are not convinced:

On the biological side things are not so rosy. The question of atmospheric composition may prove more vexing than O'Neill imagines, and the problems of maintaining complex artificial ecosystems within the capsule are far from solved. The micro-organisms necessary for the nitrogen-cycle and the diverse organisms involved in decay food chains would have to be established, as would a variety of other micro-organisms necessary to the flourishing of some plants. . . . Whatever type of system were introduced there would almost certainly be serious problems with its stability—even if every effort were made to include many co-evolved elements. We simply have no idea how to create a large stable artificial ecosystem.[65]

Another skeptical commentator is John Todd, a distinguished design ecologist who in 1976 already has considerable practical and experimental experience on a research question that has occupied ecosystem ecology since the 1950s—the composition, construction, and testing of materially closed environments (obviously, they remain open for energy flow) creating artificial replicas of natural ecosystems. Articulating the epistemological humility appropriate to the planetary imaginary, his lengthy critique notes:

After a decade of living intimately with designed ecosystems I am coming to know that nature is the result of several billion years of evolution, and that our understanding of whole systems is primitive. There are sensitive, unknown and unpredictable ecological regulating mechanisms far beyond the most exotic mathematical formulations of ecologists. When I read of schemes to create living spaces from scratch upon which human lives will be dependent for the air they breathe, for extrinsic protection from pathogens and for biopurification of wastes and food culture, I begin to visualize a titanic-like folly born of an engineering world view.[66]

In other words, we might well build and launch a titanic space colony only to see it shipwrecked almost immediately upon the hidden iceberg of its own unpredictable ecological evolution toward a nonviable state.[67] In *Neuromancer*'s not-so-distant near-future fantasy, only Zion with its "cooked vegetables . . . and ganga" has succeeded in becoming what any space colony with staying power will need to be, a miniature Gaia, a self-sustaining, self-maintaining home for permanent residents, "a closed system, capable of cycling for years without the introduction of external materials."

Inside the Villa Straylight, the Tessier-Ashpool clan aspires to be a closed system, but it has its ecology all wrong: It cannot recycle; it can only consume others as well as itself. Their generations alternately awaken and hibernate, while pursuing an exclusive purification of their own genetic line through cloning, cryogenics, inbreeding, and murderous infighting. They are a fit parody of Western humanity en masse when it dismisses its own coevolutionary embeddedness within diverse living environments that transcend human understanding and control. But, as Zion intimates, a space colony could be viable if it can learn how to take Gaia with it. That is, if it can recreate a fully interpenetrated coevolutionary consortium of living systems, including human beings, resting on virtuous microbial foundations and environmentally interlaced with the necessary suite of elemental geochemical cycles, it will be able grow its own ecosystem. Placing Zion on the sidelines as a tenuous but possible lifeline, the text then projects its otherwise flagrant organic deficits elsewhere, upon Marie-France, Ashpool, and 3Jane's incestuous inward spiral. So the Tessier-Ashpool clan at the tip of the spindle also epitomizes both the detachment and the exclusivity of the space-colony idea when it is rendered as a purely human engineering challenge and not a needfully Gaian phenomenon, that is, rendered as a global but not a planetary phenomenon.

Finally, it is the same Tessier-Ashpool clan that has constructed the sentient artificial intelligences, the AI's that are pulling the puppet strings of Case and his crew. The success of one of these cyber-entities, Wintermute, in stage-managing its own digital evolution through the overcoming of human prohibitions against its intercourse and merger with its reluctant counterpart AI, named Neuromancer, is a rich scenario full of trenchant ironies. But in the end, the AI element cannot be taken any more seriously than can the novel's visions of cyberspace, which realm is specified as the virtual environment within which the AI's carry out the term of their artificial being. And even granting these transhuman systems some future-cosmic plausibility, they too will still depend upon viable environments, no matter how galactic, planets of some sort, material and living environments that transcend the ultimate finitude of the AI's own systemic complexity. Thus, they too must find themselves embedded within some more diverse ecology that will determine the limits of their current and future possibilities. This take on the novel's most sublime, disembodied, digital, and cosmic apocalypse, then, leads us back home once again to the planetary imaginary *Neuromancer* keeps mostly under wraps—the one incubated by the systems counterculture of the 1960s and '70s. This Gaian ethic is anything but escapist in its recognition that, no matter how high-orbital or cosmic we may go, we will remain systemically coupled to the ecological conditions of our own mundane and Earthbound situation.

7.

BRINGING CELL ACTION INTO EVOLUTION

JAMES A. SHAPIRO

Watching cells in action was always one of the joys of attending a lecture by Lynn Margulis. Often she showed videos of eukaryotic microbes and bacteria from an exotic environment, like the termite intestine, moving with extraordinary synchrony. Typically, the protist would be coated with thousands of cells from one or more bacterial species providing the motive force for the larger eukaryotic cell. Lynn saw these intimate associations as clues to the evolution of eukaryotic organelles.[1] What I learned from Lynn's fascination with symbiogenesis and the intimate associations between cells from different biological kingdoms was the largely unacknowledged power of both prokaryotic and eukaryotic cells to control their activities and to coordinate with other cells. It is impossible to imagine how a symbiogenetic relationship could succeed without both metabolic and genomic integration. How did the merging cells synchronize their cell cycles so that one did not outgrow and lose or overgrow and destroy the other? We do not have the answer to such questions, but merely posing them opens the door to a realm of active cell control regimes that, I believe, will revolutionize twenty-first-century biology.

LIVING ORGANISMS IN ACTION

I wish to discuss how much of genome evolution results from active cell processes. The focus on genome evolution reflects my personal experience as a bacterial geneticist, repeatedly surprised by how successfully *E. coli* managed to engineer its own genome.[2] The idea of genome change as a result of natural genetic engineering is a distinct departure from the conventional wisdom developed in the middle of the twentieth century, which posited a stochastic succession of accidental changes as the ultimate sources of evolutionary variation.[3] If we remember that conventional evolutionary wisdom was formulated before the discovery of DNA as a major carrier of inherited information, then we will be less resistant to thinking that assumptions about the nature of genome change were destined to change as our knowledge of and our technology for reading DNA sequences advanced in the succeeding decades. Lynn, of course, realized that the early symbiogenetic thinkers were correct in seeing active cell processes at work in evolution making rapid major changes.[4] She correctly emphasized two aspects of the evolutionary process that have been amply confirmed by DNA sequence evidence:

1. The origin of the eukaryotic cell results from one or more symbiogenetic mergers. The genomic evidence is overwhelming in showing that mitochondria and plastids are descended from endosymbiotic prokaryotes and that eukaryotic cells share many features descended from both bacterial and archaeal ancestors.[5]
2. Symbiogenesis and genome mergers continued throughout eukaryotic history. The genomic evidence has confirmed many secondary, tertiary, and higher-order symbioses since the origins of the first eukaryotic and first photosynthetic eukaryotic cells.[6]

In my own work on bacterial multicellularity, I came to recognize the underappreciated ubiquity of symbiotic relationships.[7] We have only to think about the recent excitement over the human microbiome to understand how important symbiosis is for growth, health, and disease of the host organism.[8] Moreover, a large number of symbiotic associations involve the gonads or other pregerminal tissues and so will have a potential evolutionary effect on the genome passed down to future generations.[9] It is also significant that sexual reproduction involves a merger between two distinct cells. This basic fact means that the evolutionary history of all sexually reproducing organisms involves coordinated cell-cell interactions. We generally take these interactions for granted, but we will see later that sex between different populations or species lacks the usual smooth intercellular coordination and often unleashes events of profound evolutionary importance.

DNA AND THE IRONY OF MOLECULAR BIOLOGY
Crick's Central Dogma

The demonstrations that DNA carries inherited information and the elucidation of its self-complementary double helical structure initially offered the possibility of a purely physicochemical explanation of heredity.[10] The idea of coded information in DNA sequences determining the functional proteins and thereby the properties of cells found its sharpest exposition in Francis Crick's "Central Dogma of Molecular Biology."[11] The central dogma fit well with the idea of the gene as a basic unit of heredity. But the unitary, elemental notion of the gene began to dissolve as molecular geneticists probed how cells utilize DNA information. Starting in the 1950s, the atomistic gene was deconstructed into complex clusters of coding sequences and regulatory sites needed for expression and control.[12] In the 1970s, many coding sequences were found to be split into separate exons that had to be spliced together, adding another layer of control and flexibility.[13]

Cell and Genome Networks

As molecular genetic analysis dissected various aspects of cell and developmental biology, the abundance of interacting execution and regulatory factors came to represent a basic principle of biological action.[14] By the beginning of this century, the notion of individual cell or organismal characters encoded by single genes gave way to the "systems biology" concept of all traits determined by the coordinated action of molecular networks.[15] These networks contain components both at the genomic level and outside the genome in proteins and regulatory RNA molecules.

Bacterial Cognition in Metabolism

Molecular studies produced another unanticipated insight into biological function when they revealed the central role of intracellular and extracellular sensory processes. These cognitive functionalities were apparent from the earliest days of pre-molecular biology. In 1942, in Nazi-occupied Paris, Jacques Monod completed his doctoral research on quantitative measurements of bacterial nutrition and growth.[16] By asking what happened when he presented the bacteria with a mixture of sugars to consume, Monod made the striking discovery that they consumed them in two stages. Using the simple but elegant method of changing the relative proportions of the two sugars, he demonstrated that the bacteria completely consumed the preferred sugar before taking a pause to adjust their metabolism and then consume the other sugar. This discovery was the start of Monod's pioneering work with his Institut Pasteur colleagues on the *lac* (lactose) operon of *E. coli*.[17] Together with other work at the Institut Pasteur on viral control, Monod's *lac* operon research was to form the basis of our current understanding of genome regulation. The key concepts of interactive communication between regulatory proteins, specialized recognition sites in the DNA, and signal molecules all arose from this research.[18] From later studies, we learned that bacterial sugar sensing utilizes molecules that have other tasks as well, such as DNA binding and sugar transport across the cell membrane.[19] From the molecular evidence on *lac* and many other systems, we can conclude that there is no Cartesian dualism in bacteria. Sensing is built into the basic fabric of the cell.

Bacterial Cognition in Proofreading DNA Replication

The role of sensory molecules also emerged from studies of how cells protect their genomes. The accuracy of *E. coli* DNA replication is such that there is fewer than one mistake per billion nucleotides incorporated.[20] This precision is more than five orders of magnitude (100,000-fold) better than the replication polymerase achieves on its own. All the extra precision comes from two overlapping proofreading systems that detect and then correct mistakes after they occur.[21] When errors are found, the proofreading apparatus can distinguish the old (presumably correct) strand from the newly synthesized (presumably mistaken) strand and fix only the latter. Assuring the accuracy of DNA replication in this way has the same cognitive basis as quality-control processes in human manufacturing.

Bacterial Cognition in Responding to UV Damage

When genomes suffer damage from radiation and chemicals, the results include cell death and increased frequency of genetic changes (mutations). Decades of research have shown that cells play an active role in repairing both lethal and mutagenic damage—and also, surprisingly, an active role in producing mutations in response to damage. The inducible nature of the repair and mutagenesis phenomena was clearly shown in some highly original experiments during the 1950s by the Swiss physicist-turned-molecular geneticist, Jean Weigle.[22] In order to distinguish between the genetic target of UV irradiation and the cell environment, Weigle used the bacterial virus, lambda, as his test system. In the viral particles, lambda DNA could be treated independently of the host cell, while the host cell could likewise be treated independently of the target DNA prior to infection. Weigle obtained the following set of results from the four possible combinations of independently irradiating virus and host cell:

a. Untreated virus infecting untreated cells → no killing, basal virus mutations

b. Irradiated virus infecting untreated cells → killing, elevated virus mutations

c. Irradiated virus infecting irradiated cells → less killing, even more virus mutations

d. Untreated virus infecting irradiated cells → no killing, elevated virus mutations

The baseline control situation was unirradiated viruses infecting unirradiated cells. The greater lethality but reduced frequency of mutations when irradiated viruses infected unirradiated cells showed that radiation of the cells prior to infection induced both the capacity to repair lethal damage ("Weigle repair") and also stimulated the ability to produce mutations ("Weigle mutagenesis"). The unexpected mutation frequency when unirradiated virus infected irradiated cells confirmed that UV irradiation induced mutagenesis capacity in the cells prior to infection and showed that mutational activity could occur on untreated DNA.

A couple decades' research revealed the bacterial reaction to UV irradiation and chemical mutagenesis to be an integrated "SOS" response to

DNA damage.[23] A multifunctional sensor protein, RecA, binds to regions of single-stranded DNA that accumulate in cells with damaged genomes. This binding activates the expression of a suite of protein functions that include repair and recombination proteins, several of which are "mutator" DNA polymerase activities that actively introduce the incorrect nucleotides, which register as mutations at the end of the experiment.[24]

Cell Cognition in Checkpoint Control of the Cell Cycle

Among the SOS functions is an intriguing protein that inhibits bacterial cell division until DNA repair has been completed.[25] This constitutes what we now recognize as a "checkpoint" function that holds up the cell division cycle until the genome and other necessary components are in the proper state to proceed.[26] Checkpoints are ubiquitous in biology and represent an inherently cognitive set of controls that ensure the reliability of cell division, a process that involves hundreds of millions of biochemical and biomechanical events.[27] The cell monitors its own internal status and adjusts the progress of different subroutines so that none outruns the coordination needed to produce intact daughter cells with complete genomes. Among the most impressive is the "spindle checkpoint," which senses and guarantees that all the chromosomes are properly aligned along the cell division spindle apparatus so that each daughter cell receives one, and only one, copy of each duplicated chromosome.[28] As with the super-low error frequencies in DNA replication, the remarkable success of cell division depends upon cognition rather than mechanical precision.

Sensing, Communication, and Response in Programmed Cell Death

Not only cell survival and reproduction but also programmed cell death depends on cognitive processes. All cells, from bacteria to plants and animals, have systems for a regular cell death process.[29] These cell suicide routines are useful for defense against pathogens, modeling multicellular development, establishing symbioses, and eliminating in a predictable manner cells that cannot repair serious damage. Cell death decisions are typically contingent upon the receipt of intercellular signals. A good example of how cognition operates in cell death decisions can be seen with mammalian cells exposed to DNA damaging agents, like ionizing radiation. If the cell receptors detect antisurvival signals, such as tumor necrosis factor (TNF), they tend to suffer higher levels of programmed cell death, whereas cells that detect proliferative signals, such as extracellular matrix proteins or tumor growth factor beta, undergo more frequent genome repair and experience much less lethality for the same amount of damage.[30]

By the beginning of this century, the results of molecular cell and developmental biology made it clear that new basic concepts were necessary to explain how the properties of living organisms came to be and were passed on from one generation to the next. The one gene—one character paradigm and the unidirectional flow of information in Crick's central dogma were no longer tenable.[31] Complexity, networks, signaling, and cognition (i.e., sensing and decision making) had emerged as essential features of living cells.

GENOME REPAIR AND GENOME RESTRUCTURING

An independent chain of events leading to our modern understanding of cell action on the genome dates back to the 1930s, when Barbara McClintock began her studies on the chromosomal basis of X-ray mutagenesis in maize. She took pains to recount this history in her Nobel Prize lecture.[32]

Repair of Chromosomes Broken by X-Rays

Herman J. Muller discovered "X-ray mutagenesis" in the 1920s.[33] X-rays were the first of many external agents shown to induce mutations, and thus hereditary changes, in living organisms. However, it is not very well known that these X-rays also triggered cell repair functions and sophisticated chromosome rearrangements, not just simple "gene mutations." McClintock set out in 1931 to analyze the X-ray mutants Louis Stadler had isolated from maize plants at the University of Missouri. She was the right person to do this. As a Cornell graduate student in the 1920s, she personally developed the microscopic methods that allow scientists to visualize the ten maize chromosomes. Some X-ray mutants were unstable and "variegated" in their inherited properties. We can see similar variegating instabilities in spotted or striped kernels on ears of Indian corn. Based on her understanding of chromosome behavior, McClintock was able to explain the "variegating" mutants.

She proposed that variegating plants carried ring chromosomes. Such rings formed by the fusion of two broken ends near the telomeres of a previously linear chromosome. Ring chromosomes sometimes failed to distribute properly to daughter cells as the plant grew, thus producing the variegated patterns. Although colleagues scoffed at McClintock's idea, she went on to identify the predicted ring chromosomes in the variegating mutants.[34] Other mutants induced by X-ray treatment also carried chromosome rearrangements. Sections of chromosomes were deleted, translocated, inverted, and duplicated. All of these rearrangements could result from breaking chromosomes at two sites and then rejoining the broken ends to build brand new chromosome structures. These chromosome rearrangements were quite different from the "gene mutations" imagined by Muller and his colleagues.

The Cognitive Nature of Broken Chromosome Repair

McClintock reasoned that maize cells must have an inherent capacity to join broken chromosome ends when two of them are present in the cell. In the later 1930s, she devised experimental methods to induce new chromosome arrangements. Using experimentally generated breaks, she demonstrated conclusively that maize cells can detect, juxtapose, and fuse broken chromosome ends.[35] McClintock realized that the X-rays broke chromosomes wherever they happened to strike. Breakage alone, however, was insufficient to generate a mutant chromosome. Broken chromosomes would get lost. The cell's ability to repair the damage by fusing broken ends was essential. In other words, X-ray mutagenesis provoked cell action. McClintock had started down a novel path of thinking. As she wrote in her Nobel lecture:

> The conclusion seems inescapable that cells are able to sense the presence in their nuclei of ruptured ends of chromosomes and then to activate a mechanism that will bring together and then unite these ends, one with another. . . . The ability of a cell to sense these broken ends, to direct them toward each other, and then to unite them so that the union of the two DNA strands is correctly oriented, is a particularly

revealing example of the sensitivity of cells to all that is going on within them. . . . There must be numerous homeostatic adjustments required of cells. The sensing devices and the signals that initiate these adjustments are beyond our present ability to fathom. A goal for the future would be to determine the extent of knowledge the cell has of itself and how it utilizes this knowledge in a "thoughtful" manner when challenged.[36]

By raising the idea of cells operating in a "thoughtful manner," McClintock was self-consciously trying to point biological research in a revolutionary new direction. This comment led Dennis Bray to point out that she was the first scientist to ask what a cell knows about itself.[37]

Transposable Element Activity and "Genome Shock"

The studies on chromosome breakage and rejoining led directly to McClintock's unexpected discovery of mobile genetic elements.[38] They prepared her for analyzing variegation and chromosome breakage, and they led her to explore the consequences of introducing broken ends into maize cells at fertilization. Rather than explain in her Nobel lecture how she demonstrated genetic mobility, McClintock chose to relate experiments confirming that the "shock" from receiving a single broken chromosome end had the extraordinary effect of awakening previously latent transposable elements in the genome: "It seemed clear that these elements must have been present in the genome, and in a silent state previous to an event that activated one or another of them. . . . It was concluded that some traumatic event was responsible for these activations." Most of the remainder of McClintock's lecture provided "further examples of response of genomes to stress." She wished to emphasize the generality of her observations across the living world, taking examples from both plants and animals. With respect to evolutionary change, she emphasized the "shock" of interspecific hybridization, a process that leads to whole genome doubling, widespread genome reorganization, and formation of novel species.[39] For McClintock, genome change was not accidental. Change was a response to life history challenges.

Why did McClintock focus so much attention on cell sensing and not on research that provided molecular evidence in support of her previously heretical views?[40] From the way she ends her lecture, we can appreciate that McClintock had her perspective directed toward the years ahead, not those behind:

> In the future attention undoubtedly will be centered on the genome, and with greater appreciation of its significance as a highly sensitive organ of the cell, monitoring genomic activities and correcting common errors, sensing the unusual and unexpected events, and responding to them, often by restructuring the genome. We know about the components of genomes that could be made available for such restructuring. We know nothing, however, about how the cell senses danger and instigates responses to it that often are truly remarkable.[41]

McClintock wanted to draw special attention to what she saw as the most challenging problems in biology: cognition and purposeful action by living cells. As she knew well from her long experience with twentieth-century genetics and cell biology, whether life has special "vital" properties that separate it from inorganic matter has been among the most fiercely disputed topics in the history of science. In its early days, molecular biology promised to provide us with an explanation of life in terms of physics and chemistry. However, as we saw previously, it has succeeded instead in amazing us with the richness and sophistication of intra- and intercellular control and communication networks.

McClintock's Preview of Molecular Discoveries

In three important ways, McClintock's work anticipated the molecular rediscovery of mobile genetic elements in the 1960s and 1970s.[42]

1. It demonstrated that genome repair and restructuring are regulated and responsive processes the cell executes with a range of dedicated functions;
2. It showed how mobile fragments of genetic material could alter the expression regime of virtually any coding region or genetic locus throughout the genome.[43]
3. It illustrated the ability of these mobile "controlling elements," as she named them, to generate genome networks coordinating the expression of unlinked loci through insertion of related elements.[44]

McClintock laid the groundwork for a revolution in our understanding of the possibilities for genome variation in evolution.

THE MOLECULAR TOOLBOX FOR GENOME RESTRUCTURING

While McClintock was working out many details of action involving her "controlling elements," the pioneers of molecular genetics were discovering the many ways that cells acquire, transfer, and restructure DNA molecules.[45] Multiple seemingly independent lines of research supplied abundant evidence of these capabilities. Although this section will seem rather technical to some readers, it is essential to mention the details and thus emphasize the point that DNA change is not a series of accidents. On the contrary, all DNA changes result from dedicated cell functions doing their jobs ("the molecular toolbox"). The sheer variety of cellular DNA restructuring functions is unknown to all but a small number of specialists in the field. The existence of these functions needs to be more widely appreciated because they are such basic components of cell action in evolution.

DNA Uptake from the Environment

Studies of bacterial transformation, first identified in 1928, provided the first evidence that DNA was the "transforming principle" and therefore carried hereditary information.[46] More recent studies have identified "competence" for DNA uptake from the external environment in many diverse groups of bacteria.[47] Analysis of the genetic control of competence has revealed many different modes of regulation, including control by stress responses, and intercellular signaling molecules, and activation by metabolic signals in the environment. Later studies, mostly for genetic engineering purposes, using cultured plant and animal cells have shown that the ability to take up and integrate DNA from the environment is widespread among living cells. Unfortunately, little is known about the molecular apparatus that enables these examples of DNA incorporation by eukaryotic cells.

Viruses, Proviruses, and Site-Specific Recombination

By following single dividing bacteria through hours of patient microscopic examination, André Lwoff, Monod's doctoral supervisor, convincingly

documented that some bacteria possessed the hereditary capacity to produce viruses.[48] This lysis-causing property was named "lysogeny" because
the liberated viruses could infect and lyse closely related bacteria. The
study of lysogeny in the following years provided many important insights
into the relationships of virus and cell genomes, a major source of evolutionary novelty.[49] Viruses can infect cells and establish their own genome
as part of the cell genome, called a "provirus."[50] In the quiescent provirus
state, parts of the viral DNA can encode functions that alter the properties of the host. For example, the bacteria that cause diphtheria do so only
when they harbor a provirus that encodes the diphtheria toxin.[51]

Viruses can also mobilize DNA sequences from one cell to another.
This DNA "transduction" capacity was discovered in the 1950s with bacteria, but it appears to have few limits.[52] So-called metagenomic analysis
of unpurified environmental viruses has shown that some contain combinations of sequences from all three domains of life: Bacteria, Archaea,
and Eukarya.[53] This means that the progenitors of these viruses passed
through cells from all domains and thus can carry sequences from one
to another. The giant DNA viruses that infect amoebae are particularly
noteworthy in this regard.[54] So it appears as though amoebae and other
protists may constitute an evolutionary "melting pot" for mixing DNA
coming from different sources.[55] Some amoebal viruses are also known
to infect other kinds of cell, including plants and animals.

The study of how provirus genomes enter and leave host cell chromosomes has revealed a number of different mechanisms for mobilizing DNA
molecules. In bacteria, one of the most important mechanisms is the process called "site-specific recombination," in which specialized "recombinase" proteins act on particular short sequences to carry out reciprocal
exchanges.[56] This form of specialized recombination has been adapted for
many additional purposes in bacterial cells—for example, turning protein synthesis on and off, reducing double circles of replicated DNA to two
single circles for equal transmission to daughter cells, excision of DNA
introns during cell differentiation, and modification of protein structures
by inverting segments within the coding sequence.[57] At a low but significant frequency (about .001 percent), the provirus exits from the bacterial
chromosome by an aberrant event rather than precisely targeted site-
specific recombination. In these aberrant events, the viral genome can ac-

quire an adjacent fragment of the bacterial chromosome.[58] The bacterial fragment is effectively cloned into the virus genome by cell action and reproduces together with the virus. This kind of in vivo cloning can be applied to many regions of the bacterial chromosome. It served as the cloning process used to obtain the specific DNA sequences for the first isolation of a defined genetic locus.[59]

Antibiotic Resistance and Horizontal DNA Transfer in Bacteria

The spread of antibiotic resistance among bacterial pathogens starting in the 1950s and 1960s was a real-world stimulus to research in molecular genetics. The story of how we came to understand this major evolutionary episode that took place in real time subject to detailed investigation is a fascinating and instructive chapter in the history of science. It illuminates the insight that scientific reality consists of more than experimentally confirming hypothetical predictions.

In the early days of molecular biology, bacterial geneticists applied conventional evolutionary concepts to explain the evolution of antibiotic resistance. The theory was that mutations could alter the structure of cell components and either block entry of the drugs into the bacteria or prevent their action on cellular targets, such as the enzymes essential to cell wall synthesis. Even if the initial mutation did not confer a high degree of resistance, accumulation of several sequential changes would result in resistance to the antibiotic levels used in clinical medicine. Indeed, a wide variety of laboratory experiments confirmed this theory, and bacterial geneticists isolated the predicted mutant strains.[60] In virtually all cases, the resistant mutants grew less well than the parental sensitive bacteria, leading to the comforting conclusion that resistant bacteria would not significantly accumulate in nature.[61] The degree of confidence was so great that the US Surgeon General in 1967 declared that "the war against infectious diseases has been won."[62] However, there were problems both with the science and the new public health policy based on it. The Surgeon General "misunderestimated" bacteria, which followed their own evolutionary rules and did not listen to what the scientists said they should do. Although experimentally confirmed, the mutation theory of antibiotic resistance failed to account for most cases in the real world. Resistance continued to spread among bacteria isolated in clinics around the globe. Even more ominously, different strains of pathogenic bacteria increasingly displayed resistance to more than one antibiotic at a time.

Research pioneered in Japan found that multiple antibiotic resistances could be transferred simultaneously from one bacterial species to another.[63] The DNA agents responsible for this transfer are circular molecules that are called multidrug resistance plasmids, which can move from one cell to another.[64] Multiply resistant bacteria were not altered in their cellular structures or inhibited in their growth properties. Rather, they had acquired new biochemical activities that could destroy or inactivate the antibiotics, chemically alter their targets, or remove them from the bacterial cell.[65] Multiple antibiotic resistance clearly represented genome change and evolution of a type unimagined in the pre-DNA period. DNA molecules were transferred "horizontally" between unrelated cells rather than descending from ancestral cells. Horizontally transferred DNA could carry complex sets of genetic information encoding multiple distinct biochemical activities. Evolutionary leaps involving several characteristics at once could occur through horizontal DNA transfer.

In the early 1980s, two obscure microbiologists published a book called *A New Bacteriology*, postulating a radically different approach to thinking about bacterial evolution.[66] Sonea and Panisset argued that bacteria have a huge collective genome distributed throughout nature in different kinds of cells, in viruses and latent in the environment. When a new ecological niche appears, bacteria can assemble the genomic assets they need to exploit the opportunity. Subsequent research has bolstered Sonea and Panisset's initially outlandish idea. As we have seen, bacteria have all the abilities they need to acquire DNA from the environment, from viruses and from other cells.[67] Viruses are the most abundant biological entities in the environment and carry a multitude of sequences encoding important cellular capabilities, such as photosynthesis.[68] Detailed study of many bacterial characteristics, especially pathogenicity and virulence, indicate that they are encoded by plasmids or by critical segments of the DNA, so-called "genomic islands."[69] The sequences of genomic islands show that they have been acquired from unrelated organisms and often integrated into the cellular genome by site-specific recombination.

Among the DNA sequences most often found on plasmids and in genomic islands are those encoding molecules needed for virulence in pathogenic bacteria.[70] Among the most important virulence molecules are those that form intricate complexes in the cell envelope and allow the bacterial pathogens to inject protein and RNA molecules into cells of the host organism, whether animal or plant. In so doing, they subvert host cell regulatory circuits in a way that meets the invading bacterium's needs. Similar macromolecular injection systems are involved in established symbiotic relationships. Bacteria must be the smartest cell biologists on the planet; they control events in cells of higher organisms in a way that mere human scientists can only dream of imitating. The macromolecular transport structures bacteria use to acquire DNA from the environment and transfer plasmids and other DNA molecules between cells are very similar.[71] These DNA mobilization structures, in turn, are closely related to other envelope-spanning structures involved in the synthesis of high-energy storage molecules and rotation of bacterial flagella (literally, "whips") for swimming through fluids.[72] They are even used as push-pull apparatus for so-called "twitching" movement across solid surfaces. Clearly, there has been wide-ranging use and reuse of these elaborate systems for multiple functions in the course of bacterial evolution.[73]

Horizontal Transfer Outside Bacteria

We have learned through genome sequences that horizontal DNA transfer is not limited to the prokaryotic domains of bacteria and archaea.[74] There is well-documented DNA transfer from bacteria to plants in nature and, in the laboratory, to yeast and animal cells.[75] The genomes of endosymbiotic bacteria are found in the nuclear genomes of host invertebrates. There are DNA sequences that have moved between unrelated plants, animals, fungi, and protists. Both microbial kingdoms have contributed functional coding regions to multicellular plants and animals, and genomic systems permitting certain key ecological adaptations include horizontally acquired sequences. In addition to these examples of intercellular horizontal transfer, it is highly appropriate in a volume honoring Lynn Margulis to mention the sort of intracellular horizontal transfer following symbiotic events that she closely studied. Most of the proteins originally encoded in the genomes of bacterial ancestors of mitochondria and plastids are now encoded in the nuclear genomes of eukaryotic cells because of postendosymbiosis organelle to nucleus DNA transfer.[76] In cases where algae have formed secondary endosymbioses in other eukaryotic lineages, there are four or more genome compartments: nucleus, mitochondrion, nucleomorph (former algal nucleus) and plastid.[77] Comparing the genomes of free-living algae with their symbiogenetic descendants, it is evident that DNA sequences migrated from the nucleomorph and plastid genomes to the nuclear chromosomes.

Transposable Elements in Bacteria

McClintock's discovery of mobile elements in the 1940s was extended by bacterial geneticists in the 1960s and 1970s studying "spontaneous" mutations (i.e., genetic changes not induced by mutagenic treatment, whose nature and origins were unknown).[78] A number of these mutations had properties that were not explicable by the recognized types of DNA changes: nucleotide substitutions, frameshifts, and deletions. They could terminate or activate transcription.[79] The anomalous mutations proved to be the result of insertion of specific DNA sequences that occurred at many locations in the genome. The mobile genetic elements responsible for the insertions came to be called "insertion sequences" or IS elements. Later, certain viruses were found to integrate their proviruses into the bacterial chromosome and replicate their genomes by the same processes that mobilized a subset of IS elements from one genomic location to another. Some bacteria contained dozens or hundreds of IS elements in their genomes. The majority of the cellular and viral elements encoded proteins, named "transposases," that mediated their movement (transposition) through the genome.[80]

The discovery of IS elements was quickly followed by the observation that many antibiotic resistance determinants are also mobile.[81] The ability to move DNA to new locations in the genome had evolved as a powerful means to combine different antibiotic resistances into a single plasmid DNA molecule. The mobile elements that encoded a specific phenotype, like resistance, were called "transposons," and that has become the generic term for mobile DNA elements. As studies of mobile DNA in bacteria continued, naturally occurring and experimentally derived transposons were shown to be able to include any segment of the bacterial genome.

Integrons and the Rapid Evolution of Multiple
Antibiotic Resistance

In the 1990s, as they analyzed multiple antibiotic resistance plasmids and transposons, Ruth Hall and her colleagues recognized an additional mode of DNA mobility used to construct complex genomic determinants.[82] They discovered that many plasmids and transposons acquired additional individual coding sequences one at a time by the same kind of site-specific recombination mechanism that inserted proviral genomes into bacterial chromosomes. The insertion site was part of a structure she called an "integron," which encoded the recombinase activity necessary to insert the cassettes downstream of an active transcription signal, where they would be expressed.[83] Where the circular cassette structures originate remains a mystery. But we now know that integrons and larger "super-integrons" have acquired a large number of different kinds of coding sequences, including those for bacterial virulence, control of transcription, synthesis of various biochemicals, motility, and intercellular signaling.

"Legitimate" and "Illegitimate" Recombination

As we continue our review of how cells actively restructure and assemble their genomes, it is interesting to remark that it was a commonplace in the later twentieth century to call site-specific recombination, transposition, and other kinds of DNA rearrangements "illegitimate recombination," and the practice continues.[84] This exclusionary term was meant to distinguish these "exceptional" DNA exchange processes from the homology-dependent recombination that had been used to construct genetic linkage maps since the early twentieth century.[85]

We can understand the initial surprise that geneticists comfortable with assumptions about the rules of meiotic crossing-over experienced when they encountered first McClintock's work and then bacterial research on DNA transfer and rearrangements involving plasmids, transposons, integrons, and viruses. Clearly, these genome-restructuring events would appear strange to scientists used to thinking of the Constant Genome that changes only by accidents or the homologous recombination of allelic differences. Nonetheless, it is satisfyingly ironic to note that this century has found the most abundant DNA coding sequences in the genome databases to be those for transposases.[86] The reason for this abundance is that transposons and other mobile elements are the most common sequences in many genomes. For example, in the draft human genome revealed at the start of the twenty-first century, our DNA is less than 1.5 percent protein coding sequences and over 40 percent dispersed repeats consisting of various mobile genetic elements.[87]

In addition to playing an essential role in meiosis, homologous ("legitimate") recombination has been adapted for specialized targeted ("illegitimate") functions in microbes.[88] Among yeasts and other fungi that reproduce both as diploids and haploids, growth as diploids is more stable against inevitable DNA damage. Diploids have two copies of the genome, and one can serve as a template for repairing breakage in the other.[89] In order to form diploid cells following meiosis and sporulation, the haploid progeny of both budding and fission yeast spores have adapted homologous DNA breakage repair to undergo self-mediated sex-change operations.[90] When a haploid cell has changed its mating type, it can fuse with a cell having the opposite parental mating type and form a diploid.

By using a sequence-specific DNA break at the mating-type locus, both yeasts direct a process of homology-dependent recombination that changes genetic content and cellular mating type. Budding yeast use the HO endonuclease to make the targeted break, while fission yeast use a modified transposase protein.[91] So this process of directed recombination must have undergone at least two independent evolutionary steps. Once the break has occurred, it is repaired by recombination with a silent DNA cassette containing opposite mating type information surrounded by homologous sequences. The result is a unidirectional gene conversion and transfer of new mating-type information to an expression site, generating the phenotypic sex change.[92]

A similar strategy of directed gene conversion of new information from silent cassettes to expression sites is used for the process of "antigenic variation" by numerous prokaryote and eukaryote microbes.[93] The directed gene conversions change the structures of cell surface proteins and help these microbes escape the defenses of the host immune system. While the phenomenon of cassette exchange and the role of homologous recombination systems are well documented, these antigen variation events are not as thoroughly understood as the yeast mating-type switches.

Retroviruses

Among the major shocks to conventional reductionist thinking in the 1960s was Howard Temin's assertion about certain RNA viruses. He claimed they reproduced their genomes by inserting a DNA copy into the host cell chromosomes and then transcribing the proviral DNA into more viral RNA.[94] The connection with Lwoff's pioneering work with bacteria is clear. Temin's arguments were widely ridiculed until he demonstrated an activity in the viral particles that "reverse transcribed" RNA into DNA.[95] This activity came to be called "reverse transcriptase" and viruses that employed this strategy "retroviruses."[96] Reverse transcription disrupted the neat picture of Crick's central dogma, and Crick was quick to add an exception for RNA to his scheme of one-way information transfer from the genome.[97] But the genie was out of the bottle. A major new capacity for genome modification had appeared on the scene.

Retroviruses have many parallels with bacterial viruses, including the use of site-specific recombinases or, more commonly, transposase-like proteins for proviral integration.[98] We now recognize retroviruses as the prototype of a family of "retrotransposable elements," that is, mobile elements that move through the genome by virtue of an RNA intermediate.[99] Retroviruses have a complex reproductive cycle that includes the formation of DNA proviruses at many different sites in the host cell genome. It can be thought of as an infectious process of transposition from one genomic site in the first cell to a different site in a second cell.

Occasionally, a cellular RNA sequence can be incorporated into the retroviral genome. This happened in the case of the first retrovirus to be discovered, in 1910 (!), the Rous sarcoma virus that induces a transmissible cancer in chickens.[100] Because of the connection with cancer, these incorporated cell sequences are often called "oncogenes."[101] But we now understand that the product of an oncogene is simply a cell function that disrupts normal proliferation when produced in an aberrant manner. Even though we can reproduce the process in the laboratory, we do not know how retroviruses incorporate the cellular RNA sequences.[102] But the ability to incorporate and reverse transcribe cell RNAs greatly extends the genome restructuring capability of this group of mobile elements. Since retroviruses carry powerful signals for activating transcription, they can also induce tumors by inserting in the genome near coding sequences for a cell function whose aberrant expression leads to oncogenic changes in cell behavior.[103] Activating expression by mobile element insertions had been well documented in both bacteria and yeast.[104]

Endogenous retroviruses (ERVs)

Many genomes, including our own, harbor endogenous retroviruses (ERVs).[105] ERV proviruses display regulated expression patterns during development and produce viral particles in certain tissues. It is now clear that ERVs have played several important roles in human and mammalian evolution. They have contributed to functional human coding sequences, genome variability, and chromosome rearrangements in primate evolution. They have provided mobile expression signals for the generation of genomic networks active in embryonic development and the cell cycle. Most notably, ERVs were central to the evolution of the placenta, a key event leading to the radiation of eutherian mammals: Placental tissues have high levels of ERV expression and contain numerous retroviral particles, and placenta formation requires proteins (syncytins) evolved from ERV precursors.[106] In addition, ERVs play important roles in stress responses and disease.

Retrovirus-Related LTR Retrotransposons

Many mobile elements resemble retroviruses but lack sequences for essential cell infection steps. These noninfectious elements are called "LTR retrotransposons" and are found in many, if not all, groups of eukaryotes.[107] LTR stands for the "long terminal repeats" characteristic of provirus structure. LTRs are the products of the specific way that retroviruses and related retrotransposons convert their RNA into proviral DNA.[108] LTR retrotransposons are capable of intracellular retrotransposition and provide a major source of genome variability in virtually all eukaryotes, from yeast and protists through higher plants and animals. All LTR retrotransposons carry powerful signals stimulating genome expression, and genome sequence data show that many played important roles in the evolution of genome regulatory circuits.[109] Amplification of LTR retrotransposons is also a major contributor to DNA expansion in both plants and animals with large genomes.[110]

Non-LTR Retrotransposons and Reverse-Transcribed RNA

In addition to the retrotransposable elements that share the LTR structure with proviruses, there is another class of "non-LTR retrotransposons" which utilize an entirely distinct process for reverse transcription and integration of DNA copies.[111] Genome analysis identified some of these non-LTR elements as dispersed repeat sequences and gave them the names SINEs (short interspersed nucleotide elements) and LINEs (long interspersed nucleotide elements). LINE elements encode their own reverse transcription and DNA integration functions, while SINE elements are too short for such coding and depend on LINE-encoded activities for retrotransposition to new locations in the genome. In some genomes, non-LTR elements make up the most abundant classes of DNA. In our own, for example, there are 850,000 LINE elements comprising 21 percent of the sequenced genome and about 1,500,000 SINE elements that account for 13 percent of the total DNA.[112] Intriguingly, SINE elements tend to be extremely taxon-specific. When Rick Sternberg and I examined SINEs in mammals, we found that each order had its own special group of elements.[113] It appears that these elements originated and expanded rapidly when a new mammalian order emerged.

One of the most versatile features of non-LTR retrotransposition is its lack of sequence specificity for reverse-transcribing RNA and integrating the DNA copy. This major difference from the LTR elements (which use specific LTR sequences for integration) means that LINE-encoded activities can reverse transcribe and integrate the DNA copy of virtually any RNA sequence into the genome.[114] This molecular promiscuity adds greatly to a cell's genome restructuring capabilities.[115] All kinds of processed RNA molecules can be immortalized. This is the way that intron-free copies of spliced RNAs enter the genome as additional copies of a coding sequence ("retrogenes") or as "pseudogenes" that may lack full coding capacity but often play regulatory roles. Sometimes the insertions produce novel exons or fusion sequences encoding novel proteins.

Another way that LINEs and SINEs mediate genome creativity is by picking up adjacent sequences and moving them to new locations. LINEs and some SINEs frequently include downstream sequences when transcribed, and these external sequences are incorporated into the genome together with the non-LTR element.[116] There are also SINEs that similarly mobilize upstream sequences to new locations. Laboratory studies and genome sequencing have shown that non-LTR elements can insert either upstream or downstream exons into a distant coding region. This exon mobilization is significant in evolution because it mediates protein innovation by "domain shuffling."[117] DNA transposons have the same innovative exon shuffling capacity.

Nonhomologous End-Joining in Genome Rearrangements

Although haploid cells cannot use homologous exchanges to restore double-strand DNA breaks in unreplicated genomes, repair still takes place by a process known as "nonhomologous end-joining" (NHEJ).[118] In NHEJ, the ends of broken DNA molecules are processed and joined together. There are multiple ways these coordinated events can occur and multiple possible outcomes. In some cases, the original DNA structures can be restored precisely. In other cases, mutations or deletions occur at the site of joining broken ends of the same molecule. When the ends of different molecules are joined together, of course, various types of chromosome rearrangements occur.[119] Since chromosome rearrangements frequently accompany taxonomic divergence, NHEJ is an important evolutionary process.[120] For example, we differ from chimpanzees and gorillas by the fusion of higher primate chromosomes 12 and 13 to form human chromosome 2.[121]

There are two important evolutionary consequences of chromosome changes. One consequence is the potential effect of location at different chromosome positions on the expression of coding regions. This can have a direct input into physiological or morphological alterations. The second evolutionary consequence is that chromosome differences serve as a barrier to sexual reproduction between two recently diverged species. Most mating events will be sterile without genome doubling. In this way, chromosome changes separate the reproductive destinies of species that may still be quite similar to each other under normal conditions. At the same time, as we shall discuss below, chromosome restructuring and accumulated epigenetic changes set up the potential for unleashing genome variability after interspecific matings under crisis conditions.[122]

Natural Genetic Engineering as Evolutionary Process

When we consider all the active cell processes that alter DNA molecules discussed earlier (and this account is far from comprehensive), we have to conclude that living cells possess all the tools they need to restructure their genomes in any fashion compatible with the basic chemistry of DNA. I have called this capacity "natural genetic engineering" (NGE) because it is analogous to the kind of genetic engineering we carry out in the laboratory or for biotechnology.[123] Like its human analogue, NGE is not a random or accidental process. Each process makes predictable types of changes in DNA molecules. For example, the movement of a transposon or retrotransposon to a new location always introduces the same constellation of coding and regulatory and DNA restructuring signals. This kind of reproducibility is key to cells having the ability to construct coordinated networks rapidly throughout the genome. We know from genome sequence data that this NGE capacity has been utilized in evolution.[124]

Recognizing natural genetic engineering as a normal and legitimate cellular capability makes a radical change in our understanding of genome variation in evolution. Rather than a series of random, accidental, independent events, we have the documented ability of cells to make non-random, reproducible and concerted changes to the genome. This widens our ability to think about the sources of biologically useful novelty in evolution. Rather than selection slowly molding a gradual process of (in Darwin's words) "slight, successive modifications" toward adaptive improvements, we can envisage functional systems arising by a process analogous to human engineering, where defined sets of components are assembled in new combinations to accomplish novel tasks. Such outcomes are not deterministic because even the conscious, intelligent example of human engineering occurs by trial and error, with the failures typically outnumbering the successes.

In order to consider genome variation as a serious candidate to be the major innovative force in evolutionary change, we need to ask two questions:

1. Is there a known evolutionary system where novelty clearly results from natural genetic engineering?
2. Do cells have the ability to control natural genetic engineering in ways that would be necessary for concerted functional innovation?

The answer to both questions is "Yes." Pursuing these questions will help us define unknown issues that remain to be explored in evolution science.

EVOLVED EVOLUTION BY NGE: THE ADAPTIVE IMMUNE SYSTEM

The adaptive immune system provides a positive answer to the first of our two questions about NGE and evolution. Adaptive immunity is a naturally evolved system for rapid protein evolution. It solves the seemingly intractable problem of how to generate a limitless variety of properly structured antigen-recognition molecules while using only finite DNA coding capacity. There is not space here to cover the details of how rapid evolution by NGE has been adapted for antibody production, but I have described this process in an online appendix to my book and in a pair of well-referenced blog postings.[125] Here, let us simply review some conclusions we can draw from the remarkable evolutionary capacity of lymphocytes, the cells of the immune system. Lymphocytes teach us three important lessons about evolution that are well outside orthodox thinking:

1. They demonstrate the virtually unlimited creativity of NGE processes. In generating endless pairs of antibody chains, B lymphocytes show that DNA change can be simultaneously targeted and highly flexible.[126] Determinism and randomness are not the only possible outcomes.

2. In switching selected antibodies from one class to another by targeted exon swapping, lymphocytes demonstrate that cells can signal one another and direct the outcome of an NGE process.[127]

3. By undergoing a well-defined series of steps before and after contact with the invading antigen, lymphocytes illustrate how evolutionary NGE proceeds in a stepwise fashion. The first step (VDJ, or variable diversity joining) produces a molecule that works at a basal level. The next step (targeted somatic hypermutation) fine-tunes that initial product.[128] And the third step (class switching) adapts the fine-tuned product to a particular distribution in the body.

If we keep the immune system example in mind, we can distinguish parallel cases of sequential multistep NGE events with varying degrees of precision, communication, and control in a wide range of evolutionary scenarios. If we think about the evolution of new body parts, for example, we can trace their initial appearances, subsequent refinements, and further adaptations in the fossil record.[129]

REGULATING AND TARGETING NATURAL GENETIC ENGINEERING FUNCTIONS

A principle philosophical tenet of the neo-Darwinian Modern Synthesis is that there can be no biological input to the process of hereditary variation. Once we realize that this variation results from dedicated cell functions, as detailed earlier, our experience with all aspects of cell biology teaches us that regulation and control must play a central role in their operations. From a molecular perspective, Weismann's idea of a germ plasm segregated from the rest of the cellular system makes no sense.[130] There is no reason to separate the biochemistry of DNA molecules from other biochemical functions.

Activating and Targeting Natural Genetic Engineering

The answer to our second question is definitely "Yes." Cells can control and target NGE. The first way we can document these assertions is to tabulate the many stimuli that activate episodes of genome change and the many different ways that NGE events are targeted within the genome.[131] With respect to cell NGE targeting capabilities, it is important to note that they utilize the same basic molecular interaction specificities as other genome control functions: protein recognition of DNA structure or sequence, DNA-DNA and RNA-DNA sequence pairing, and protein-protein binding.[132]

The Effects of Hybridization

From an ecological, population, and evolutionary perspective, among the most important stimuli for genome restructuring by NGE are mating events between individuals from different populations or different species.[133] The work of Stebbins on abrupt speciation by interspecific hybridization, mentioned earlier, is but one example of many where interspecific hybridization leads to rapid changes in genome structure and phenotype that generate new species of plants or animals.[134]

Within-species genome destabilization goes by the name "hybrid dysgenesis." Hybrid dysgenesis has been documented in insects and mammals; it reflects the way mobile elements establish themselves in the genome of a species after horizontal transfer to the germline of an individual, presumably by some kind of viral, symbiotic or parasitic vector. Hybrid dysgenesis manifests itself when a sperm introduces chromosomes carrying transposons, retrotransposons or ERVs into an egg cell that lacks the epigenetic controls to inhibit their expression. The results of such an introduction are activation of that element and high frequencies of transposition and chromosome restructuring during germline development. Sometimes the effects of hybrid dysgenesis on germline development in the progeny are so severe that the offspring are sterile. But often, dysgenic germline development successfully produces gametes carrying a dramatically restructured genome. Since these genome changes occur before meiosis, hybrid dysgenic progeny can produce multiple gametes carrying the same restructured chromosomes. In species that produce multiple progeny from a single mating, these gametes can lead to the formation of siblings capable of interbreeding and founding a population with a new genome architecture.[135]

Interspecific hybridization has an analogous destabilizing effect on the progeny genomes, altering epigenetic modifications, transcription patterns, splicing patterns, and genome stability.[136] Sometimes these destabilization effects last two or more generations. The precise molecular basis of this destabilization is not as well defined as in hybrid dysgenesis, but we can reasonably hypothesize that species differ in their epigenetic control systems.[137] According to this hypothesis, NGE operators in chromosomes from the male parent would not be properly regulated and would initiate an episode of genome instability. In any event, interspecific hybridization as a trigger for genome change is so important in evolution because it is a causal link between failing species and episodes of genome restructuring. Individuals will normally mate with members of their own species. But when populations are depleted because of some ecological disruption, they will not find normal partners and consequently will mate with other species. This feedback process promotes evolutionary innovation because the hybrid progeny undergo germline destabilization for at least one and often for several generations. In other words, the ecological conditions that require genome renewal result in precisely those sexual events that stimulate the needed natural-genetic engineering processes.

THE GENOME AS MEMORY SYSTEM WITHIN A COGNITIVE CELL

Evolution science, like the rest of biology, will have to incorporate two new fundamental concepts:

1. The genome is a RW memory device subject to modification by the cell at all biological time scales.
2. Cells are cognitive entities operating on acquired information about external conditions and internal operations.

The RW Genome

The RW nature of the genome is basically a summary statement about the last six decades of molecular discoveries about genome regulation and genome change. Regulatory studies have revealed transient modes of genome modification that operate largely within the cell cycle. These involve the formation of specific "nucleoprotein" complexes between DNA, RNA, and proteins that control the replication and expression of the genome.[138] Focused mainly on problems of multicellular development, regulatory studies have also uncovered the molecular basis of epigenetic modifications to the genome, which are heritable and operate over multiple cell generations in somatic development and sometimes over multiple organismal generations.[139] The key novelty in thinking about regulation by transient nucleoprotein complexes and epigenetic modifications as part of a RW memory system is to realize that they are more or less temporary inscriptions on the genome which do not alter DNA sequence information. The genome operates on three biological time scales:

1. Within the cell cycle, where transient nucleoprotein complexes predominate.
2. Over multiple cell cycles, as in somatic development of multicellular organisms, where epigenetic modifications are most important.[140] There are also some cases of DNA restructuring that play a role in these intermediate term inscriptions.[141]

3. Over evolutionary time, where symbiogenesis, horizontal transfer, and other genome sequence and structural changes predominate.[142] However, the role of stable, transgenerational epigenetic modifications also play an as-yet-to-be-defined role in this evolutionary form of genome inscription.[143]

As outlined earlier, the study of changes in genome sequence information have revealed a whole world of cell functions that restructure DNA molecules at all levels, from single nucleotide substitutions to massive genome rearrangements. As whole genome sequencing reveals, each new taxon has its own characteristic genome features. In other words, natural genetic engineering constitutes the active process of genome rewriting in evolution.

The Cognitive Cell and the Modern Research Agenda

The area where we have the most to learn is in what I have called cellular cognition. How do cells acquire and process information about external and internal conditions and make decisions about the appropriate biochemical and biomechanical operations to undertake? We know a lot about the different kinds of molecules that comprise cell sensory receptors, signal transduction pathways, and internal networks, as well as about their biochemical and biophysical interactions. But we have virtually no understanding of how these networks operate. What logical and informatic principles do they use? How are algorithmic computations performed? In terms of evolutionary biology, how do these cell networks use the tools available for regulating and targeting natural genetic engineering to generate useful adaptations to new conditions? These and many other questions need answers in order to take biology forward in this new century. I wish I had the answers. Nonetheless, as I used to explain to my students, and as Lynn Margulis would surely agree, you know you are truly doing science when you have questions to ask but cannot anticipate the answers. In that situation, you are guaranteed to learn something new.

8.

SUSTAINABLE DEVELOPMENT

Living with Systems

SUSAN OYAMA

DEVELOPMENT: EARTH, LIFE, AND SYSTEM

This chapter is about development and how it is understood in developmental systems theory (DST). What are its relations to contingency and evolution, and what does it take to produce and sustain it? Traditionally, development is viewed as quite autonomous—internally driven and highly regulated (even "programmed")—while conditions outside the organism are placed outside conceptually as well, taken for granted or admitted as random "influences" without being integral to formative processes themselves. In both developmental and evolutionary theory, in fact, customary distinctions between the innate and the acquired typically cast genes and environments in contrasting roles. Treating them in a systems framework, however, allows us to frame ontogeny (development) more expansively, explicitly incorporating the temporally and spatially extended ensembles involved in bringing an organism into being, helping it grow, change, make its way (which is also, in several senses, to make its worlds), and perhaps reproduce. Indeed, in DST, the plasticity, redundancy, reciprocal dependencies and integrated functioning of multiscaled developmental complexes rework gene-environment relationships, along with those between ontogeny and phylogeny, and even among disciplines.

Understood in the inclusive manner advocated here, in fact, development connects the major themes of this volume: Earth, life, and system, environment and evolution. Our planet is as we know it because it has entered into eons of ever more numerous interwoven developmental pathways: organisms develop *with* and *through*, as well as *in*, environments, including living ones, altering them as they are themselves transformed. Narrow definitions of evolution can thus be enlarged and enlivened by adopting a deeply contextual (socially/ecologically embedded) notion of development in which phylogenies—evolutionary histories—trace flexible continuity and change in ways of changing, with all the bodily and worldly involvement that entails. Freed from traditional internalist assumptions (innateness, genetic "control," etc.), this is an ample concept, its temporal and spatial compass extended to take in the whole life span, incorporating into its very workings (endosymbiotically!) the environments that help constitute it. For development to be sustainable, of course, its ever-changing internal and external conditions must themselves be sustained, or otherwise made appropriately available. A developmental system (DS) is an organism (or other developing entity) *and its developmentally relevant environments*, all approachable on multiple scales of space and time: for instance, we can study within the same frame somatic cells and their immediate surrounds, or organism-environment encounters as diverse as wind desiccation or sibling play.[1]

To some, *system* implies homeostatically regulated stability, and as we shall see, organisms themselves have frequently been considered to *be* such systems, fairly self-sufficient and distinct from their surroundings. This is consistent with the convention just adverted to, of treating development as internally controlled. And depending on conditions, including those of observation, DSs can indeed demonstrate great regularity (say, certain metabolic cycles and motor routines). But such orderliness need not be criterial, and processes and constituents can be more variable than is immediately evident. A DS's workings run the gamut from robust reliability to extremely loose, variable associations among its *interactants* (abiotic and biotic participants in a DS, including the changing organism). Such changing complexes are "systems" in the sense that the focus is on the shifting relationships among the many factors that influence development. Traditional understandings of organisms as enclosed and autonomous, then, should not be transferred to DSs; rather, the heterogeneity, openness, and fluid interdependency so often found in development should alter our view of the organisms that affect, and are continuously made by, the constructive processes[2] that produce life courses in all their variety and regularity. Later we will return to this view of systems as not necessarily well behaved. For now, let us note that what enters into a developmental interaction, and the results of that interaction, are *systemically contingent*: they depend on the rest of the system.

In developmental systems theory, then, environments are not external or secondary to development, supporting it (or not), even competing with it for explanatory dominance. Far from being "nurture" to the genes' "nature," these developmental contexts, organic and inorganic, multifarious and in transition, are implicated in the ontogenetic formation of natures. DSs are open-ended (they do not terminate at adulthood, say, or reproduction, and their courses and possibilities are not uniform or preordained), yet constrained and constraining (not just *anything* can happen). In such contingent, often permeable systems, organisms come into being and live out their brief or protracted lives. Conceived in these terms, an ontogeny—a developmental course—integrates organisms with their developmental worlds, pressing academic specialties toward the sorts of collaboration that do justice to those relations. In fact, by challenging some of the ways that scholars have traditionally distinguished themselves and their fields (sociologists questioning psychology's preoccupation with individuals, biologists and anthropologists lining up along nature/nurture divisions), DSs virtually demand a transdisciplinary approach.

Developmental systems give rise to the kind of reliability that can suggest internal control *as well as* to the responsive variability that some associate with shaping from without. As we will see later, this versatility still leads many to classify traits as innate or acquired, but DST does not decompose organisms according to contrasting sources of developmental input (e.g., bodies mostly genetic, behavior less so). Dispensing with such *developmental dualism*, it directs attention to concrete interactive construction of organisms and their surrounds, including much that goes unexamined in more conventional approaches.

In many circles, conceptual dualisms have properly fallen into disrepute, but one finds scant agreement about alternatives, or about how far the rethinkings should reach. Accordingly, I will sketch DST's (potentially misunderstood) emphasis on contingency and its distinctive reworking of nature/nurture (innate/acquired) contrasts, showing how these previous views have helped to perpetuate problematic notions. I will offer a reconstruction of development and evolution that relates them more intimately and links them to diverse disciplinary concerns. Recent work on developmental plasticity will come into play, and I will disambiguate my intentionally multivocal title, to relate sustainability in development to its more familiar ecological and economic senses; living beings, after all, are always integrated with their environments, though not always to their benefit.[3] This chapter will conclude with some thoughts on sustaining communication about the systems with and within which we live.

All this asks us to disregard some influential classificatory and inferential conventions. Lynn Margulis, of course, was known for such scandalous breaches. She and I traveled different roads, which seldom crossed. Yet our approaches had telling commonalities: a respect for complex dynamics at diverse scales, an affection for systems of wildly heterogeneous parts, and an irreverent attitude toward certain time-honored distinctions.[4] In addition, motlies of great and small environments were important to Lynn's work on Gaia and symbiosis, as they have been to my own writings on ontogenetic embeddedness. Flexibility and inventiveness over developmental and evolutionary time have been pivotal for us both. Her single-celled protagonists endured engulfment and invasion; often under duress, each participant in these interactions became another's immediate environment (part of its DS), the one adjusting, and adjusting to, the other. Blithely miscegenistic, oblivious to species essences and individual identities, they cobbled together their ingenious modes of cohabitation, helping to make life's sprawling history. Lynn and I never compared notes on any of this, yet my regard for her always carried, in addition to astonishment and admiration, a bit of fellow-feeling. Perhaps I recognized in her a bracing appreciation of the rigors, hazards, and satisfactions of life on the edges.

CONTINGENCY: CHANCE AND CAUSAL DEPENDENCY

The notion of contingency is pivotal to my project. Some caution is in order, however, for its usual sense, of chance or unpredictability, can overshadow a second one: causal dependency.[5] We see the first (chancy) meaning in biology's association of contingency with largely capricious phylogenetic and ontogenetic environments. The vagaries of evolutionary milieus are said to shape species' genomes, while accidents of development add acquired detail, insult, or quirk to the resulting innate base. Evolutionary happenstance, then, supposedly produces genetically encoded "natures" (instincts, biological bases), leaving development to supply their variable "nurture." What is ontogenetically acquired, however, is barred from *evolutionary* significance, because, in this view, only genes are passed on.

A related definition of contingency points to the causal dependencies that presumably *explain* the particular events considered predictable or not. An outcome contingent in this sense is not inherently chancy; it is as probable as that on which it depends—the complex of factors giving rise to it. An interaction in a DS can involve many such contingencies (more than are usually considered, let alone explored). These depend in part on the investigator's interests and breadth or depth of focus, a matter to which we will return. For now, a few straightforward examples: a drug may have a quite reliable average effect in many studies, but finer or broader investigation could well show that men and women react differently, or that the effect varies with the time of day, and there is always a social context. Factors may interact statistically, the impact of one depending on another. Sea turtles reared at constant temperature produce offspring of only one sex, and amphibians are less susceptible to damage from pesticides in the lab than in the "wild," where pesticide toxicity is heightened by chemical cues from predators.[6] Predictability in DSs is correspondingly contingent, but sufficiently regular circumstances can fade into the background, lending outcomes such an air of inevitability and self-sufficiency that their contributions may be written out of an account altogether.[7] DST draws attention to the specific events and relations that constitute development, summoning them downstage and giving formative credit where it is due. Identifying oft-ignored generative backgrounds discourages premature inferences while widening the domain of the conceivable. What uncertainty is found in such developmental descriptions does not necessarily attach to external factors, any more than regularity is conferred by internal ones; both arise from contingent interactions in developmental systems.

In 1994, Jan Sapp observed that the notion of interdependence posed major difficulties for biology.[8] As we shall see, it still does, and not only for biologists. Interdependencies allow for analysis of variation in outcome, but they prevent us from dividing organisms or their features—phenotypes themselves—along nature/nurture fault lines.[9] It is not that "both nature and nurture are needed" to make an organism, as popular wisdom has it, but that it is a mistake to depict them as contrasting causes to begin with. The picture of development being presented here thus undermines many of the presuppositions of practices that classify ontogenetic processes and products—including their relative predictability—according to the relative contributions of genes and environments. Natures cannot be identified with genes, whose various developmental influences are manifested only through mutual dependencies with their own immediate or more distant environments. "Nurture," meanwhile, is just what I would term "development" or "ontogeny," and thus applies to all features. The moth-eaten oppositions themselves must go.

Despite the internalist connotations of *system* alluded to earlier, I believe systems-talk of a certain sort can be valuable for addressing these topics. Peter Taylor writes of approaching systems either as well-bounded, largely autonomous and self-regulating (the "strong" sense), or more loosely, as characterized by "unruly complexity," in which boundaries, scales, and levels are unclear, external relations are complicated, and structures can change as heterogeneous and mutable components interact. His subject is ecosystems, but he applies his framework broadly.[10] In many respects, the complexes I describe lend themselves better to his unruly complexity stance (and his related notion of "intersecting processes") than to traditional systems perspectives, which, he observes, tend to be associated with fairly simple causal schemes and prescriptions.[11]

Despite the sometimes impressive variation in developmental processes, furthermore, degrees and kinds of order can certainly be observed, in both their gross and fine features. That these can be analytically tractable, and can appear to be highly regular, even across generations,[12] invites a seamless but mischievous slide between ideas of gene transmission and trait transmission. But however similar they may be to parental ones, traits must develop. By suggesting genes being "mapped" onto/into organisms, transmission-talk reduces periodic ontogenies to DNA transfer, while obliterating the distance between molecular and organismic levels.[13] The protean spatial and temporal relations of development are lost; if organisms must be explicable in terms of genetic "information," DNA transmission becomes the needle's eye through which "innate" characters must pass.[14]

Diagrams of heredity tend, in fact, to feature genetic arrows between generations, with dead-end Ps (phenotypes) thrown off at an angle from the Gs' (genes') resolute trajectory through time.[15] The collapse of (p)henotypes into (g)enes "for" them may seem an innocent synecdoche, with parts standing for wholes, or vice versa, but it confers on DNA molecules the generative powers of entire developmental systems: if only genes are passed on, they must be able to make the next round of phenotypes.[16] But as the discussion of plasticity will make clear, there is more to intergenerational relations than such point-to-point correspondences, even allowing for substantial "chance" variation. In the language of coding and mapping, *redundancy* is observed if several pathways converge on the same trait; different genotypes, for instance, can be equivalent with respect to that result. Then there is *ambiguity*: one genotype involved in multiple phenotypes.[17] As I did with the mapping metaphor, I am tracing the genotype-to-phenotype relationship to connect with conventional presentations, but similar points can be made for other interactants. It is now common knowledge that a gene may be linked to divergent developmental outcomes, and similar ambiguity is possible with other factors (food for one is toxin to another, a stimulus has different effects at different ages or in different settings). In phenocopies, meanwhile, a given trait can be redundantly associated with particular genes *or* impinging environments. Once stark G → P arrows are supplemented by converging and diverging paths reflecting even a few developmental contingencies, the treachery of the synecdoche is evident: it makes ontogeny and environments disappear, a sleight/slight whose import will become evident.

ENVIRONMENTS, SYSTEMS, AND NATURES

When environmental influences are seen to be deeply implicated in the very processes of formation, nature/nurture dichotomies dissolve, along with genetic programs, recipes, and other such representations. It is not traits that are inherited, I have argued, but the developmental means or resources for their construction. These become available for developmental interaction—are constituted *as interactants*—throughout life, and equivalent interactions may eventually affect another generation (if this is an organism with generations). Inheritance is generally treated as the receiving end of transmission, but I often find such vocabulary too mischievous to use, not least because, as many other commentators on heritability have observed, it tends to confuse population patterns with descriptions of development.

Even if they are closely correlated with lineage, traits are not literally transmitted ("given" to offspring, as "information" or "instructions"). Neither, therefore, are they inherited ("gotten") as such. They must develop from available means. A retooled notion of heredity would point in other directions.[18] The ears you "got" from your grandmother, the fact that you have two of them, and the piety encouraged by what they take in, would all be seen as the results of development from a vast number of interactants, some tracing to your parents, some not, some widely distributed among organisms, others not. Whatever their degree of predictability, the developmental significance of these contributions would not be prespecified but would emerge in constructive interaction.[19] Under different circumstances your ears might not have resembled Granny's, or they might have been like hers even with different conditions, genetic and otherwise. All this goes far beyond conventional interactionism, which masks its dualisms with sanitized vocabulary: genetic potentials or propensities, say, "interacting" with learning or culture.[20] But what if there are not two kinds of causal pathways to begin with?

Natures are phenotypic, mutable products of ongoing processes of development. A useful frame-destabilizer, this formulation itself eventually needs contextualization, lest *product* be read as *completed*, or as *effect* but not *cause*. Neither prefigured nor necessarily fixed, natures are just (!) organisms in their worlds, influencing their own development, interactants in their (and others') DSs. What a nature is at one moment influences what it may become in the next, in part by influencing its (living and nonliving) surrounds. Yet our languages seem better at distinguishing causes from effects, or agents from patients, than at capturing processes in which factors depend on each other and changing products enter into their own transformation.

There is more to nature/nurture distinctions, however, than causal attribution. An obscure notion of genetic/biological essence helps explain why controversies recur.[21] When people ask whether sex differences, say, are "real," I doubt they have in mind statistical reliability (one meaning of *real* to scientists). They might mean "irreducibly different," but that is unanswerable, for its implied range is all possible conditions. I submit that this query is at heart a demand for the *really* real, beyond mere observables. Its para-empirical spin suggests an evolutionary essence: what is meant to be, what Nature intended, what really is. This normative loading would explain why discussions of "nature" are so often politically and morally fraught, and indeed, why feminists, among others, have been so alert to their undertones.[22] Hence the intense public interest in reports of murderous, empathetic, or homosexual birds, rodents, or chimps. Nonhumans have long been asked to reveal us to ourselves, showing us the beasts we are not, or displaying what we—deep down—most truly are. I do not minimize the variety or subtlety of other creatures' behavior, but I do question some of the meanings scientists and others seek to derive from it. The suggestion of cross-species (or, in other contexts, cross-cultural) resemblances seems to encourage a sense of revelation, of glimpsing a cryptic truth whose existence and import for our own lives transcend real-life particulars. What is "innate" in this sense is already there, defining us as the "acquired" does not. Often assumed to be necessary for characterizing organisms or species, these preformations and essences continue to be widely accepted inside and outside science, however hedged or denied.[23]

Like scientific consensus, popular understanding increasingly allows that some parts of biological "nature" can be suppressed. What neither imagines is that there is no *there* there. Or, to think it another way, that there is *always* a there there, and it is phenotypic: changing organisms, with changing possibilities, always in their changing surroundings. These are interrelated. No snippet of DNA enters into developmental construction unless it is made relevant by—in this case, constituted as chemical environment to—other interactants. It may seem odd to refer to DNA as "environment," but surrounds have scale, and interactants are environment to each other (consider symbiosis, or epistatic gene interactions, in which one gene's impact depends on another's). If a segment of DNA is in the effective neighborhood of certain other molecules, it may be involved in making a product. A transcribable sequence, that is, *becomes* a gene, or part of one, only in context, often after much snipping and stitching. Under different circumstances it could become part of another one—or not be engaged at all. Geneticist Wes Jackson defines a gene as one or more stretches of DNA with the context that allows transcription.[24] Usually considered the prime movers of development, genes begin to sound like its products, arising through interactions of diverse factors that may produce a complex that might assemble a protein. Odd things happen to master molecules when viewed through a DS lens. It is not that the gene actually *is* an effect and not a cause, but that, as was the case with the phenotype as a whole, seeing it as part of a DS makes the search for original causes less compelling, and the distinction between causes and their effects less than simple.

Developmental significance is constructed over time. Far from being given once and for all, natures are continually being "given." To conclude from this construction-on-the-fly that everything is up for grabs, though, is to deploy exactly the wrong inferential habits.[25]

DEVELOPMENTAL DUALISM, PLASTICITY, AND EVOLUTION

The consequences of viewing organisms as always in DSs extend to evolutionary theory, where the blackboxing of development is under pressure from hybrid fields like evo-devo and eco-evo-devo,[26] and from the burgeoning interest in developmental plasticity.[27] Yet dualisms still exert drag, for they are built into theory itself. Insofar as developmentalists rely on a binary frame, they reinforce the very structures that evolutionists feed back to them as the distinction between innate and acquired characters. We find, then, a reverberating circuit of genecentrisms: (1) heredity is narrowly defined as transmission of genes; (2) development is their "expression"; (3) evolution is change in their frequencies. In this narrative, experiential traces on/in the organism lack evolutionary import; only genes' imperturbable "flow" through organisms counts.[28]

John Maynard Smith defends this orthodoxy in a paper notable for its unapologetic bluntness: "biologists draw a distinction between two types of causal chain, genetic and environmental" and "evolutionary changes are changes in nature, not nurture."[29] Genetic information is *intentional*, he explains, and this semanticity or meaningfulness derives from a history of natural selection. Developmentalists study its translation into adult organisms (making the formation of juveniles puzzling, but never mind), while evolutionists explain the *origin* of the information: selection designs the genome that "generates an adapted organism."[30] This tidy division of labor pretty much excludes developmental considerations from evolutionary studies (an exclusion strenuously contested by the developmentally inflected approaches just mentioned). Surely, one might think, blackboxing does no harm; if ontogeny is just the translation of preexisting information (G "mapping" to P), why should it command evolutionists' attention? (Apologies to actual translators of actual texts.)

Given such mutual conceptual support, it is hard to envision a description of development or evolution, let alone their relations, without the developmental dualism of those discrete causal chains. The genecentric circuit *makes it make sense* that selection should be of genes for traits, development should be their readout, and evolution should be change in their relative numbers. Maynard Smith does not deny developmental plasticity. He does not have to: Although environmental "fluctuations" can leave traces, he assures us, they are *noise*, not information, and their effects are confined to one lifetime (192). (Notice again the association of environments with random perturbation—noise, not signal.) Having excluded environmental influences from the transmitted signal, he deems it "reasonable" to think "the genome contains enough information to specify the form of the adult," because it is hard to imagine where else it could come from (186). Alternatives become unimaginable when their possibilities are precluded beforehand (e.g., developmental role of environments, evolutionary role of development). Theoretical frameworks too have their systemic qualities, and his puts plasticity firmly beside the point: Whatever an organism's ontogenetic fortunes, it passes on only its DNA, so evolution is about nature, not nurture.

Now contemplate the evolutionary process as a succession of diverse developmental systems. *Plasticity* can refer to change in individuals (by learning, muscle strengthening, etc.) or to the developmental ambiguity demonstrated by variation across otherwise similar individuals exposed to different conditions. As we saw, though, developmental outcomes can be less variable than their ensembles of interactants, and recent writings on plasticity and evolution highlight canalization or buffering—redundant mapping to a phenotype. For any of these comparisons to be made among individuals, though, developmental changes must occur in each one. Recent plasticity work has centered on environmental changes that perturb development, unmasking (exposing to interaction, or in the perspective being suggested here, constituting as interactants) hitherto cryptic or silent genetic variants, sometimes influencing evolutionary change.

This idea, like the whole topic of development in evolution, has fairly deep historical roots but is still controversial; it is perhaps best viewed against a backdrop of expectations about determinate gene-character mapping and genes as guardians of species type. The claim is that canalization can produce a species-common phenotype despite substantial environmental or genetic variation.[31] Genetic variants lacking correlated phenotypic variants are not "selected out," and they accumulate silently in the population, available for eventual engagement in novel responses to new circumstances.[32] As noted, plastic or ambiguous genotype-phenotype relations are implicated when a different organism is produced under changed conditions.[33] If a previously cryptic genetic variant is now correlated with a new phenotype, it might be selected; a process starting with developmental flexibility could lead to genetic change in the population.

When the "same" gross phenotypic feature persists over long stretches of evolutionary time, biologists may speak of conservation or stasis, but its developmental interactants and construction may have shifted considerably.[34] To this literature's focus on masked genetic variation, I would add previously uninvolved environmental factors, internal and external, which could also facilitate developmental innovations. Altered intracellular processes tweak immediate-neighbor relations, modifying biochemical possibilities, including gene interactions. Changed organisms may have changed effective external environments as well, for relevant environments depend on the organism (just as relevant aspects of the latter depend on the former).[35] In ontogeny's progressive unmasking and obscuring of interactants, the ground for further events moves: Constraints and possibilities arise as previously absent or unengaged factors come into play and others shift roles or drop out. Development just *is* more or less ordered plasticity: tissues grow, cells die, bones remodel, memories form and reform—the very dynamics of developmental construction suppressed by the transmission synecdoche. It is interactive construction all the way down.

Evolutionary theorists have tended to consider development an "internal constraint" on natural selection, a conservative brake on change and variation.[36] In plasticity studies, though, developmental robustness participates in the *generation* of evolutionary diversity. Let us return to that circuit of genecentric (or infocentric) definitions. Its linked construals feed, and feed on, developmental dualism, maintaining a landscape on which Maynard Smith plausibly falters at the very idea that organism-specifying "information" could come from anywhere but the genome.[37] But if (1) an organism inherits a mobile set of intra- and extraorganismic interactants—developmental resources it "gets" over time and from which it is made and remade; (2) development becomes their participation in interactive construction, from the molecular level on up, as discarnate "information" is replaced by temporally and spatially extended beings in their changing worlds; and (3) evolution expands as well, from naked alleles bobbing in gene pools to the derivational history of overlapping, interacting developmental systems.[38]

No coherent line of inquiry is excluded by this reworked DS circuit, though some questions and inferences may not make the transition unscathed. One can still track alleles and/or other interactants across generations, study multigenerational niche construction,[39] or compare phylogenetic relatives, assuming rigorous identification methods are available.[40] Traditional distinctions between inherited and acquired characters are, *pace* Maynard Smith, unnecessary, although degrees and kinds of intra- and intergenerational similarity can be explored.

SUSTAINING DEVELOPMENT, SUSTAINING ENVIRONMENTS

My ambiguous chapter title, then: *Sustainable Development*. So far we have considered the contingent ontogenies of phenotypic natures, species-common and otherwise, and the evolution of the systems that produce them. Once developmentally consequential surrounds are acknowledged as full players in formative dynamics, DS sustainability is seen to be inseparable from more familiar senses of the word in ecology or economics.[41] This fits with both the interdisciplinary reach of the present collection and its systems orientation. My discussion moves, accordingly, to that systemic continuity, then closes with the challenges of contending with our worlds' multiscaled complexity.

For a population to persist, some significant proportion of its members' development must be sustainable, across not one but many life spans, and it is both possible and satisfying to see such continuity in terms of inter-connected ontogenetic contingencies within, across, and beyond body limits, knitting developmental and ecological/economic systems together. Recall that many accounts of plasticity in evolution begin with just the dependable phenotypes that inspire talk of transmitted, conserved, or even static traits. Yet it is their developmental heterogeneity that is key, their resilient responses that start the ontogenetic and phylogenetic cascades. Genetic and environmental variations concealed beneath canalization's cloak *become* relevant (enter into developmental interactions) in altered circumstances, and new phenotypes change the patterns of selection, with or without subsequent genetic mutation.[42] Alongside their regularly cultivated potatoes, traditional Andean farmers maintain other varieties, pressing them into service in hard times.[43] Farmers and (most of) their practices persist as the makeup of harvestable crop varies. Inside or outside the skin, standby diversity can buffer some changes while opening ways for others to take hold. Notions of difference and persistence, though, depend on focus. What is "sustained" depends on what we are tracking, and at what level, so that different metrics support quite different narratives (conserved phenotype or shifting developmental processes; particular potato crop or family/community; "the Church" or the practices/experiences of its members). Deviations can lead back to (redundantly converge upon) the usual path, or diverge ambiguously into the new. Recall the way development—life—compromised the spare linearity of the transmission metaphor.

Conservationists often speak of stewarding forests and wildlife for their grandchildren. Meanwhile, Richard Levins and Richard Lewontin tell of a question asked by one of their offspring upon encountering any new animal: "What does it do to children?"[44] Some adults' concern with legacy and a child's trepidation give us two faces of sustainability: what we do to/for/with our surroundings, and what they do to/for/with us. To oversimplify, these are the directions of influence in a DS, and the adults clearly know that causal arrows can veer and return.[45] To nurture a child is to tend its relevant surround. Mutually constituting organisms and environments bring to the fore geography, ecology, the social sciences, and other areas of study, as seemingly remote factors (foreign economies, oceanic pH) are seen to be part of organisms' fluctuating, recurrent, or standing developmental conditions: their means of continued life. Sustaining de-

velopment, then, entails sustaining developmental environments, not because the latter contain or support the former, though they may, but
because they are *part* of it. Development thus connects to considerations
of environmental sustainability in myriad ways.

In this volume, Sankar Chatterjee describes early Earth's basins, temperature gradients, convection currents, and sticky substrates, a kind of
planetary morphology and protophysiology for emerging life. Similarly,
what psychobiologists call "nonobvious" factors can be subtle developmental circumstances, sometimes found or provided by conspecifics, even by
the organism itself.[46] The two disparate fields connect in that an ecological feature that helps constitute a life course, species-common or not, is
an interactant in that developmental system, and DSs are not just *like* ecosystems; they are finally the same systems.[47] Yet, ecological structure and
dynamics are not generally worked into developmental and evolutionary
theories.[48] DST can gather these threads into a coherent but flexible framework, showing the interdigitation of the developmental, evolutionary, and
ecological while insisting on distinctions appropriate to particular foci:
organism or community scale, metabolic or geological time, and so on.
Constructive interactions undo visions of self-sufficient, prespecified ontogenies and constitutively unpredictable environments, while an expan-
sive framing and recognition of systemic contingency raise just the
questions a researcher needs to ask about organisms' relations with their
worlds, without separate inheritance channels or other nature/nurture
impedimenta prestructuring the analytic field: what conditions are needed
for this feature to appear; how direct, redundant, and reliable are the relationships; with what do they vary; are there compensatory mechanisms;
on what further conditions do these depend; what kind of feedback is there
to other interactants?

Observations of ecosystemic or even planetary integration have led
some to liken such large-scaled entities to (unified) organisms. Organisms'
heterogeneous composition, meanwhile, points others in the opposite direction, toward viewing "individuals" as ecosystems themselves, for we
are legion, largely composed of microbial multitudes within us and on
us (not an easy distinction, what/whoever "us" is—consider the human
gut, and our doughnut topology).[49]

I likened the flexible Andean potato-growing system to a developmental system, but finally the crop is *part* of these people's DSs, as are their

resources for cultivation, including those standby potato strains, and perhaps, further out, their urban markets, where nonstandard tubers may now be valued as specialties. Any population's persistence in its region hinges on such interactants, which can themselves be affected by larger-scale (and smaller-scale) changes.[50] Although sustainability often sports a (greenish) halo, furthermore, the global slave trade endured a long time, maintained by dense economic, geopolitical, cultural, and other linkages; it took major upheavals to interrupt that incarnation of it, and various kinds of trafficking in persons go on today. This is not pure snark on my part; sometimes the challenge is to *disrupt* a state, pattern, or trend; consider invasive species (including us) and the feedback loops that so frequently stabilize endemic corruption, social and economic inequality, or rampant ecosystem damage. These last are dire precisely because of the systemic continuity of development with the ecological and economic realms. No longer can the environment be regarded as just the place where we do our thing. Popular imagination now pulses with malign molecules passing from pesticides to pollinators, plastics to puddings, with a corresponding rise in demand for "healthy" environments.[51] Drilling towns and redlined neighborhoods are at once developmental, medical, ecological, and economic/political/cultural phenomena.

Many organisms termed "ecosystem engineers," such as kelp seaweed, lack the glamour of top predators, but they create habitat and other resources for diverse organisms—until they don't.[52] Sea urchins eat ecosystem-engineering kelp. Sea otters eat urchins, preventing them from destroying kelp forests. Killer whales deprived of their usual prey by overfishing hunt otters, with downstream effects on the massive and usually dependable kelp habitats and their innumerable beneficiaries.[53] Thus, otters indirectly tend kelp, part of their own standing developmental conditions, while the whales that eat them can, even more indirectly but potentially ruinously, compromise that nurturing. Widespread and/or prolonged damage to such rich systems, of course, can change the size and distribution of populations, even extinguish them, by altering their developmental environments. Given such conjunctions of development and ecology, tales of natural selection can be retold in terms of the relative sustainability and resilience of developmental systems, variant organisms making their livings (or not) with/in their varying worlds. Ecological, developmental, and evolutionary questions are mutually implicating:

the parasites, climates, and food sources frequently cited as selection "pressures" are so deemed because they affect organisms' developmental (including reproductive) fortunes. One generation influences what happens in the next, not by "transmitting" traits, but by affecting the interactants (symbionts, DNA, status, epigenetic marks, wealth) that will be available for construction, at the moment or down the line. Kelp forests, for instance, sequester carbon dioxide. Do "developmental context" in one case and "selective pressures" in the other refer to different environments or to different scales and questions? The constructed niches, multiple replicators and information channels in diverse evolutionary models are largely about cross-generational availability of DS interactants. These may affect organisms' life courses, and hence their contributions to later ones.[54] Not all factors are equally significant over every timespan: many will damp out rapidly, while some will recur or endure. Yet as the symbiosis, plasticity, and developmental systems literatures suggest, phylogenetic change need not always be extremely gradual and slow, and complex multiscaled systems afford many opportunities for the kinds of cascades that can move DSs onto different evolutionary paths.[55]

This section links developmental and environmental sustainability. Their joining follows from the conceptions of development as interactive construction, and of inheritance as the "getting"—the developmental engagement—of the contributors (resources or interactants, internal or external to the organism) to the constructive processes. In these processes traits come into being, are altered, maintained, and degraded. Widely distributed in time and space, interactants arise repeatedly, and their role is a partial function of the organism's features, state, and activities. The disciplinary division of labor that so neatly served Maynard Smith's argument in the previous section precludes the very crosstalk that this alternative vision calls for.

SUSTAINING CONVERSATIONS

Crashing kelp forests harm many other organisms, those whales, otters, urchins, and numerous fish species presumably among them. But tolls are taken when they are functioning fine, too, and policies to halt their destruction would produce losers, plausibly including urchins (increased otter populations) and fisherpersons (tighter regulations). Short- and long-term interests of diverse parties are bound to collide, sometimes with grim consequences for individuals or even whole populations. To navigate the plethora of viewpoints, conversations must be sustained across many borders. There will be talk about the workings of these systems, of course, but surely attention must be paid to what kinds of talk will keep (which) conversations going, what perspective will be adopted, what boundaries respected, and whose voices will be heard.

Much of biology has been marked by genecentrism, for instance, and one response to gene talk is to challenge the supposed omnipotence and autonomy of the DNA. James Shapiro in this volume deploys cognitive terms for his cells' astonishing doings. Toward the other end of the size continuum, planetary systems shrink humans to motes: comparing the impact of Margulis' symbiotic vision to the Copernican revolution, Peter Westbroek in this volume describes civilization as something the Earth does through us.[56] The language of agency can be an effective tool in theoretical exposition and dispute, especially when different levels of explanation are being contrasted. It can direct attention and create engaging and accessible narratives while showcasing the marvels of the phenomena thus described. These are legitimate aims, and when faced with what seems an overblown agential account, one is tempted to invoke a countervailing one. Yet dueling agencies risk perpetuating some of the less useful aspects of traditional disciplinary competition, including attempts to identify *the* causes or operative units of change. Such descriptions, that is, can exacerbate the already difficult task of managing diverse system levels (with their associated scholarly allegiances), by attributing the dynamics of the whole to particular components, or privileging one level over others. They can also propagate unreconstructed notions of choice and action.[57] It is through causal embeddedness that things happen, of course, so the systemic interdependencies we have been reviewing are in tension with at least some agent-talk.

It is thus convenient for my purposes that Westbroek[58] describes his remark about the civilizing Earth as made from a particular point of view. He refers to conflicts between the human and natural sciences, and warns against deeming ourselves helpless pawns in the face of scary threats. These look like nimble moves by someone not glued to one spot, for the sense of diminished effectivity that so often accompanies confrontations with "the system," whether bureaucratic, economic, or climatic, is one of the perils of treating humans and systems as alternative agents.[59] It raises the worry that people, overwhelmed, will give up. Calling us "the principal players in the civilizing process," Westbroek seems to deny that agency resides at just one level, and the action that animates his tale about surviving a maelstrom depends critically on engagement in a causal world. Such gestures permit the sort of movement within a heterogeneous system that cross-scale integration would appear to call for.[60] Left for other days and others' labors, though, is the demanding work of reconstructing the agent, of describing activity and possibility without committing to any of the mind-matter, subject-object dualisms lying about like landmines.[61]

Multiple perspectives have been important in this discussion: choice of scale, focal phenomenon, or object—and one reason for the conceptual reconfiguring I am urging is that framing a process amply rather than restrictively can extend the perceptual field, increasing the number of noticeable phenomena and potential perturbations. Think of the psychobiologists' nonobvious developmental factors mentioned earlier, many of which would have been invisible to more traditional researchers. Intervention sites do not have to be *called* parts of a DS to be efficacious, but perspective works in part by bounding the realm of the seeable (and thinkable: recall Maynard Smith's constriction of the notion of information, and his subsequent inability to imagine beyond it). At the same time, increasing our choices puts pressure on us to take responsibility for including, excluding, ranking, deferring. No detached scientists here, peering at nature from no-place. One kind of organism-environment interaction, after all, is research, and professional lookers-at-the-world can deny neither their parts in the exchange nor their embeddedness in their contexts.[62] Squier in this volume tells of theoretical biologist Conrad Waddington's queasy dismay that a chick might, God forbid, eye, approach, even imprint on him. Is this what happens when the one-way mirror comes down and nature looks back?

Boundaries have figured prominently as well—when they are invoked and what they are supposed to signify, how porous they can be, what conceptual barriers they can pose. The absence of determinate ones in my account of DSs will no doubt alarm some, but neither absence nor alarm need halt investigation. In any case, constraints on actual research tend to hang on theoretical considerations, funding, investigative momentum, available tools and skills, and the like. Economic and other (unruly, complex) social systems are notoriously leaky, as are ecosystems—and organisms, of course, if you look closely. People are accustomed to thinking of the last as very self-contained, and part of the anxiety that fuzzily bounded DSs provoke in some could be contagion from conventional internalism about organisms and their ontogenies.[63] Zolli and Healy note that disasters and large world events often uncover previously unrecognized dependencies, some of them crossing political, ecological, and economic divides—contingencies, once again, that can multiply the sites of feasible collaboration and action.[64]

I am convinced that a whole range of conflations, inflations, and conceptual dead-ends can be avoided by paying closer attention to the framings and assumptions that affect the questions we ask and the answers we accept. This may necessitate some recalibration of expectations: fewer quests for foundations and unitary causes, say, or for just the right statistical techniques for uncovering biological bedrock. Eventually the search for the preexisting representations and essences that will provide the "real" explanations of living beings can yield to the everyday (hard) business of discovering what makes a difference in a given system, under what circumstances. However modest this may sound, it brings with it enough complication and subtlety to keep us very busy: what kinds of differences, with what sorts of temporal or other limitations, what kinds of immediate and longer-term consequences? The scale, indeterminate boundaries, and complexity of ecosystems like kelp forests or coral reefs render human decision-making fraught: not all participants can profit from change, whatever their species and conditions, however profit is defined. These systems frequently span national, occupational, cultural, and other borders; keeping communication productively open can test even the most dedicated.[65] Yet the conjunction of the scientific and the political that makes these confrontations so combustible is very much to the point. Sustaining particular ecological relationships requires defining, adjusting, or sustaining social ones as well, but recalling systemic ambiguity and redundancy might keep us at the bargaining table; many routes seem blocked, but there may be more ways forward than we think, and there are sometimes alternate paths to a given end.

A vital part of these negotiations is the airing of economic externalities. I am often sympathetic to demands that "true" or "full" costs of some policy or technology be divulged,[66] but surely a definitive account is no more necessary, or likely, than the true catalog of interactants in a DS. The plea is rather that consequences for easily ignored populations, human and otherwise, be recognized, counted, heard. Like those nonobvious influences that turn out to be developmentally significant, ramifying externalities are by definition beyond the field of focus. But there is more than analogy here. Externalized costs are often, sometimes in quite straightforward ways, developmental influences . . . on *somebody*. Pressure thus builds to bring such externalities to light, these consequential but as yet unnamed impacts and interests: life—and death—off the books.

Much of science, indeed, of living, is about judicious effort, or lack thereof, in the face of uncertainty, and the systems in question are moving targets. The characteristics that make them interesting and important render once-and-for-all descriptions or prescriptions unlikely. Like the primary DST move of honoring/recognizing environments' varied participation in ontogeny, indicating the continuity of DSs with ecological and other systems should widen our field of vision, keeping alive a perimeter of possibility. Not everything will be significant, and we can never capture it all (what could that mean?), but I have suggested that an advantage of taking these relations seriously is that their very multiplicity allows the detection of hitherto unsuspected sites for monitoring, research, and intervention. The counterfactuals with which Taylor probes causality and difference-making are just the kinds of what-ifs that open up an investigation to novel contingencies and consequences. This sensitive *what-iffery* shows scientists as, among other things, engaged and imaginative beings.[67] Indeed, Bruce Clarke in this volume invokes the significance of the "planetary imaginary" in their productions. Certainly the present collection is in part about the disciplining, the forming and reforming, of imaginative vision. By asking for more inclusive accounting, but without hope of a final one, furthermore, we must face our own investments and sensitivities, limits and distinctions.

In a short story called "The Ones Who Walk Away from Omelas," Ursula Le Guin told of a seemingly utopian city of light and contentment. Its inhabitants were happy, but their happiness was not simple. For reasons not made clear, the blessings they enjoyed depended on the abject misery of a single child, whom citizens could and sometimes did view as it cowered in its filthy locked cellar. Le Guin's title refers to those few who, at odd intervals, quietly departed their beautiful city. The story tends to persist in memory, a shadow drawing the eye to another perspective on the contingencies I have referred to.[68] In a DS, outcomes for our chosen entity (citizens?) are foregrounded, reverse effects typically described when they further that inquiry. If we instead consider the spectrum of impacts implicated in sustaining our own lives, we find a growing preoccupation with carbon footprints, organic agriculture, and the like, and such concerns are obviously informed by an awareness of short- and longer-term feedbacks. But once you start looking upstream and downstream, tracing the paths of contributions (not only positive) to our lives, and the consequences (not only negative) of living them, it seems hard to stop short of the utterly horrific.

Maybe the conditions of possibility for our lives recede so easily into the background not just because they are reliable, indirect, subtle (nonobvious?), distant, at an inconvenient scale, or even because the dimensions of our discourse can be so constricted. Perhaps some conditions are too discomfiting, even shameful, to confront. A liberal education teaches about microbes, strangler fig trees, and assorted creatures crowding out, poisoning, and otherwise wreaking havoc on fellow organisms. These too are developmental contingencies. We are all consumers, and our skeins of dependencies tend to circle back, like Waddington's Ouroboros.[69] It is no simple thing to say across what spatial and temporal reaches any of us should attempt to trace the lines of connection, but nobody/thing exists outside them. There are no clean hands. No roads leave our Omelas, though perhaps we can reposition ourselves within it, and while we cannot avoid making choices, we can try to own them, think honestly, and communicate about them, even revise them.

Children seem to be popping up in this narrative, so we might as well attend to them as we form some questions about these interdependencies, for it appears they impose more than cognitive burdens: If we find ourselves at the cellar door, we could ask, "are gazing or turning away our only choices?" Told that it takes a village to sustain a child, we could wonder, "And what does it take to sustain a village?" If we reflect on the production of the things we value, or just take for granted, we could inquire, with the wary youngster we encountered pages back, "What does it do to children?"

PLATE I. Cosmic evolution as depicted by the Exobiology program at NASA Ames Research Center, 1986.

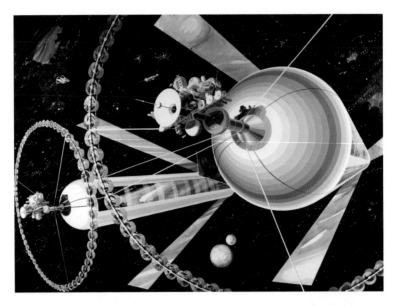

PLATE II. Exterior view of a double cylinder colony. Artwork: Rick Guidice. NASA ID Number AC75-1085.

PLATE III. Interior view of a cylindrical colony, looking out through large windows. Artwork: Rick Guidice. NASA ID Number AC75-1086.

PLATE IV. Endcap view of a cylindrical colony with suspension bridge. Artwork: Don Davis. NASA ID Number AC75-1883.

PLATE V. Portrait of Michiel de Ruyter by Ferdinant Bol (1667). Collection Rijksmuseum Amsterdam.

PLATE VI. Earthrise seen from Apollo 8, December 24, 1968, in its original orientation. NASA.

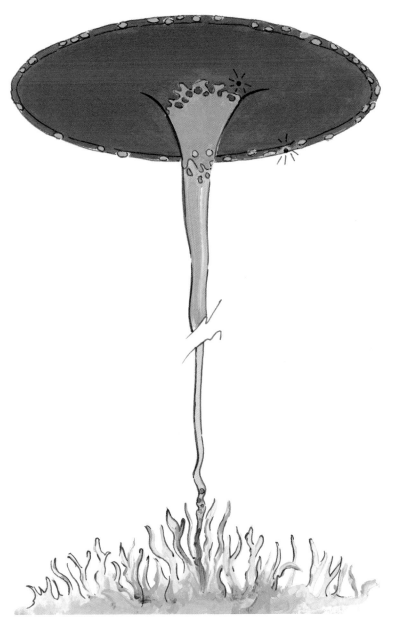

PLATE VII. Artistic impression of the civilizing process in humans (the giant mushroom). The timing of civilization is on the vertical axis and its influence on Earth dynamics at any moment in time is indicated by the size of the corresponding horizontal cross-section. Green, below: biological roots; pink: fire regime; purple: agrarian regime; deep red: industrialization or modernism; green specks on the rim of the toadstool: emerging symbiotism. Illustration by Cees van Nieuwburg.

PLATE VIII. Contemporary modernism from Earth's point of view: a magical slot-machine. Illustration by Cees van Nieuwburg.

9.

BOVINE URBANISM

The Ecological Corpulence of Bos urbanus

CHRISTOPHER WITMORE

> Where there was once a "lifeworld," there must now be
> air-conditioning.
> —PETER SLOTERDIJK

The top of the mill offers an elevated view that encompasses something
approaching the full expanse of a dense city. To the west, a regular grid
of streets extends north for a mile. From foreground to background, thousands of the city's residents stand shoulder to shoulder in silence. Block
after block, their fenced yards are devoid of greenery. To the north, sheet
metal buildings line the main thoroughfare—horse stables followed by
an office building, a hospital followed by what looks like a warehouse;
and beyond, other sundry facilities. To the right and parallel to the main
street, linear mounds of black waste stretch for two-thirds of a mile. Beyond these shadowy grounds lie fields, some green, some fallow. And
somewhere out of sight, a main office building sits by the port of entry,
where every rig must stop to be weighed. All the while, there is the constant smell of manure. Welcome to Cattle City, Texas, population 58,000.

This essay begins the task of understanding Cattle City, what the meat
industry calls a "concentrated animal feeding operation" or CAFO, as an

object of both ecological and archaeological concern.[1] A city of any sort is not an obvious object of ecological thought in the ways that the Amazon rainforest or gray wolves in Montana are, but it is an ecological object nonetheless.[2] My thought here is aimed at understanding the conditions under which residents in such an urban habitat do or do not live and thrive; the dependencies and rapports between various Cattle City interlocutors; the loops and transformations involved in its energy captures and feedbacks. What work goes into sustaining such a city, into upholding its order? What does it take to maintain 58,000 cattle inside a contrived envelope of less than a square mile?[3] I tease out some of the artificial atmospheric conditions (in Peter Sloterdijk's sense) necessary for bovine populaces to be "housed" within their own cities. My ecological thought recognizes systems of interaction that expand in weight and distance to the point of being obese. How many heterogeneous interlocutors does this city draw together or weigh upon? How far does this city extend its reach? Ecological thought reminds us to make explicit from that assemblage of enclosures the spillover that occurs through the maintenance of a life-support system geared toward lives soon to be shortened.

My ecological thought is mindful of the derivation of ecology from the ancient Greek *oikos*, specifically, the corporeal house and the common household, its management and maintenance, its locality and linkages. I will not be so bold as to stray fully into the nearby realms of symbiogenesis or the coevolution of species. Instead, I will look at a situation that arises through the pervasive divergence of species, where more and more humans unhook themselves from an agrarian or pastoral existence, no longer to share lives with their former labors' companions or the living wealth that underwrites their status; where more and more cattle are removed from familiar pastures to become urban residents. Nonetheless, from my angle, Earth, life, and system remain key ingredients in understanding Cattle City: soil and metal, steers and microbes, humans and horses, feedback loops and energy flows. Here, my ecological thought strives to recognize cattle, humans, or horses not as insulated or immunized beings thrust into corporate relationships, but rather as complex material and ecological assemblages, as integrated organisms, which regularly exchange microbes, thus shaping one another in symbiotic codevelopment.[4]

With regard to archaeological thought, long-term human interactions with our fellow creatures and their encompassing milieu have been an abiding concern. One might profitably approach Cattle City as an outcome of a longstanding generalized immunology where living is, according to Peter Sloterdijk, "a measure of defense through which a zone of well-being is walled off against invaders and other bearers of illness."[5] This is a matter not exclusive to history, that is, to a continuity of human experience over the millennia. Rather, it is predominantly one of material genealogy, of both radical discontinuity and the contingency of connections as much through mundane and silent things as through urbane and talkative ones.[6] Thus, my archaeological thought aims to learn from situations lost to practical experience or recall, but whose memory remains in things and the rapports between them.[7] In so doing, I treat Cattle City as a matter that goes beyond archaeology as an academic discipline. I approach this bovine living situation with an archaeological attitude that encompasses ecological concerns. With an aim to incite the imagination in an era of the very short term, an era when immediate reckonings count for more than the experiences accumulated by those who have lived with animals and land for centuries, I take every opportunity to connect present matters to erstwhile scenarios.[8]

To approach Cattle City both ecologically and archaeologically, one can place the conditions required for cattle to be "housed" in their own cities under three categories: collection, accommodation, and sustenance. Collection attends to the basic material conditions that make it possible to gather together and contain 58,000 cattle in one place; accommodation enumerates the ingredients necessary for making a city suitable and providing for bovine-being-toward-slaughter; sustenance delves deeper into those life-supports needed to maintain, to preserve, to uphold, to prop up cattle in a crowded, urban setting. In approaching Cattle City, I take quite seriously the argument that to understand the dynamics of a complex system, it matters where one stands, or sits. For instance, the quadruped bodies of cattle would never gain access to the oblique and elevated view with which we began.

From Childe to Mumford to Cronon, no archaeologist, no philosopher of technology, no historian has looked beyond a city as a human-designed ecology *for* human populations. With Cattle City we encounter a city designed *for containing* animals, though it is still largely *for* humans (one could acknowledge the benefits to other species in the form of manure as fertilizer, animal byproducts for pet food, etc.). Through the essays in this volume runs a shared concern over anthropocentrism along with a counteremphasis on a more modest, less masterful, understanding of the place of humans on the planet. I will add to this thread by placing rapports between steer and steel enclosure, rumen gut bacteria and corn— between the things that are, and make up, Cattle City—on the same footing as that between cowhand and horse, driver and truck. In all of this, I touch lightly on the exceedingly rich sociology of cattle, albeit one that is now heavily mediated by urban architecture as well as endocrinological and microbial intervention.

Cattle City and CAFOs more generally are beset by a litany of problems: pollution/depletion of ground water and climate change (a much-cited study[9] suggests that livestock are responsible for 18 percent of greenhouse gas emissions in terms of CO_2 equivalence), animal exploitation (comparisons of CAFOs to concentration camps are pervasive[10]) and the reductionism that regards living animals as proto-meat, use of antibiotics and the development of antibiotic resistant bacteria, biodiversity loss and genetic manipulation.[11] These problems will escalate in even more catastrophic proportions if global meat production reaches 465 million tons by 2050, as projected by the United Nations' Food and Agriculture Organization (FAO).[12] Such Atlas-like responsibilities situate Cattle City at the heart of struggle for a workable agropastoralism against an overweight meat-industrial complex.[13] Strange though it may seem, Cattle City operates under the guidance of familiar values derived from those long-term practices associated with farmers and herdsmen, two helmsman of an erstwhile agrarian world. Here, I question whether the mixed pragmatics of herdsmen and farmers can provide sound guidance for something larger than monstrous, a bovine urbanism beyond leviathan. I will press such matters further by pointing to considerations of what I call "ecological corpulence," a notion that draws together a raw measure of material weight and energy expenditure with the image of gluttony.

I touch briefly at the end on the question of options. In an era of technical intensification, Cattle City will eventually fall short, not only by failing to silence mounting discontent, not only by failing to mollify those ecological, biopolitical, energy, and ethical burdens that hang over its enclosures, but also by failing to provide the most optimal conditions for exerting control over flesh. The laboratories of meat will usurp bovine urbanisms. Relieved of the ecological emergencies and the ethical alarms raised by those who link human rights with animal suffering (a debate I will refrain from entering here), laboratory meat, in decoupling the breathing animal from edible flesh, will fully put an end to farmers and herders as helmsmen in a new sector of food production in all ways but metaphorically. Still, in passing from a fossil-fuel economy to a new one based on solar, old ways of the past gain new meanings and here we will find new, "lighter" rationales for keeping cows, as fellow symbiotic ensembles, around, perhaps in closer proximity, definitely in lesser numbers.

COLLECTION

Cattle are gregarious animals that form groups, herds. Cattle City collects cattle as herds in containers where every resident (each "head") is allotted a patch of ground, the size of which varies by season—150 square feet in winter, 135 square feet in summer. Fenced lots, open to the sky, are the constitutive element in a vast "life support" system. Divided block by block into a grid of cells, well over 300 animal pens cover more than half of the square mile, upward of 320 acres of Cattle City.[14] Organized from south to north beginning with line A, pen A1, A2, A3, all the way to Z and beyond, expanded capacity equates to multiplying the number of cells. The first section, was constructed in 1969, the second in 1985, and a third in 1997. With each new district, slight design alterations occurred by virtue of what bovine relations were taken into explicit consideration: changes in the grade of slope encouraged the flow of standing water, changes in the width and overall shape of the trough (locally known as the "bunk") followed on the movement of heads in confined spaces, and other changes were made to the fences. Cable, which succumbs under the weight of quadruped bodies clamoring for a slot at the bunk, was abandoned in favor of steel pipe, which refuses to give way under the mass of a 1,400-pound animal. This pipe was also raised in order to provide more depth for consuming feed at the trough.

Movement between fenced cells occurs via a system of alternating roads. One road acts as a drovers' alley. A "lot" of cattle or individual, sick, or wayward steers or heifers can be moved to any pen along these narrow

FIGURE 9-1. Cattle at the bunk. Photo: C. Witmore.

corridors that add up into a network of alleys, which is manipulated by the use of swing gates. Even the arms of these gates are devised so to be opened by a person mounted on a horse, thus their design also takes into account the sociology of cattle, which perceive a peripatetic human as a threat. Lined with troughs, the other road is for distributing feed from large container trucks. Hence, this space is organized: food in the front, animals and waste out the back.

My Cattle City guide describes this system as the materialization of a Cartesian divide, which he frames in management terms—the design of cattle containers follows on a philosophy of "managed-being." On one side, along the feed road, we encounter an ordered, systemized and scheduled world, on the other side, in the drovers' alley, moves a world of biology, animal health, behavior, horses, and handlers. Clashing across this bifurcated space are cultures, values, styles, sensibilities, and practical knowledges. Each side deals with matter of animals in very different ways and where they sit or stand makes a difference—in the pen, on a horse, next to a steer in squeezebox or at a desk, at the conference table, in a control room. The matter of animals for food requires both care and estrangement; this is manifest in, and thus reinforced by, urban design, information structures, and labor management.

In order to specify the relevance of Cattle City for long-term, archaeological understandings of bovine ecologies, and to provide more examples that run against the popular belief that this "space-saving" system of cattle compressed into feedlots arose in 1950s Kansas,[15] allow me to touch lightly upon a far deeper history. From Greece to Rome to colonial America, herdsmen and farmers have followed a round that involved fattening animals for market or sacrifice. A fundamental distinction exists between moving with animals to food versus bringing fodder to livestock within enclosures. Seasonality and locality, property and law, appetite and community, all conspire with respect to enclosing animals. Importantly, keeping animals alive until there was a need for the meat was the norm before refrigeration. It therefore follows that precedents concerning the provision of animals in holding cells are beyond pervasive: consider popular examples such as the Forum Boarium (the cattle market) in the Rome of the Emperors or the ruins-turned-animal-pens in Pannini's eighteenth-century Rome. It should be noted that our ability to get at what has become of these corrals archaeologically is hampered more by an expectation that the erstwhile existence of ruins, of structural remains, was centered upon human rather than animal lives.[16] Indeed, the need for bounded animal space lies behind the transformation of many abandoned buildings into enclosures.[17]

Closer to Texas, James Whitaker traces the roots of a round from pasture to feedlot to market among herdsmen in the backcountries of colonial America.[18] By 1860, a new mass, density, and reach are achieved in the shadows of the Chicago meatpacking industry, with the help of railroads and refrigeration. "Grassland gave way to pasture, and pasture to feedlot," as William Cronon puts it: "The general tendency was for people to replace natural systems with systems regulated principally by the human economy."[19] Placing aside the question of the overly dramatized separation between what is natural and what is human, the Texas Panhandle, one might then suppose, witnessed a transformation of rangeland into ranch, ranch into feedlot.[20] Indeed, across the flat expanse of these high plains, the famous XIT ranch once provided three million acres for its cattle herd, which numbered 150,000 head by 1887.[21]

William Cronon regards the pen as the historical outcome of ever-shrinking containers for grazing, as a reduction in the square footage of the fenced pasture—from the XIT to today's Texas Panhandle, cows and cattle lose almost 558,000,000 square feet of residential space.[22] However, from a genealogical angle, the constitutive element of the Cattle City cattle pen is not a fenced pasture, but the corral; it is an enclosure for the herdsman, which concentrates the dispersed crowd and offers different living conditions. There are, to be sure, numerous differences in corral materials, in corral practices—a series of wooden fence pens controlled by wranglers in the Chicago Stockyards is different from an enclosure of stacked fieldstone maintained by a fifth-generation cowherd in North Yorkshire, England, a rope circle cordoned by cowboys on the XIT is different from a round of interlaced maquis rebuilt each spring by a shepherd in the southern Argolid region of Greece. All corrals, nonetheless, may be seen to share some rudimentary attributes.

The corral demarcates inside from outside. As a container with defined limits, it boxes and orders. As a container, it constrains kinetic life and facilitates management, whether for purposes of care, protection, or manipulation. The differences with respect to pasture, whether enclosed or not, are far from incidental. The pasture has grass, the corral is stripped bare of green; the pasture has space to roam with every animal moving over their personal patch of gramineous ground, the corral is about confining and controlling movement; the pasture allows bovine-sanctioned social relations and frolicking, within the corral a herdsman aims to suppress these activities in favor of calm animals. The genealogy of Cattle City is traced not through a reduction of pasture but the multiplication of the corral and this longer sojourn in the enclosure involves a daily round between water tub, raised mound, fence line (though mediated by steel pipe, and in some cases cable, socializing occurs between cattle in adjacent pens), and feed trough, while congregating within bunches among a couple hundred lot mates. While subtle, this nuance is nonetheless fundamental, for the confines of a Cattle City pen are not about restricting the comfortable, roaming space of cattle but rather replicating the artificial, controlled space and increasing the length of an animal's residency within. And this space, while far from new, is now far more prevalent. A herd grows into an urban population.[23]

Collected into these regular, redundant spaces, which make for the ease of synthetic routine and procedure, immigrants from California and Oregon, Montana and Oklahoma, South Carolina and Florida join together as very specific breeds, as residents among themselves. This selection rests upon the choice of the right kinds of bovine masses. There are "red cattle, black, white, and yellow cattle," but most are young Angus crosses, animals little more than a year old.[24] There are no fierce Texas Longhorns, no "meek-eyed milch cows."[25] There are no great bellowing bulls—some breeds can be too large for slaughterhouse infrastructures. There are no little calves of recent birth. Here and there one finds one or two *Bos indicus*, cattle originally brought from India—these are filtered into lots clandestinely to make up for price shortfalls. Importantly, there are no other livestock species such as swine—a common companion in feedlots of the nineteenth century.[26] Here the cattle reside for a period of 100 to 360 days, depending on their entry weight, a period that culminates in the death of the animal and packaging of its flesh. The silent residents of Cattle City lots exist unawares in a state of being-toward-slaughter.

ACCOMMODATION

To make these three hundred and more collecting spaces "livable," cattle require shelter, food, water, air, and waste removal. Shelter: the pens are without roofs. In lieu of cover, residents are provided with a few square feet upon an earthen mound raised above the muck. In the parched air of the Panhandle, raised mounds dry out faster than flat surfaces; thus, they provide an island in the midst of mud. (Anyone familiar with paths and areas of gathering in pastures will note the bare patches under trees, in byres and by barns—cattle like their shade and often prefer a roof to enduring sun and rain.) These pens also shelter cattle as herds, thus they form and maintain their own groups. So deeply seated is the psychosocial commitment to the herd that even an animal that tastes the sweet air of freedom by being pushed out of the cable pens when competing for space at the trough will not stray far from the group with which it entered. And when handlers move about in search of sick animals, cattle will feign normal behavior in an effort to mask their illness. Whether for group solidarity or against predation, no steer, no heifer, wishes to be singled out through isolation.

In sheltering the herd, group dynamics require constant attention. Reducing stress on animals in this crowded urban environment is key to their coexistence as "citizens," by which I mean recognized dwellers of Cattle City, behind shared fences.[27] A measure of the success of this endeavor is found in the absence of lowing residents—where one might expect a sound "as of all the barnyards of the universe" there is but a strange silence. In order to maintain calm masses, daily feeding schedules are kept consistent, drugs suppress the estrous cycle, noise out-of-doors is kept to a minimum, and handlers on horses or observers within raised-viewing platforms undertake routine surveillance in search of sick animals and wayward "bullers."[28] A buller is an animal that exhibits some form of aggression toward others. Usually a sign of domination, an action such as mounting a fellow resident is taken as a sign of stress within Cattle City. Wayward animals are reshuffled with new pen mates and regularly monitored—an activity sustained through the night by a well-placed lighting system.

Food: The pivot around which Cattle City revolves is feed. Accommodating the appetites of the quadruped throng demands a mountain of mixed victuals. With each spin of the Earth, trucks capable of holding twenty-five tons distribute 1.7 million pounds of feed into over eight miles of concrete trough—a Cattle City inhabitant consumes 25 to 30 pounds every day. A steady stream of big rigs, thirty a day, supplies a large feed mill, a food factory, which stores the constituents of the feed and mixes them in 8,000-pound batches every 102 seconds, twelve hours a day, seven days a week.

In order for Cattle City to continually house 58,000 cattle, the mill must be kept in solid working order. The well-being of the mill comes about through an assembly of components operating in harmony: two

FIGURE 9-2. Composite photograph of mounds of silage and corn-gluten feed adjacent to the feed mill. Photo: C. Witmore.

400-horsepower boilers, screw augurs for horizontal conveyance, bucket belts for vertical movement, conveyor belts for inclines, variable speed drives, rub sensors, large storage silos and holding tanks, a front-end loader (a Deere 544J dumps bulk ingredients into boxes which supply the conveyor belts), a New Generation 668 grain cleaner, eight feed-flaker mills, a steam cooker and tempering system, a series of scales (large, small, and micro), a chain-drive mixer, and a control center at the heart of the facility.

Designed to attain the right combination of protein and fat, bovine diet varies with how long an animal has dwelled within Cattle City and their overall weight. From starter to finish rations, four different feed combinations are provided over the course of an animal's residency—a slow shift from a high-roughage diet to a highly concentrated one aimed at adding weight is necessary to control acidosis. The feed is a combination of silage (fodder from corn stalks or cotton burs), flaked corn (corn heated by steam and compressed in a roller mill), corn-gluten feed (a by-product from the manufacture of cornstarch and corn syrup, this provides a kind of medium level protein), and tallow (rendered animal fat)—the constituents are determined by market price.[29] Accommodation also carries connotations of adjusting or adapting. Food is not so much adapted to the ruminant; rather, with the addition of other ingredients, it is the other way around. Added into the mix is an ionophore, which suppresses the appetite in the animal but improves their "efficiency" (the argument being that one achieves the same rate of gain with a little less feed), Tylan 100, an antibiotic used to reduce liver abscesses that result from a high-concentrate diet, MGA 500, an estrous suppressant, and a probiotic.

Water: every pen contains a round water tub into which water is automatically released. On average, each heifer or steer consumes ten gallons of water per day—well over half a million gallons collectively.[30] Cattle City currently owns water rights to 1,100 surface acres, with no restrictions on the amount of water they can pump.[31] This does not account for the water that goes into other processes, such as growing the food necessary for feeding its population. Air: in the dry lands of West Texas hoof action generates huge amounts of dust—billions of particles of cattle dung. In making the air breathable, Cattle City spends $20,000 to $30,000 a month in maintenance costs, which includes spraying even more water over the pens in order to keep the dust down. Waste removal: pens are cleaned twice a year in step with the seasonal demands of regional agriculture. Large loaders clean the manure from pens and transfer it to fields, where it is spread out into linear mounds covering fifty acres. A sewage system drains liquid runoff into a retention pond, which is high in salts. High in nitrogen, phosphorus, and potassium, most of this wastewater is repurposed in the composting of manure. Beyond adding nutrients, the use of this water helps maintain a higher moisture level for generating compost.

To accommodate its residents, Cattle City provides the subsequent conditions: a bunk of milled feed, a tub of water, an envelope of conditioned air, a patch of seasonally cleaned ground, and a guarantee of social interaction among a circumscribed number of similar lot mates with cordoned mobility.[32] In Cattle City, residents must live in a bubble of life over which they have little control. The "lifeworld," as Sloterdijk states in the epigraph, gives way to air-conditioning during a circumscribed time on a bovine island where every resident lives in being-toward-slaughter, the final stage of its life, however brief.

This being-toward-slaughter characterizes the lot of cattle as proto-meat. It bears distinguishing this concept from the more familiar, Heideggerian notion of being-toward-death, where humans exist both with the certainty of death and uncertainty of when it will come. Throughout Cattle City, living residents know nothing of their impending end. This secret of the looming demise of the entire populace is a specter that hangs over the calm masses of Cattle City, which every bipedal worker, every cloths-wearing visitor, carries with them. Being-toward-slaughter affects the human sense of being-toward-death in this sense and another, not necessarily through the certainty of death, but through an ironic lack of somatic kinship with respect to both ownership and end. While bovine lives exist in far greater numbers for far less time than their ancestors, these mammalian bodies do not exist for themselves. Ultimately, the cattle of Cattle City are accommodated toward an end, not in death per se, but as meat-within-plastic-and-Styrofoam-packaging-toward-consumption. Death, which comes at the end of a captive bolt gun held by a "knocker" on the kill floor of an industrial slaughterhouse a hundred miles away,[33] is but a stop on a "disassembly line."

SUSTAINING BEEF CATTLE

The existence of Cattle City is an effect of packaged meat. Its aim is to produce what its proprietors describe as high-quality beef by adding weight in the form of muscle and fat to feeder cattle that began their lives in pasture. A key step toward the realization of this aim occurs in a mandatory destination for every resident, a metal building on the north side of Cattle City. It is here that the measure of each body is taken. Inside, cattle that have been on the feed for seventy or eighty days are herded into what the industry refers to aptly as a "cattle-sorting system." Coaxed along by paddle or prod, whistles and shouts, residents are driven into a curved corridor at the end of a metal crowding tub. A Temple Grandin design, this ordering helper isolates each animal into a single-file arrangement and presents them but one option—to move forward. Filled by quadruped bodies muzzle-to-tail, this restricted space, enclosed by solid-steel panels to a height just above the withers, insures that each animal can see nothing of what they can hear before them. Turning from the now silent handler, eyes and ears focus ahead. Amidst a whining machinic cacophony, which drowns out even the most tenacious bellows, a squeeze chute declares its presence. This steel box compresses each heifer or steer about the shoulders, ribs, and flanks. A head restraint holds the neck from the point of the shoulder to the base of the skull. While the animal's movement is constricted, a worker shaves off a section of hair from the back and applies oil onto the bare patch. At the end of a second enclosed channel, cattle enter the grasping, hydraulic, steel hand of another squeeze chute, where an ultrasound is taken.

An image of intramuscular fat, marbling, and muscle depth of the rib eye appears on a screen. This image is all that is seen by those who work in office building at the edge of the city, a mode of engagement that renders two very different views of a heifer or steer and these contribute to two very different styles of understanding the matter of animals for the forty-five paid employees of Cattle City and beyond. One is a living animal clasped firmly within a squeezebox; another is an image on a screen of what will become a steak.[34] A visualization that can be reshuffled on computer screen in offices anywhere is a very different thing from a bawling, 1,000-plus-pound animal besieged in the unforgiving clutches of a metal squeeze chute somewhere in Cattle City. As with urban design, these

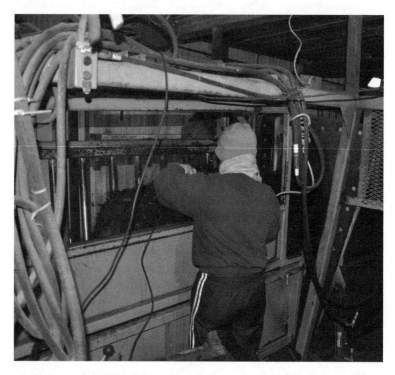

FIGURE 9-3. A worker prepares a heifer in the squeeze chute for an ultrasound.

things are operative in the formation of very different "practical identities" within the various cultures of Cattle City.

Meanwhile, a hormone implant, trenbolone acetate 200 mg with tylosin tartrate 29 mg, a local antibiotic, is injected beneath the skin just behind the right ear. Subsequently sorted by virtue of the right combination of fat and marbling into five classes, cattle are provided with color-coded tags and induced to exit through one of five automatic gates. In this way, "citizens" are ordered into divisions, synchronized with their fellow urbanites at similar stages in their bodily development. Every week, three thousand cattle go through this rite of passage, which is paramount to determining the length of one's residency in Cattle City.

Operating costs for Cattle City are calculated in terms of "head day." That is, costs are distributed across each inhabitant for each day of their residency in Cattle City. With 1.8 million head days per month and 21–22 million head days per year, Cattle City proprietors aim to increase weight gain while reducing the length of residency. Such orientations have given rise to the synthetic manipulation of endocrinological and immunological systems. Over the last fifty years and more, the formerly latent worlds of bovine biochemical processes and microbial symbioses have become matters of conscious concern and routine management. Endocrinological manipulation through the use of both trenbolone-acetate hormone implants and ionophore feed additives increases the rate of weight gain while transforming metabolic cycles. Immunological management through the use of tylosin in feed or other antibiotics aims to minimize the collateral damage of a concentrated diet, which can destroy the rumen digestive system (Cattle City loses one to two residents a day from diet-related issues), and underwrites the expansion of the bovine populace to an urban level. Let us compare Cattle City to Sloterdijk's generalized immunology: "It is not just in their intricacy that immune systems confuse security demands of their owners; they irritate even more through the immanent paradox that their successes, if they turn out to be too thorough, can turn into a cause for their own kind of sickness: The growing universe of auto-immune pathologies illustrates the dangerous tendency of the self to be victorious to the death in the fight against the other."[35]

The ease with which industrial science is integrated into routine processes follows on the insurance of expedited growth. Indeed, among many herdsmen runs the belief that microbial management reduces the "competition" for nutrients—more calories go to muscle building and less to microbes. These seeming gains are not without their concomitant, often unrecognized, losses. Akin to sleepwalking in a minefield, the use of antibiotics gives rise to transformations in the composition of the bovine intestinal microbiome that offset processes themselves the outcome of long-term symbiotic rapports. The familiar logic of competition favors insularity at the expense of the symbiotic—some gut bacteria salvage calories from otherwise indigestible carbohydrates, and cattle can use these reclaimed nutrients as fuel.[36] As a symbiotic consortium, the cow is an organism whose anatomy, physiology, and behavior is informed or regulated by a rich community of cellulose-digesting bacteria, ciliated protists, and anaerobic fungi.[37] Looming over the varying degrees of short-term success in microbial management are potentially catastrophic failures.

Antibiotics, deployed as a means of reducing the number of abscesses, are recognized as increasing their frequency, which is detectable mostly on the viscera table—toxicology and residue chemistry research has yet to make explicit the effects of antibiotics upon edible muscle.[38] Useful for a relatively narrow spectrum of bacteria, tylosin is a macrolide, which increases the frequency of antibiotic resistance in bovine gut bacteria. Thus, it escalates the pace of microbial evolution: The "hygienic" gut becomes an incubator for new pathogens that can be transmitted among cattle and humans. Use of tylosin can also lead to cross-resistance to other members of the macrolide group of antibiotics, including erythromycin. Poorly absorbed, oral antibiotics like tylosin are largely excreted in urine and feces. These residues enter the soil and water supply and thus adversely transform the microbiome beneath the hooves of cattle.

While, after years of suspicion as pathogenic sympathizers, good microbes are now welcome members of the shared human/bovine *oikos* with the recent introduction of probiotics, bad microbes remain suspect and continue to be blacklisted. Adept at recognizing the subtle indications of microbial imbalance, handlers continually search for the presence of these harmful agents or the absence of friendly and benign bacteria in sick animals. Once found, these animals are taken to the hospital and treated. Common ailments include pneumonia, diphtheria, and acidosis; the latter, which is relieved by a stomach tube three-quarters of an inch or so in diameter and eight to ten feet long that delivers a defoaming agent, results from an imbalance within the rumen microbiome.

On one level, Cattle City presents a model of husbandry less controlled by the physiological processes of cattle than by an industrial assemblage of urban infrastructures, machinic manipulators, pharmaceutical companies, and a knowledge economy that divides up the undividable; ever more management of bovine biological functions, ever more power over the formation of tissue and even over the bovine microbiome. Still, how many times have we encountered this story of expanding control? Charging into the fray between antibodies and microbes comes with an experiential blindness for those herdsmen who are now expected to do so by proxy. Sloterdijk is right to speak of "symbiosis with the invisible"—herdsmen have never had the benefit of the microbiologist's microscope or their ways of knowing. Nonetheless, management of biochemical processes and the bovine microbiome extends the concept of enclosure into the interior world of bovine bodies. Thus, it would be more appropriate to think of Cattle City as a metacollector, which attempts to maintain the integrity of the herds despite the increasing protests now mounting within bovine guts, the soil, and the surrounding atmosphere.

Incidentally, one would be unwise to ignore the vast implications of detaching cows from grasses, from contact with fellow pasture mates—pigs, sheep, farming or herding families, and the majority of humanity—on a daily basis. Ten thousand years of living together came with the requirement that one stayed in place or moved together to care for each other. Recent estimates suggest that more than 90 percent of human lives are now spent indoors, and cattle now live as much as half of their lives within their own cities.[39] Beyond the issue of biochemical and microbial management and manipulation, the consequences of living apart for microbial exchange and the human microbiome are immediately prevalent. While a far more pervasive issue with farms in general, children of hyperhygienic parents who develop allergies in air-conditioned, extremely controlled, microbially impoverished atmospheres are now encouraged to return to the farm, which they and their parents have never known, and play in the mud. A day out may be good for both the human and bovine microbiome—such a situation is mutually beneficial for the cows—but it does little to renew estranged relationships, at least in the way that we all once lived, or counter the collateral damage of antibiotic use in Cattle City. And while we may never return to our erstwhile relationships, at least in the way that they were, they have perhaps never been more relevant for so many.

OLD WAYS AND ECOLOGICAL CORPULENCE

Morals and appetites, it has been pointed out, parted ways early in human relations with animals.[40] From Hesiod, at a beginning for talkative history within Greece, we learn for the farmer, as Victor Davis Hanson writes, that "raising food is not a clean experience; living by toil is not necessarily always moral."[41] Hanson, a fervent advocate for the long-term traditions of otherwise silent independent farmers, reminds us of the hard work, pain, and sacrifice that come with working one's own land. He also reminds us how passionate yet naïve pictures of all-caring farmers, romantic and detached images of the natural living conditions of cows as ecstatic beings in green pastures are self-delusions, oblivious to the realities of small, independent agrarianism; such pleasant fictions are ignorant of history.

Insofar as animals raised for food are concerned, one should look not only to the farmer, but also to the herdsman. An agrarian morality, tied to a patch of earth, deviates from a nomadic morality, which, in order to move with animals as their desire leads them, refuses to root the self. Aristotle regarded pastoralists and agriculturalists as completely different social entities with radically different styles of life, sensibilities, and accumulated wisdoms. Varro, despite his estranged aristocratic background, reiterates this difference in Book II of *De re rustica* in the following way: "The farmer's object being that what ever may be produced by cultivating the land should yield a profit; that of the shepherd to make his profit from the increase of his flock." Here Varro lightly touches on an antithesis, which, as Sloterdijk has argued, "sets the stage for the contest between rootedness and speculation, between a spirit of preservation and blithe wastefulness, between vegetable provisions and animal assets."[42]

The agrarian notion of self is tied to a patch of land. It is around this enclosed area of terra firma that the farmer's commitment has revolved for millennia with the protection of plants, both those that are the fruits of his labors, those of his neighbor, and those of his labor's companion, yoked oxen. The value of land grows through the hard work of breaking earth and raising a crop. The herdsman, on the other hand, is willing to live out his life, the life of his family, around the wandering herd. His lot revolves around animals, paths of transhumance, seasonal movements among pastures and temporary camps, not fixity upon a patch of land. Thus, long-term cohabitation creates a unique assemblage of cowherd and cow herds. The herdsman, robust in body, knows how to both sustain and consume the herds by maintaining the right combination of breeding and feeder stock. He watches and learns from cattle and knows when to defer to their leadership in moving the herd. The herdsman, however, does not know how to make the best use of land in the way of the farmer. Over the millennia, agrarian economy has repeatedly usurped nomadic ecology through the expansion and efficient use of valuable farmland. While the careful historian may be stuck, quite rightly, by an overwhelming and uncanny sense of déjà-vu associated with a historical cycle between unrooted herdsmen, independent small farmers, and absentee landlords witnessed many times since the agricultural revolution, we also run the risk of missing some subtle, yet profound, differences.[43]

In Cattle City, raising animals as a profit-driven enterprise is melded with the ambition to expand the herd to the greatest extent. Here we encounter a break down in the age-old antithesis of herdsman and farmer. Multiplying familiar collecting spaces, replicating familiar collecting practices, extending familiar collecting knowledges, all these transformations, more pragmatic than ethical, mask the unfamiliar as they push into the realm of the formerly latent. Atlas-like responsibilities, endocrinological and microbial management—no farmer, no herdsman ever lived in such a world. One wonders what the old Greek pastoralists, those who held their cattle in such high esteem as to bestow upon their daughters names such as Euboia ("good for cows"), Phereboia ("bringer of cows"), Polyboia ("worth many cows"), or Steneboia ("strong in cows"), would make of Cattle City.[44] The notion of enclosure works well for wolves and dangerous pathogens, but not for commensals and symbionts.

The mingling of agrarian and nomadic aims can occur because they are subsumed by that fourfold of graces—economy, technology, science, law—with their larger loyalties to profitability, efficiency, objectivity, and formalism.[45] For ten thousand years and more the helmsman of society was the farmer, the herdsman, the sailor, the fisherman. Too few now live out in the weather with the populace of Cattle City. Too many meet what becomes of this population in the grocery store and nowhere else. Even I, as one who grew up on a small farm with cows[46] and whose observations follow from an on-site visit, engagements with articles, books, aerial photographs, childhood memories, and subsequent correspondence, have written the majority of this essay indoors at a distance from Cattle City, far from its residents, from those who live and work there. What right do I have to make decisions concerning Cattle City? Indeed, who decides? The silent farmer and herdsman with their old traditions are not at the helm. Legislators armed with the latest figures and projections, administrators following budgetary considerations, quarterly profits, and shortfalls, scientists working in air-conditioned laboratories with the latest genetic research: all are absentee decision makers who divide up their areas of knowing. The Earth with its ever-changing microbial foundations and climatic fluctuations is also deciding. Who is listening?

Cattle City, as a heterogeneous assemblage, involves many different interlocutors—handlers and horses, veterinarians and their treatment regimes, drivers and trucks, engineers and natural-gas boilers, office workers and computer screens; it involves many different ways to relate to animals—drovers and swing gates, care givers and squeeze boxes, corporate managers with flat images and quarterly margins; it draws together many different, and distant, locales—corn fields, cattle pens, administrative offices, and feed mills; it has many different ways to calculate how to best turn a profit. Urban infrastructure and "global" reach, explication and management of bovine internal ecosystems, radically modify the rapports between cattle and herdsmen, cattle and enclosures, cattle and packaged meat. Indeed, the urban multiplication of enclosures and the scientific manipulation of interiors, both managed through synthetic routines and propped up by fossil fuels, expand these styles of living to the point that they become grossly overweight.

Can the rich, accreted knowledge of agrarian and herdsman traditions—ways derived from ineffable relations with locality and animal idiosyncrasy, ways regulated to the margins, scrapped for their usable aspects—deal with the world that now lies before us? A farmer's self-certainty lies in generating the greatest yield, in making the most "efficient" use of resources, in maximizing improvements with an aim toward profitability. Such plans work well for a couple of acres of olives or several acres of vines, but what happens when this sound and familiar farming logic concerns a biomaterial assemblage that grows into something larger-than-monstrous? A herdsman's wisdom accumulates by living with animals season after season and with those who have lived with them; it selects for the best stock and centers the self on the mobile herd. Such ways care well for the health of the herd, discern the best attributes for the locale, and even keep the wolves and jackals at bay, but quality packaged meat does not necessarily require healthy bovine lives; Cattle City residents are not selected for living on West Texas pasture; microbes are neither wolves nor jackals. Oriented toward a herd of a few dozen, can these pragmatics hold together with herds in excess of 58,000? Can these pragmatics cope with a populace that adds up to more than 60,000,000 pounds in animals, plus an equal weight in urban infrastructure? Can they cope with a vast energy flow derived from fossil fuels? Can the associated ethics rooted in locality, crafted in corporeal proximity, provide guidance for a bovine urbanity (with due irony, given the Latin connotations of *urbanitas*) that grows even larger through a network of other cattle or cow cities, some twice as large and through which some 700,000 to 800,000 residents move each year?[47] Can such wisdom, born of corporeal experience with animals conceived as insular entities, operate in the realm of the invisible with microbial ecologies that cannot be observed from the saddle, save through the symptomatic effects of an imbalance that is managed through divided fields of knowledge? Neither ancient farmers nor XIT herdsmen had words for entities beyond leviathan; neither farmers nor herdsmen understood their relations with animals as fellow symbiotic assemblages shaped through mutual, multispecies exchanges.

Through its multiplication, the enclosure expands into an urban situation far beyond the concerns of farmers and herdsmen. The integrated yet compartmentalized city is now the model of one state in husbandry whose scope aims at gigantic proportions. Cattle City expands out as the accumulated output of thousands of farms and ranches in one urban locale. Cattle City expands into the bovine microbiome in order to supply and maintain an industrial food chain with a global distribution. Cattle City maintains the right climatic conditions for its populace to thrive in being-toward-slaughter. This weight, what we might call "ecological corpulence," should also be defined in terms of energy. From an ecological perspective, this corpulence increases with the distance from feeding upon light. Weight gain is also consequence of blocking another entity's ability to feed upon light—corpulence is also tied to a will to mastery. An entity can be said to be overweight when its distance from feeding upon light and its presence in blocking another entity's ability to feed upon light encumber wider systems of relation, whose former equilibrium resulted from a balance between what it contributed and what it took.

Cattle City exists by virtue of increasing the steer or heifer's distance from the sun. Left to their own devices, cattle would move in search of tasty greens and display a preference for particular grasses—buffalo, mesquite, and grama grasses in the Texas Panhandle, or, as in those South Carolina or Florida pastures where many Cattle City residents were born, a smorgasbord of big and little bluestems, fescues, timothy, river-oats, foxtail and various legumes. Now between sun and plant, tasty greens and ruminant enters a vast chain comprising an armada of big rigs and corn-syrup factories, grain cars and a railroad system, "shadow acres" in the

Corn Belt and a monoculture infrastructure, a bovine city and a food factory that consumes natural gas. A short loop has been replaced by an obese system where tens of thousands of residents are shipped in from all over the United States; where mountains of feed are brought to cattle rather than allowing quadrupeds to move around and fertilize thousands of fields with their own manure. Out the other end, Cattle City's manure is composted in mounds covering fifty acres, which will supply most farms within a sixty-mile radius; an area encompassing 11,310 square miles. Between the anus and the ground enters a dozen rigs, a hammer mill, irrigation lines, fifty acres of manure, half a dozen loaders, numerous spreaders, and thousands of gallons of diesel.

From the fertilizers used in Iowa fields to the diesel used in transporting feed constituents, cattle and meat products to the plastics used in packaging these products, fossil fuels underwrite every stage in this system. Recent calculations suggest that it takes three-quarters of a gallon of oil to produce one pound of beef.[48] By the end of their residency, each citizen of Cattle City would have required six barrels of oil. By some estimates industrial agriculture requires ten times more energy than lighter, local alternatives and perhaps ten times more energy than what is derived from the food that is eventually produced.[49] Feeding upon fossil fuels is several hundred million years distant from feeding upon light.

Cattle City is further encumbered by units and enclosed property; it is further impacted by the whims of the property owner; it is further burdened by local infrastructure—vast portions of the high plains are set up to produce cotton, not corn. Cattle City gains even more weight with the repetitive and pervasive model of the grid, of the city. Obligations and legal bonds are material, concrete and visible as roads, fence lines, or the local cotton gin. From the perspective of the farmer, agrarian property pushes range to the unwanted, non-cultivatable edges. Only through concentration can the herdsmen equal agrarian gains. And through concentration, Cattle City further weighs upon resources—water, soil, air, and fossil fuels. Through ways of knowing that partition and reduce, in other words, enclose even smaller parts, Cattle City operates through division, and this generates a composite husbandry whose problems arise from their disintegration. Cattle City adds even more weight through its attempts to enclose "patches" of the rich ecosystem of gut symbionts. A grossly overweight, corpulent system, Cattle City gains, not in robustness, but fragility.[50] Ever refined, the unitary lacks flexibility in absence of diversity. From the perspective of the herdsmen who once chose to live their lives in close proximity around the things that grow, this arrangement would seem to invert ten thousand-plus years of iterative practice and accumulated wisdom.

BOS URBANUS: A REQUIEM?

A painting hangs on the wall in the administrative office of Cattle City. It depicts a variety of cattle, white, brown, and black, with and without horns, large and small, within a series of pens. A few in the foreground look up from the bunk. Yellow tags dangle from their ears. In the closest pen, labeled B21, a lone cowhand rides among the cattle on horseback. In the middle ground, a sea of enclosures. In the background, a large feed mill. Beyond, the sun hangs low in the big sky, sunrise or dusk, over the high plains. A caption reads: "Feeding the World." The painting depicts a Cattle City, but its image evokes largesse. This massive existence is legitimated as a robust response to a meat-eating humanity growing beyond leviathan. Cattle City shoulders this world-responsibility like the Farnese Atlas.

Extending boundaries through enclosure, groping attempts to corral the bovine ecosystem prove to be futile in the face of present realities. Wholly unhooking humans from cows after ten thousand years in close domestic proximity (and hundreds of thousands of years in common ecological proximity) with the codevelopment and coevolution of human/ bovine microbial ensembles proves to be unwise on many fronts. We err containing them within their own cities, overconsuming cattle, reducing animals as symbiotic ensembles to proto-meat, and, in going our separate ways, by not maintaining our former diversity of rapports.

Cattle City is not about exporting the human need for interior space; rather, it is about maintaining megamachinic order within bovine-urban-ecologies-toward-consumed-meat, and in this regard, the interior world of the laboratory is far more efficient. As I write, a scientific answer to bypassing ecological catastrophe while feeding billions is well under way;[51] it aims to separate the bellowing steer from its edible muscle altogether by growing flesh-without-bodies in labs, and thus skipping the issue of high-density urban living and the ethical dilemma of eating animals altogether. In a new agrarian space where the closest approximation to growing flesh will be found in the cultivation of plants, the farmer will appear to usurp the herdsman once again, at least metaphorically. The shared affinity between growing meat in a bioreactor and vegetable provisions will ignore the fact that even more control over physiological processes has been exerted. The continued, and pervasive, replacement of sound ways of living with readily manipulable, mechanical substitutes will be ambivalent to the fact that, within the far more controlled, air-conditioned space of the numbered bioreactor, life continues to succumb to the universal. This masterful realization of animal reductionism will blithely disregard the fact that we humans continue to remake the world to fit an image that suits our own needs.[52] For how long can we hold forth with such immediate reckonings? What other alternatives could possibly shoulder such responsibility?

Passing from a situation where the land available for herding existed in abundance to one where the majority of fertile land was enclosed and worked by yeoman farmers, many ancient Greeks cultivated a different cultural attitude with respect to the consumption of meat. By the late sixth century BCE, cattle as consumed flesh were associated predominantly with the sacred economy and within the agricultural economy, cows were largely used as draft animals. As one's labor's companions, oxen were fellow participants within an agrarian existence.[53] For the Greeks, meat for three meals a day came to be associated with heroes, tyrants and kings, not everyday citizens; it came to be associated with gluttony. Most Greeks chose to eat less meat and this choice was melded with an ideology of modesty, where eating meat was a shared, communal endeavor. In the midst of a burgeoning population with less high-quality agricultural land to go around, Greek standards of living changed, their ideologies changed, as did their relationships to consumed flesh. I am not idealizing the Greeks here: not all of the ensuing responses to these transformations were necessarily admirable. However, they listened and worked with land and animal in a different way.

Putting ecology in conversation with archaeology allows us to look elsewhere, to the memories held in soil and stone, in old crop plants in fallow or abandoned fields, to the old infrastructures of derelict farms, for viable alternatives. The ontological grounds for the experiences of farmers and herdsmen gain new significance and potency when melded with a multispecies mode of existence. Living before Leviathan, living without an understanding of the microbial world, living without fear of the planet striking back, even so, the past does not count for nothing. With a planet under protest, the past becomes a locus for biodiverse, ecologically rich ways to live with animals, now recognized as fellow symbiotic ensembles constituted by many species dwelling together in coexistence.

Living outdoors, the farmer listened to the interactions of soil and wheat, bees and olive blossoms. Living in the weather, the pastoralist listened to the animal and the grass, and recognized the time to move on without exhausting the greens of an area. These ancient relations were guided by these seemingly mundane things. Such interactions are now modified by new actors—Earth, microbial life, and overweight systems. Nonetheless, modest acts such as planting a garden, eating less meat, listening to other rapports in the world, cultivating an ideology based on multispecies understandings and humility could rival these more masterful, self-centered reckonings which blithely attempt to wall off, to immunize, the self against animals and microbes, which aim to increase our distance from living off the sun for economic gain. Can we not trust that these small acts will add up for more people than those who will be able to afford it?

10.

SYMBIOTISM

Earth and the Greening of Civilization

PETER WESTBROEK

The Rijksmuseum in Amsterdam has a portrait on display of a national hero in the Netherlands, Admiral Michiel Adriaenszoon de Ruijter (1607–76). Ferdinand Bol, a highly successful pupil of Rembrandt, painted it in 1667, at the height of the Dutch Golden Age. We see the imposing personality of de Ruijter, dressed up with abundant gold, brocade, and jewelry (Plate V). He casually rests his right arm on a globe, in his hand the admiral's baton. The golden hilt of a sword is just visible on his left side, and a half-opened curtain in the background unveils his flagship, the *Seven Provinces*, from which he has just brought the Second Anglo-Dutch War to a glorious end. His employers, the merchants of Amsterdam, had commissioned de Ruijter to conquer the world. All they wanted was power and wealth. The inexhaustible riches of the Earth waited to be taken. The goal was straightforward, the conviction unshakable. Here was a man in one piece, not hampered by doubt about the rightfulness of his mission. Significantly, the globe in the painting does not represent the Earth itself but the heavens with the zodiac. The implication is that even plunder, murder, and repression could count on God's blessing, provided they served the glory of Holland. Something great was wrought in what the Dutch call Frogland!

Alas, as times go by, old dreams melt away. In our day and age, it is not pride but doubt that prevails. The issue of *Time* of October 1, 2012, has Bill Clinton on the cover, lovingly clasping a globe.[1] The retired world leader unfolds some promising plans, good for business and good for Earth as well. Good for Earth? Here, a new notion enters the scene. De Ruijter may have been concerned with the fortune of Holland, but the well-being of the entire planet was not on his mind. Clinton, conversely, is filled with the best of intentions. The truth about this planet may be inconvenient, he seems to think, but there always is hope as long as we as respectable stewards take good care of this place.

One wonders why Clinton hangs on to a dusty old globe while NASA's unforgettable photographs are on everybody's mind. Did not the 1968 Earthrise picture, with Earth hovering over the lunar surface, herald the glory of God's own country, the United States of America (Plate VI)? It suffices to imagine Neil Armstrong replacing Bill Clinton to see the absurdity of this cover image. This cannot be an accident; it must be a Freudian slip of the mind. But how could it happen? The answer is that this particular image eloquently expresses hybrid feelings toward the Earth, feelings that all of us share. Clinton is eager to hold the planet in his hands, while he knows that he can't. The *Earthrise* picture turns out to be a symbol both of glory and of humiliation. We should be grateful to Clinton for bringing our trauma into the open.

At first sight, the contrast between de Ruijter's confidence and Clinton's doubt may appear incidental, but when we try to place their convictions into the context of Earth system science they acquire a highly significant meaning. What we then perceive is the dawn of a global metamorphosis that may lead to a new relationship between the worlds of nature and civilization.

METHODOLOGY

In the following, I shall abundantly refer to "Earth system science" and "civilization." These terms may be unfamiliar, so let me explain from the outset what I mean by them. Earth system science is the relatively new scientific approach that regards our planet as a coherent entity with its component parts interacting—the solid Earth under our feet, the waters, the air, the biota, and culture. The scope of this field embraces the full forty-five million centuries of the Earth's existence. Radioactive decay in the deep earth and solar radiation are the main sources of energy that keep the system away from thermodynamic equilibrium. The Gaia theory advanced by Lovelock and Margulis is a development parallel to Earth system science.[2] The phrase "civilizing process" was introduced by sociologist Norbert Elias to designate the full gamut of cultural evolution, from its biological roots to the present.

Placing the personalities of de Ruijter and Clinton right away into the overwhelming perspective of Earth system science is asking too much; we are forced to proceed in two steps. In the first part of this essay, we expand our view from the century scale of conventional history to the entire process of civilization. As Elias observed, the short-scale study of human history often reveals a chaotic world of contingent events, while his own large-scale approach to civilization brings to light an overall sequence of societal configurations. Thus, although the first step in our exercise reduces Ruijter and Clinton to almost imperceptible vestiges, it allows us at least to determine their position in this large-scale patterning.

However, it is important to realize that the process of civilization does not evolve in isolation, but is part and parcel of an encompassing system—Earth. Unlike the planet, civilization is not a system, as its dynamics do not follow from the interaction of its component parts. Despite his bold way of thinking, Elias remains in line with the tradition in all the human sciences by perceiving the civilization process primarily as a human phenomenon. This has the annoying consequence that our newly attained overview is not free of anthropocentric overtones. It is this weakness that forces us to make the second step in our procedure—to further extend our perspective from civilization per se to the level of the entire Earth. Now, from this vantage point, we can leave behind the idea of a central role for humanity and regard civilization as a late acquisition of this planet's long-term dynamics. System Earth and not humanity becomes the foundation of civilization. It goes without saying that this second step in our procedure is enabled by the advent of Earth system science. Adding a third step, all the way up to the cosmos, would not significantly help our understanding of civilization, as this phenomenon is sufficiently defined in the context of Earth dynamics. So, in this chapter I essentially take the cosmic viewpoint, but regard Earth as a part representing the whole of the cosmos.

STEP 1: DE RUIJTER, CLINTON, AND THE CIVILIZATION PROCESS
Civilization

Elias selected a seemingly insignificant detail of social development—changes in eating habits in European courts since the fourteenth century—as a model system representing the entire civilization process. This approach reminds one of the practice in biochemistry where a single bacterial strain, *Escherichia coli*, gave access to the molecular organization of all living organisms. Thus, for Elias, rude, ill-mannered warriors who adapted themselves to the strict rules of life at the court became the *E. coli* of civilization. Not that these people liked it, but forced by changes in the prevailing power structure, they learned to repress their impulses and got access to the courts in return. It did not take long before they had internalized the new behavioral patterns, so that they were no longer aware of submitting to the artificial straightjacket.

This model system allowed Elias to study in detail how the civilizing process manifested itself both in the human individuals and in society at large and to show how these two levels of organization were connected. He also demonstrated in detail how new patterns of self-restraint spread from the courts, where they had their origins, through society at large. Elias was able to draw yet another major conclusion from the study of his model system. He showed the civilization process to be of such magnitude that individual people could hardly influence its course. It just occurred, and people were largely unconscious of the huge transformations in which they were entangled.

Following the theory of Elias, Joop Goudsblom formulated a comprehensive subdivision of the overall civilization process.[3] The mushroom-like image of Plate VII is an artistic impression, representing the time scale along which civilization proceeds vertically and the size of the human footprint on Earth dynamics horizontally. The colored segments illustrate the major stages of civilization. The green stem below stands for our biological roots, whereas subsequent socioeconomic regimes are indicated in shades of red, with pink for fire domestication, light red for agriculture and animal husbandry (agrarianization), and dark red for industrialization. The human footprint grew almost imperceptibly throughout the phase of fire domestication (from before 200,000 to 10,000 years ago), accelerated during agrarianization (10,000 to 400 years ago) and exploded during the industrialization period, from about 1600 CE until the present. Two dots indicate the approximate positions of de Ruijter and Clinton. They both appear in the dark red field, the former with this regime in its blossoming youth, close to the transition from agrarianization, while the latter is at the edge of the mushroom, where modernism appears to shift into a subsequent regime, shown as green dots. The latter transition is the main subject matter of this essay.

The major socioeconomic regimes comprise, in addition to the dominating technologies of subsistence, characteristic power structures as well as major worldviews known as animism, geocentrism, and heliocentrism, respectively.[4] It should be noted that in the course of the four centuries of its existence, the heliocentric worldview itself underwent profound transformations, mainly resulting from scientific and technological innovations. This is why I prefer the term "modernism" to heliocentrism as an umbrella concept covering the entire industrial regime, including its worldview and power structure.

An important characteristic of the overall civilization process is its ratcheting type of development, or growth by accumulation, whereby new elements are added to the already existing configuration. Thus, emerging regimes incorporate their predecessors by modification rather than obliteration. For example, the biological past and the regimes of fire and agriculture persist under the present modernist order, but their configurations are curtailed to suit the prevailing modernist requirements. This ratcheting mode of development suggests that a formidable memory underlies the entire civilization process. Goudsblom has also observed that the transitions between subsequent regimes are never sharp, but gradual.[5] New regimes appear locally at first, as islands amid the dominating regime; these islands then enlarge and coalesce, until they take over the entire system. In Plate VII, these transitions are indicated as mottled zones, with the competing regimes showing up in the appropriate colors.

It is worth noticing that over time the productive regimes, power structures, and worldviews do not evolve independently, but in conjunction.

As long as they agree with or even support the prevailing power structure, the worldviews form islands of mental detachment and stability amid an ocean of involvement and chaos. However, as people accumulate more experiences with their environments, their ideas will eventually exceed the prevailing worldview's capacity for accommodation, so that a conflict with the associated regime and power structure becomes apparent. This conflicting situation may lead to the emergence of a new worldview, with an associated regime plus power structure, together providing a new island of detachment.

Involvement and Detachment

Elias sees the dynamics of worldviews as resulting from a subtle interplay between involvement and detachment.[6] People are "involved" when, overwhelmed by their fears, they are left to the dynamics of their surroundings, unable to act in any purposeful way. They are disoriented, escape into emotions and give their fantasies free rein. This attitude makes them all the more vulnerable. "Detachment" refers to the reverse situation: People with minds open to current reality can acquire an understanding that allows them to adjust challenging situations to their advantage. Detachment requires the suppression of fear, the acquisition of orientation and a balanced state of mind. Involvement does not disappear but is curbed to become the driving force of purposeful action.

The emergence of modernism, the currently dominating worldview, may serve as an illustration. It chased humans out of the central position they had occupied under the preceding feudal regime with its geocentric worldview, and placed them on the margins of reality, together with their newly marginalized planet. For the first time, nature emerged in the conscious mind as an independent, indifferent world. This marginalization of humans increased during the following centuries. Earth was mapped out, the human body was analyzed, geological time was charted, humanity was reduced to an animal species, our emotional world was shown to be the scene of subconscious drives, and astronomy drove even the sun out of its central position.

Step by step, all vestiges of a center were removed from the modernist worldview. This change of mind went hand in hand with a rapidly intensifying exploitation of the natural world by science, technology, and industry. Thus, the unfolding of modernism brought a growing level of mental detachment among the intellectual and economic elites. But it also provoked a backlash of increased involvement and fear. The old feudal establishment had to endure the loss of its power, and, later on, a diverse but persistent anti-Enlightenment movement sought to reinstall the lost mystery in an otherwise demystified world.[7] Only one stronghold of anthropocentrism persisted. *Humans remained the masters of the Earth.* The conquest of new land was the key business of de Ruijter, and we have only to look at the globe Bill Clinton holds in his hands to see the vivid colors on its surface, showing how the elites divided the planet for their own convenience.

An Emergent Worldview: Transgressing the Bounds of Modernism

Plate VII sketches the beginning of a mottled zone with green specks at the very edge of the red segment. The suggestion is that symptoms of a new regime and its accompanying worldview are popping up right now, although they have not yet acquired a dominant position. If this is true, we may expect this emerging configuration to contradict the basic assumptions of modernism, raising conflict, anxiety and hope. This situation would explain the difference between de Ruijter's self-confidence and Clinton's incertitude.

Ironically, it was the most explicit manifestation of modernism—space travel—that brought the new worldview into the open. Halfway into the twentieth century, the modernistic icons of Earth, the globes, lacking any more mysterious territories, had lost most of their original attraction. The appropriation of space would further extend the modernist ambitions. In December 1968, astronauts Frank Borman, Jim Lovell, and Bill Anders were circling the moon in their Apollo 8 capsule. After the third orbit, upon making a course correction, they witnessed a breathtaking spectacle: from behind the barren lunar surface appeared a mysterious globe with subtle hues of blue, green, brown, and white (Plate VI). And they instantly knew: This enigmatic phenomenon was home, far more so than God's own country. We didn't own it but were destined to obey its unpredictable dynamics. This observation transgressed the bounds of modernism and heralded a new phase in the history of civilization and of Earth. Earthrise was a giant step in human consciousness that would erase the globes of plaster and cardboard from memory.

Alas, it took more than a lunar voyage to get the new regime under way. If the emergence of modernism had been received with the mixed responses of Enlightenment and anti-Enlightenment, it was the turn now for the new worldview to arouse mixed feelings of hope and fear. The prospect of a new relationship between the civilizing process and its home planet seemed to mark the end of petty problems of human strife, the Cold War in particular. But at the same time, Earthrise removed the last remaining bastion of anthropocentrism. Public awareness of global issues regarding the environment, population density, climate change, and sustainability emerged as early signs of the new worldview. Soon afterward, policymakers responded to the inconvenient environmental truths with the international and interdisciplinary research programs known as Global Change. The goal was to help the politicians with their decision making by predicting the future of the Earth on the century scale.

What is common to these developments is that they stretch the modernist order beyond its limits and can thus be regarded also as incipient manifestations of a new regime. They respond to challenges that are essentially revealed by the Earthrise image, and break through the fragmentation of science. This is why I made them show up as green specks in Plate VII. Yet, not unlike Clinton's attitude, their motivation is hybrid. Global Change research is a case in point. It has been and continues to be a gigantic and visionary enterprise. Suddenly, thousands of researchers all over the world began to break through the confines of scientific fragmentation that for so long had been enforced by the modernistic order. They started mutual collaborations embracing wider and wider circles of expertise, in an attempt to understand the infinitely complex web of interactions between the many component parts of the planet. And they generated a flood of new data and staggering insights into issues such as global climate, overpopulation, and urbanization. On the other hand, Global Change does not study the evolution of Earth as a whole, but only how it currently operates at the century scale. The goal is managerial and technological rather than fundamental. As a result, it does not diminish, but inflates public anxiety. The ambition is not to prepare for a new future, but to keep modernism in charge. As to the methodology of this paper, we sense the advent of a new regime, but remain entrapped within the anthropocentric view of civilization. To take the second step in our procedure, a higher level of mental detachment is required. So let's now move to Earth system science.

STEP 2: CIVILIZATION AS SEEN FROM THE EARTH

Earth system science was born in the aftermath of the Earthrise photograph that showed the Earth and the Moon in a single view. This research reveals that in the whole universe there is nothing more underrated than planet Earth. In everyday life, we tend to regard our surroundings as normal and stable. But the viewpoint of deep space and time uncovers an unusual and unfixed world. Earth is strange, astonishing, and even absurd to a degree that widely surpasses the capacity of our intellect. Only rarely, as in a flash, may we succeed in grasping some of that singular strangeness. This planet slowly turns itself inside out, and it has done so without interruption, throughout its existence. With the exception of a few particularly resistant fragments of the crust, everything in the outer Earth— the crusts of the oceans and the continents, mountain belts, water, air and biota, including everything of beauty and discomfort—is swallowed, demolished, and even melted. This is why Earth has always been a place of death and ruin. Meanwhile, the building blocks reappear and reassemble spontaneously into a new world, teeming with youth and vigor, but otherwise an almost exact copy of the one just vanished. Earth must be endowed with a memory of astronomical proportions for being able to recreate the bygone global order in such exquisite detail, over and over again.

But there is more. The new order emerging from any particular round of the global geological cycle is never a *precise* copy of its predecessor. The long-term view reveals an overall pattern: an ongoing increase in complexity, organization, and connectivity, and a persistent differentiation of the planetary configuration. We know that Earth began as a glowing sphere in bewildering turmoil, with showers of meteorites splashing into the oceans of magma. And look at it now! Only energy is allowed to exchange with the surrounding space, while the Earth has become a virtually closed system for matter. The boisterous currents of magma are converted into an orderly arrangement of core, mantle, crust, ocean and air, while crustal plates change their positions at the speed of a growing fingernail. We see bacteria, trees, flowers, birds, buildings, mobile phones, and works of art; we hear sounds and think thoughts that a short while ago could not exist. The Earth's cycle is no cycle at all, but rather a ratcheting spiral. As Earth turns inside out, the currents of matter split up, make detours, and diversify, creating more and more levels of organization. New emergent phenomena proliferate and they in turn combine into constructs of even more dazzling complexity. "Creative reproduction" may be an appropriate term for this dynamic behavior.

From the point of view of Earth system science, one may regard life as an emergent property of creative reproduction that speeds up the Earth's differentiation. Thus, from a systems point of view, biological evolution is not an autonomous process, but part and parcel of the creatively reproducing planet. Furthermore, there are good reasons to assume that the civilization process further increases the rate of this acceleration. The spectacular proliferation of human implements, thoughts, means of communication, works of art and institutions is adding to the differentiation of the present world at a rate unparalleled in Earth history (Plate VII).[8] Indeed, the civilization process is not an autonomous creation of humanity. Not only life but also civilization must be regarded as an emergent property of Earth's creative reproduction. We *are* Earth; the entire planet participates in the ratcheting process of civilization. Primarily it is the Earth that civilizes, not we ourselves. Recall that, when the Apollo 8 astronauts saw Earth from deep space, with the lunar surface in the foreground, it was Earth that discovered itself, for the first time in all its existence, through our eyes. Likewise, the entire process of civilization, particularly the inexorable advance of science, must be understood as the dawn of planetary consciousness.[9]

With all of this in mind, we now can fully understand why the first step of our procedure, the one from de Ruijter and Clinton toward Elias's encompassing view of civilization, could never have been allowed to break away from modernism. Elias's anthropocentrism forces us into the uneasy position of masters of the world. Even our worries about issues such as global climate, the environment, or biodiversity are anthropocentric, as our prime concern remains only what these hazards mean to us humans. Only Earth system science offers the key insights by which we can establish a proper relationship with Earth. We have to accept that we can help ourselves only by helping the planet.

A Name for a Worldview

One reason for Clinton's mixed attitude may be the lack of an appropriate name for the new, emergent worldview. The creation of names plays a crucial role in the entire civilizing process.[10] Names are labels that provide otherwise vague notions with a discrete identity in our mind, allowing us to think and communicate about what they refer to. I wrestled myself with this problem for years. In countless lectures about the new worldview, I was unable to come up with a proper name. As a result, my argument remained confused, and I could not figure out the consequences of my intuition. After Lynn Margulis's death in November 2011, I was kindly invited to write an essay to her honor.[11] I leafed again through her work, when all of a sudden my attention was drawn to her short memoir, *Symbiotic Planet*.[12] This reminded me of a subject close to Lynn's heart—bacterial ecosystems and how they are organized.

I remembered an excursion we had made in 1980 for a NASA summer course to Laguna Figueroa, a salt lagoon in Baja California, Mexico. Just under the surface of the salty mud under our feet, and extending over large areas in the lagoon, were microbial mats, laminated microbial consortia, each layer with a different color. The microbiologists told us that each cubic centimeter of the mats contained representatives of the entire bacterial world—myriad different types, each with its own physiological potentials. Out of this profusion of microbial life, a small team was recruited; it was doing all the work, while the rest slept. The active team collected the necessary energy and nutrients for the whole community out of the environment, recycled the nutrients that were in short supply and removed (or inactivated) the toxic materials. As soon as conditions changed—for instance, after rainfall or during the cycle of day and night—the active team went to sleep and a new selection showed up to take over the job. We were amazed at the infinitely complex networks of metabolic pathways that the mats automatically maintained in this way. Lynn told us that these microbial mats represent the archetypal biosphere from which we all descend. All of today's ecosystems are still organized along these same primeval principles. For Lynn, the microbial mats were the acme of symbiosis.

When I realized, upon looking through *Symbiotic Planet*, that the entire Earth system is indeed symbiotic, the pieces of the puzzle fell together, and "the symbiotic worldview" became the name of my choice. Lynn saw that the component parts of the Earth, including the rocks, the oceans, the atmosphere and the biota, together form an integrated and interactive whole, capable of surviving over the eons. Furthermore, she insisted that symbiosis is a major source of evolutionary innovation, whereas entities that fail to adapt to the symbiotic Earth will eventually be eliminated. That is why humanity is doomed, she thought, because humans behave as parasites upon the Earth.

I personally have the impression that, as a biologist amid a dominating modernist society, Lynn tended to overemphasize biological determinism while underrating the innovative potential of the civilizing process. She despised the huge mushroom of civilization and regarded modernism as a predetermined attribute of man. However, we should not forget that humans are flexible and complex—biological and cultural at once. An anecdote may suffice to make the point. During the interval of a lecture by primatologist Frans de Waal emphasizing the biological underpinnings of our behavior, we all were amused to recognize the apelike conduct of the entire audience. I then overheard sociologist Joop Goudsblom mumbling that it was time for a chimp to give the next lecture, so that we could hear his point of view. The message is that we must not ignore the immense distance separating us from our primate ancestors. In fact, this is the prime goal of Plate VII, despite its anthropocentric underpinnings: to show the grandeur of the civilization process. Humans do change, and they even do so more and more rapidly, increasingly affecting the global environment. There is no reason to believe offhand that a symbiotic relationship between human civilization and Earth dynamics could not emerge.

The Earth's View of Modernism

For our further orientation on this matter, a search for some early manifestations of symbiosis within the present modernist order seems appropriate. I therefore propose to sketch a simplified picture of the modernist world as it functions today, excluding any signs of an emergent symbiosis. This mental picture may then serve as a negative control, allowing the new symbiotic sprouts, like the green specks standing out in the red field of Plate VII, to be readily identified.

The cartoon in Plate VIII symbolizes today's modernist configuration. Please note first of all that this image differs fundamentally from anthropocentric perceptions of our society. What we find in the textbooks is the bottom-up view with all cultural paraphernalia placed in the center as causative factors—industry, enterprise, labor, class contradictions, trade, justice, education, war, love, religion, and so on. My cartoon attempts a top-down perception, showing modernism from the vantage point of system Earth. Here, the human paraphernalia are locked up inside a slot machine. All we can see of them are a few bits and pieces that are relevant from Earth's point of view. At the scale of Earth, even the great conflicts of interest—between social classes or nations—are no more than a shifting of funds. We look at the deep structure of civilization: the modernist world does whatever it does, but what counts is its relationship with the Earth.

This slot machine is also unusual for another reason. We all know that in ordinary one-armed bandits, more money goes in than comes out, so that at the end of the day the owner can cash in the difference. In the modernist slot machine, however, there is more money coming out than going in, and the difference between these two fluxes goes to the owner (i.e., the ruling modernistic society). The inflowing money represents the investment required to secure the proper operation of the slot machine, that is, the amount of money required for the machine to reproduce the delivered products and services. The origin of the excess outflow remains a mystery to modernism, but, as we shall see, is clarified from the viewpoint of Earth system science. All in all, we should not be surprised that it took civilization a great many years to get this magic contraption up and running.

And magic it is! This slot machine reminds one of a *perpetuum mobile*, an apparatus that, once activated, runs forever. As we all know, however, the notion of a perpetual-motion machine is incompatible with the second law of thermodynamics: such contraptions are simply impossible. As a result, faith in modernism is based on the persistent and widespread superstition that value (or money) falls out of the blue sky. Thus, over and above the investment, a suitable medium is needed for the system to work: the pink cloud surrounding the slot machine. This cloud represents the dominant, modernist worldview, i.e., the mental orientation that allows the superstition to be generally accepted. This public orientation can take on many different forms; in principle, any myth that does not interfere with the slot machine and the money flow will do.

Before giving away the secret of the magic slot machine, we must address the question what all the excess money is used for. The answer is that ultimately it is spent on improving the slot machine, so that it generates even more money than before. This is why behind the little window we can see a dual spiral of science and technology. In this contraption, teams of scientists and engineers collaborate to develop new applications for industry and the marketplace. This science-technology spiral is the motor of modernist innovation and growth. Yet, the spiral is also a vicious threat to the integrity of the modernist regime. Science has the inconvenient property to unmask superstitions, so that for modernism to flourish, it is best to lock science up inside the slot machine and exclude it from the pink cloud of public orientation.

But such an incarceration cannot last forever. Although the everyday practice of science offers a chaotic spectacle of discovery, excitement, frustration, error, and fraud, the long view is one of an autonomous process that tends to defy human regulations. It is not surprising then that with time the pink cloud of myths is increasingly contaminated with ideas derived from science. To counter this threat, the modernist regime automatically reacts by inactivating the contaminating science. All it has to do is to reinforce the ongoing tendency of science to break up into countless disciplines. This leads to a Babylonian confusion from which no coherent narrative can emanate. Thus, the science leaking out of the slot machine can do no more than counter unjustified claims of the ruling myth (e.g., that the Earth is six thousand years old, or that evolution does not exist). The isolated splinters of science in the pink cloud cannot harm the modernist worldview. They even help to lure new generations of engineers and scientists into the science-technology spiral, and remove unnecessary discrepancies between the modernist myth and the real world.

So, what is the secret of modernism? Where is the hidden source that makes up for the difference between the investment and the outflow of the slot machine? It is the Earth, of course. Our planet donates the energy it gets from the sun and from its own interior, the raw materials, the water, the air and the environment on which the entire modernistic enterprise depends. We own and plunder the Earth as we please, reduce it to a globe of plaster and hide it inside the slot machine, together with all our other paraphernalia. This is why in Plate VIII, Earth is out of sight.

Testing the Slot-Machine Model with Shale Gas

Modernism owes much of its unchallenged dynamics to the exploitation of fossil fuels—peat, coal, oil, and gas. Not long ago, we used to think that these reserves would soon be depleted, that scarcity would force us to change to alternative energy sources and so draw to an end the climate problem stemming from the emissions of greenhouse gasses. So we were told. But this perspective is now outstripped by the introduction of new technologies.

Classical drilling for oil and gas is essentially vertical; the fuels are transported straight up from the underground fields to the surface. But larger reserves than the fields can deliver are preserved in layers of fossil clay, or shale. These valuables are stuck in the sediments so that they cannot be retrieved by the classical methods. Shale gas fracking is the technology that can retrieve these riches. First, vertical wells, about a mile apart, are drilled into the target shale. From each of those points onward, a dense array of drill holes is made to extend into the shale. The actual fracking involves the injection of a hot mixture of water, sand, and a cocktail of chemicals at high pressure into the array of drill holes. This fractures the shale, the sand grains keep the cracks open, the chemicals prevent bacterial contamination and scaling, and the gas is liberated.

In the United States alone, shales distributed over large parts of the nation are suitable for gas exploitation. Many other countries, including Russia and China, are preparing to follow the American example. To all appearances, the threat of energy shortages is over, as is the Western dependency on the Middle East. A new wave of industrialization can begin. Fortunately, as the burning of gas liberates less CO_2 and other pollutants, it is less harmful to the environment and climate than coal or oil. Gas utilization delays global warming, allowing us to prepare for the ultimate step in the energy saga—solar. However, while the oil industry is flooding the media with encouraging news, we are equally exposed to more alarming information. I will just mention the reports on large-scale pollution of soils and of water reserves, including tap water, with unspecified chemicals and dangerous toxins. Wastewaters are pumped underground, spread out over the surface, or squirted into the air for evaporation. Inhabitants suffer from serious poisoning. Wildlife and vegetation are devastated. Some parts of the United States, including protected natural reserves, are becoming uninhabitable. Finally, are we to believe the promise that shale gas opens the way to solar energy? Or does the dump price of natural gas discourage the necessary investments?

When innumerable billions of dollars are at stake, reliable information is in short supply.[13] Clearly, shale gas is now at the epicenter in a heated war of propaganda. The reason why I bring it up here is not make any judgment or a prognosis of the viability of this industry, but simply to put to the test my caricature of the modernist order as seen from Earth's point of view—the miraculous slot machine with all its paraphernalia. To what extent does this apply to the practice of shale gas extraction? Let me go through the list of its component parts.

1. The slot machine itself. The investments are dwarfed by the profits (so we are told), and nobody asks what it is that makes up the difference. For the owners of the business, money spews from the ground and all they have to do is to grab it.
2. The science-technology spiral. An understanding of subsurface geology, the development of advanced drilling and fracking technology, the logistics of water management, waste processing, and transportation—all this essential expertise could not exist without the close collaboration of scientists and engineers.
3. The cloud of superstition, myth, and disorientation. I limit myself to the following points.
   a. Although it is the Earth that pays the difference between the investment and the profits, the business entirely ignores its participation in the enterprise.
   b. Although enough information is made available to give the public the broad picture, essential details, such as the composition of the fracking fluids, are kept secret.
   c. The responsible public authorities hire oil geologists and other specialists for advice and for public information. Usually, however, the majority of these experts have been involved in the business throughout their career, so that they are unable to reach impartial conclusions.

Moreover, only very few people seem to be concerned at all about the brutal destruction going on at the continental scale in the planetary crust. While the powers of Earth to cure injury inflicted to the surface environment are amazing, the deep internal scars left by fracking will be irreparable for hundreds of millions of years. Do remember that now is the time of awakening planetary consciousness through science. These shales are indispensable treasures of the geological archive, even if we can only access them by drilling and are just beginning to decipher the messages they contain. Fracking at the continental scale, as it is intended, would amount to vandalism at a scale never paralleled in history. Serious lesions in the memory of the Earth can be expected.

I conclude that my slot machine caricature of modernism reflects the shale gas scenario in striking detail. But there is more: Fracking unleashes emotions deeper than cool calculation. Instead, we witness a fanatic zeal, a savage outcry for plunder and destruction, a deep-seated hatred against the Earth. Fracking has turned modernism into a fundamentalist religion. It sheds all appearances of stewardship and moderation and brutally reduces the planet to slavery. Is fracking a signal of weakness? Is it fear for the emerging symbiotic order?

Emerging Symbiotism

Meanwhile, we see in front of us the Babylonian confusion in the disunity of science fading away, and a new convergence emerging. The slot machine degenerates, so that the globe inside comes out into the open and transforms into the real Earth. The science-technology spiral unfolds to accommodate a third partner: a science-based mental orientation. The new triple spiral now begins to demolish the modernist superstition, first of all the central dogma of that worldview—the myth that we people have free disposition of Earth, as either owners or stewards. All of this is under way, and the spearhead is fundamental science, to be precise, Earth system science. And a first intimation of the symbiotic worldview is already within reach. For instance, it gets harder to avoid a deep sense of awe and belonging with this unique planet. We become more and more aware of the depth of our roots and the special role of civilization in this gigantic unfolding. The anthropocentrism inherent to modernism declines as we begin to realize that we humans are ephemeral manifestations of the Earth's creative reproduction.

Recalling Norbert Elias's model system for the civilization process, one may compare the modernist plunder of the Earth with the coarse behavior of Western European warriors in the fourteenth century. Indeed, a mechanism similar to courtly restraint seems to apply to the present change of regime, although the issues are different and the scale is much larger. Modernism is confronted with a change in the power structure: to avoid the threat of colossal catastrophes we can no longer treat the Earth as an easy source of energy and raw materials. These circumstances urge humans to suppress their unchecked impulses. This is how the opposing worldviews are now becoming the issue of societal strife. The early signs of symbiotism that appear in the heart of the modernist order provoke defensive reactions. Shale gas exploitation is modernism's revenge!

This leaves us wondering what the symbiotic order may look like, once it is fully developed. The problem reminds one of the French writer André Maurois, who said, *"Toujours l'inattendu arrive"*—what you never expected always happens. So, let me end with an idea originally launched by R. Buckminster Fuller, renowned visionary, architect, and scientist of the last century.[14] He pointed to the fact that the amount of solar energy reaching Earth is orders of magnitude bigger than the requirements of civilization. Solar panels spread out over the planet could easily provide all the electricity we need. The problem is in the storage; during the night there would be too little power. Meanwhile, as Earth turns around its axis, one-half of its surface is always blazing in the sun. So, if we were to build a worldwide web of energy linking all solar panels on Earth, there would always be plenty for everybody. A worldwide web of renewable energy is my symbol for a civilization observing symbiotism, a worldview Lynn Margulis, despite her skepticism about humanity, inspired.

NOTES

INTRODUCTION: EARTH, LIFE, AND SYSTEM
Bruce Clarke

1. Margulis and Sagan 2000, 48.

2. That important role has been superbly met by a prior volume, *Lynn Margulis: The Life and Legacy of a Scientific Rebel*, developed from a memorial meeting at the University Massachusetts, Amherst, in March 2012. See D. Sagan (ed.) 2012.

3. Merchant 1980; Midgely 2001; Needleman 2012.

4. Waddington 1969a.

1. LIFE ON A MARGULISIAN PLANET: A SON'S PHILOSOPHICAL REFLECTIONS
Dorion Sagan

1. See D. Sagan (ed.) 2012.

2. "20 Most Influential Scientists Living Today," 2012.

3. Margulis and Cohen 1994.

4. Sapp 2012, 66–67.

5. But the problem isn't settled: see Hall 2011.

6. Two important summary papers here are Gilbert, Sapp, and Tauber 2012 and McFall-Ngai et al. 2013.

7. "Every scientific idea passes through three stages," writes William Whewell in his 1840 *Philosophy of the Inductive Sciences*: "First, it is ridiculed. Second, it is violently opposed or claimed to be of only minor importance. Third, it is accepted as self-evident." See D. Sagan 2013a and 2013c, dedicated to both parents.

8. D. Sagan 2014.

9. Lem 2013.

10. Margulis and Sagan 1997.

11. Cited in Fenster 2000, 208.

12. Gupta and Mathews 2010.

13. Wrangham 2010.

14. D. Sagan 2013e.

15. Todd et al 2012.

16. Bybee 2012.

17. Doyle 2012; D. Sagan 2013b.

18. Sapp 2012.

19. Margulis and Sagan 2000.

20. Margulis, Asikainen, and Krumbein (eds.) 2011.

21. Margulis 2007, 52.

22. "Speeches and books were assigned real authors, other than mythical or important religious figures, only when the author became subject to punishment and to the extent that his discourse was considered transgressive" (Foucault 1977, 124).

23. Shakespeare, *The Life of Timon of Athens*, Act iv, scene ii.

24. Margulis and Case 2006.

25. For instance, Gould 1997.

26. Dolan 2012.

27. "Battle of Balliol" 2009.

28. Mitteldorf and Pepper 2009; Mitteldorf 2012.

29. Wilson 2013.

30. Varela, Maturana, and Uribe 1974, 188. See also Razeto-Barry 2012. On Margulis and autopoiesis, see Clarke 2009, 295–300.

31. D. Sagan 2011; D. Sagan 2013a.

32. Abram 2012.

33. Clarke 2012a.

34. D. Sagan 2012, 2013d.

35. Mira et al 2010.

36. Zhang et al 2011.

37. Cited in Welsh 2011.

38. Shapiro 2011.

39. Atsatt 1991.

40. Pirozynski 1991.

41. Ryan 2011.

42. Sagan and Margulis 2011, xv.

43. Whitehead 1948, 91–92.

2. THE RNA/PROTEIN WORLD AND THE ENDOPREBIOTIC ORIGIN OF LIFE
Sankar Chatterjee

My thanks to Lynn Margulis, David Deamer, Christie Henry, Brendan Headd, and Richard E. Wilde for critically reviewing the manuscript and constructive suggestions, and to Jeff Martz and Bill Mueller for illustrations. The research was supported by Texas Tech University.

1. Nisbet and Sleep 2001.

2. Schopf 1999.

3. Bird 2003.

4. Chyba et al. 1990; Bernstein, Sandford, and Allamandola 1999; Kring 2000; Nisbet and Sleep 2001; Delsemme 2001; Cockell 2006; Osinski 2011.

5. Chyba and Sagan 1992.

6. Cockell 2006; Osinski 2011; Hazen 2005.

7. Palme 2004; Cohen, Swindle, and Kring 2000; Nisbet and Sleep 2001.

8. Chyba and Sagan 1992; Kring 2000.

9. Mojzsis et al. 2000; Kring 2000.

10. Kring 2000.

11. Schopf 1999.

12. Green 1972; Cockell 2006; Nisbet and Sleep 2001; Osinski 2011.

13. De Wit and Ashwal 1997.

14. Condie 1981.

15. Melosh 1989; Chatterjee et al. 2006.

16. Kring 2000; Chatterjee et al. 2006.

17. Grieve 1990; Melosh 1989.

18. Grieve 1990.

19. Glikson 2008; Lowe et al. 1989.

20. Glikson 2010.

21. Mojzsis, Harrison, and Pidgeon 1996.

22. Fedo and Whitehouse 2002.

23. Rosing 1999.

24. Schopf 1999; Wacey et al. 2011; Walsh 1992.

25. Reysenbach and Cady 2001.

26. Kring 2000; Nisbet and Sleep 2001; Cockell 2006; Osinski 2011.

27. Reysenbach and Cady 2001.

28. Cockell 2006.

29. Holm 1992; Martin et al. 2008.

30. Hazen 2001; Kring 2000.

31. Abramov and Mojzsis 2009.

32. Woese 1987; Wiegel and Adams 1998; Martin et al. 2008.

33. McCollom and Shock 1997.

34. Woese 1987.

35. Gauchy et al. 2003.

36. Alberts et al. 1994.

37. Delsemme 2001.

38. Chyba and Sagan 1992; Nisbet and Sleep 2001; Bernstein, Sandford, and Allamandola 1999; Delsemme 2001; Cockell 2006; Osinski 2011.

39. Pizzarello and Shock 2012; Deamer 2011.

40. Bernstein, Sandford, and Allamandola 1999.

41. Nisbet and Sleep 2001; Kring 2000; Cockell 2006; Wiegel and Adams 1998.

42. Deamer 2011; Deamer and Fleischaker 1994.

43. Deamer et al. 2002.

44. Schopf 1999; Delsemme 2001; Chyba and Sagan 1992.

45. Wiegel and Adams 1998.

46. Hazen 2001; Hazen 2005.

47. Alberts et al. 1994.

48. Bernstein, Sandford, and Allamandola 1999; Delsemme 2001.

49. Hazen 2001.

50. Wächtershäuser 1993.

51. Cairns-Smith 1985.

52. Hazen 2005.

53. Alberts et al. 1994.

54. Hazen 2001.

55. Cooks, Zhang, and Koch 2001.

56. Pizzarello and Shock 2012.

57. Ibid.

58. De Marcellus et al. 2011.

59. Schopf 1999.

60. Hazen 2001.

61. Bernstein, Sandford, and Allamandola 1999; Callahan et al. 2011; Deamer 2011.

62. Ferris et al. 1996; Hazen 2001; Hazen 2005.

63. Wächtershäuser 1993.

64. Hazen 2001.

65. Cairns-Smith 1988.

66. Crick 1981; Cech 1986.

67. Deamer 2011.

68. Deamer et al. 2002.

69. Ibid.

70. Cech 1986.

71. Cech 2000; Steitz and Moore 2003.

72. De Duve 1995; Shapiro 2007; Trefil, Morowitz, and Smith 2009; Dyson 2004; Hazen 2005; Deamer 2011.

73. Levy and Miller 1998.

74. Szostak, Bartel, and Luisi 2001.

75. Deamer 2011; Deamer et al 1994.

76. Cech 2000; Steitz and Moore 2013.

77. Francis 2011.

78. De Duve 1995; Shapiro 2007; Trefil, Morowitz, and Smith 2009; Dyson 2004; Hazen 2005; Deamer 2011.

79. Fox 1988; Chyba and Sagan 1992; Wiegel and Adams 1998; Nisbet and Sleep 2001.

80. Harish and Caetano-Anollés 2012.

81. Ibid.

82. Ibid.

83. Francis 2011.

84. Davis 2002.

85. Margulis 1970; Margulis 1980.

86. Alberts et al. 1994.

87. Deamer 2011.

88. Alberts et al. 1994.

89. Harish and Caetano-Anollés 2012.

90. Forterre, Filee, and Myllykallio 2004.

91. Leu et al. 2011.

92. Forterre, Filee, and Myllykallio 2004; Forterre 2005.

93. Szostak, Bartel, and Luisi 2001.

94. Woese 1987; Wiegel and Adams 1998.

95. Koonin, Senkevich, and Dolja 2006; Lupi et al. 2006.

96. Koonin, Senkevich, and Dolja 2006; Forterre 2005.

97. Prusiner 1998.

3. EXOBIOLOGY AT NASA: INCUBATOR FOR THE GAIA AND SERIAL ENDOSYMBIOSIS THEORIES
James Strick

1. One of Lovelock's discoveries with the ECD was the rising concentration of chlorofluorocarbons in the atmosphere, even far from population centers and industrial areas. Thus, he made a seminal contribution to what soon became the ozone depletion debates of the 1970s (about spray can propellants as well as supersonic transport planes). See Dotto and Schiff 1978 and Lovelock 1979.

2. Silverstein to Lovelock, 9 May 1961, Lovelock papers. My thanks to James Lovelock for access to this material.

3. NASA 1962, 181.

4. Lovelock 2000.

5. See Bruch 1966, 488–89.

6. Lovelock 1979, 1. Stapledon had enormous influence on several generations of origin of life and exobiology researchers, most notably J. B. S. Haldane. See Adams 2000.

7. For a survey of the strategies being considered at this time, see Bruch 1966.

8. It is worth noting that this basic insight of Lovelock's, seen as so challenging in 1965, has since become the new paradigm in exobiology and astrobiology. See, e.g., Conrad and Nealson 2001 and, for a well-balanced assessment of Lovelock's fundamental contributions, Schneider 2001. A recent scientific evaluation of the fruitfulness of the Gaia hypothesis is Schneider et al. 2004.

9. Norman Horowitz oral history interview (OHI), 15 January 1999.

10. Ezell and Ezell 1984, 107.

11. Lovelock 1965.

12. Lovelock 2000, 237–39.

13. Hitchcock and Lovelock 1967.

14. Lovelock and Giffin 1968.

15. Lovelock 2000, 239; Lovelock OHI, 23 March 2000; also Norman Horowitz to Leslie Orgel, 3 Feb. 1968, Horowitz papers, CalTech Archives, 5.10.

16. Lovelock OHI; see Lovelock 1979, 10.

17. Margulis OHI, 23–24 June 1998.

18. Lovelock 1986, 392.

19. Later, it turned out, the biota also regulate cloud formation and thus dramatically alter the amount of incoming solar energy reflected back to space as another powerful way of regulating temperature. See Charlson et al. 1987.

20. See Margulis (ed.) 1970, 8–13. Kramer's notes from that 1950 course are available in his papers at the University of Florida, Gainesville.

21. Lovelock 2000, 239; see also Horowitz, Sharp, and Davies 1967. Horowitz was opposed in this opinion by Sagan, Levinthal, and Lederberg 1968), 1191–1196; see also Sagan OHI.

22. Lovelock 1979b, 715.

23. Horowitz to Strick, 16 Jan. 2002.

24. Cooper 1980, 69.

25. For an excellent description of these Antarctic dry valleys, see Pyne 1986, 226–33, 312–16. In a stunning stroke of historical irony, these valleys make a spectacular reappearance in the exobiology story after Viking, as a source of meteorites, some later determined to be from Mars, most notably EETA79001 from Elephant Moraine and ALH84001 from the Allan Hills (see Dick and Strick 2004, chapter 8).

26. Horowitz, Cameron, and Hubbard 1972; Cameron 1966; and Ezell and Ezell 1984, 235–37, including errata sheet.

27. See "Can Exploration" 1966: JPL scientists Richard Davies, Roy E. Cameron, and Roy Brereton will leave May 2 on a six-week exploration trip in Chile's Atacama Desert. NASA History Office, Exobiology files.

28. Strick 2004, esp. 140–41, 155–58, 167–69.

29. Lovelock to Margulis, n.d. (~January 1973), Lynn Margulis papers, University of Massachusetts, Amherst.

30. Lovelock 1972; Margulis and Lovelock 1974; Lovelock 1975.

31. The proceedings were published as Billingham (ed.) 1981.

32. Schneider and Londer 1984. The NOVA series can be viewed in six parts on YouTube.

33. The proceedings of this conference were published as Schneider and Boston (eds.) 1991; Kerr 1988.

34. Milne et al. 1985, 24; italics in the original.

35. Ibid., 154.

36. See Young 1985.

37. Lovelock 1979.

38. Lovelock and Watson 1982; Watson and Lovelock 1983; Lovelock 1988. For a balanced retrospective on the entire controversy, see Schneider 2001.

39. Mann 1991.

40. See Wills and Bada 2001, 81–83, for the initial negative reaction of scientists to Gaia based on its "Earth Mother" aspects, and for Lovelock's response.

41. Conrad and Nealson 2001. Nealson 1999 is an excellent review of new discoveries and changed thinking in microbiology since Viking that are relevant to exobiology and astrobiology.

42. Morowitz 1992, 5–6; Morowitz OHI, 20 March 2003. For Morowitz, the "systems approach" of Gaia must have had inherent appeal early on.

43. Lovelock 2001; Lovelock OHI. Margulis has replied in numerous articles, several of them in Margulis and Sagan 1997b; see especially Margulis 1997.

44. Lovelock to Strick, 11 March 2002; see Midgley 2001.

45. See Lovelock 1979, 714–17; Lovelock 1995, xvi–xvii. He develops the discussion much further as the central focus of his autobiography, Lovelock 2000.

46. Lovelock to Strick, 10 June 2002. The articles he mentions are Lenton 1998 and Lawton 2001. The *Geophysiology of Amazonia* paper was republished as Lovelock 1986.

47. Lawton 2001, 1965.

48. Kuhn 1970.

49. Lovelock to Strick, 6 June 2002. The article to which he refers is Overbye 2002, specifically to a quote about Gaia by Kevin Zahnle. The book to which Lovelock refers is Benton 2003.

50. L. Sagan 1967. She had sent the manuscript to Lederberg, Abelson, and Bernal; Margulis OHI.

51. Margulis OHI.

52. Sapp 1987. On the development of Margulis's SET, see Sapp 2003, chap. 21. Sapp 2009, esp. 119–20, also has much of the subsequent history of the SET, and an excellent discussion of why endosymbiosis ideas were resisted in the 1920s and '30s.

53. Lynn Margulis OHI, June 23–24, 1998, 1.

54. Ibid.

55. Ibid. By 1986, Margulis's yearly funding had reached $87k; for 1987, $87,891; for 1988, $90k; for 1989, $89,850; and for 1990, $89,850 (of $125,000 requested). By 2002, however, Exobiology ceased funding her because of budget cuts; Margulis papers. In 1977 Woese was already requesting $73k; by the early 1990s he was requesting $125k and receiving most of this; Woese to Strick, 14 January 2002.

56. Sapp 2009, 120.

57. Fleck 1979, 30–31, 92–94.

58. Margulis 1997, esp. 273–80.

59. Margulis later developed the idea of symbiogenesis in much greater depth, as a possible source of entire new phyla, for example, in Margulis and Sagan 2002.

60. Steig 1987 captures well the feeling I intend here.

61. See Sapp 1994, 180–83.

62. The first collaborative papers were Lovelock and Margulis 1973 and Margulis and Lovelock 1974.

63. Brand 1980; McDermott 1980.

64. Inserted as a sidebar in McDermott 1980, 31.

65. Lovelock turned out to be mistaken about the severity of ozone depletion, as the discovery of the Antarctic ozone hole revealed a few years later. See Lovelock 1981, 63.

66. Merchant 1980.

67. See, e.g., Kite 1986.

68. Keller 1986, 48.

69. Keller 1986; Margulis 1970.

70. Margulis to Strick, 1 July 1999.

71. Lovelock 1981.

72. Lovelock 2001.

73. Midgley 2001, 172.

74. Ibid., 173.

75. Ibid.

76. Ibid., 173–74.

77. Ibid., 16–17.

78. This translation became the basis of the 1998 complete annotated English edition.

79. Vernadsky 1998, 15.

80. Needleman 2012, 20, quoting Vernadsky. Further discussion on 7–8, 18–21.

81. Callicott 2010, 186–87, discusses Ouspensky but downplays how much his ideas mattered to Leopold. That Needleman saw similar connections there suggests that the linkages are not so weak after all.

82. Needleman 2012, 114, 153.

83. Ibid., 37.

4. ON SYMBIOSIS, MICROBES, KINGDOMS, AND DOMAINS
Jan Sapp

1. Lamarck 1809.

2. Darwin 1859, 420.

3. Ibid., 479.

4. Ibid., 490.

5. Darwin Correspondence Project 2013.

6. Owen 1860.

7. Hogg 1860.

8. Haeckel 1866.

9. Sapp 2009.

10. Carter 1991.

11. Woodhead 1891, 24.

12. Breed 1928, 143.

13. Manwarring 1934, 470.

14. Huxley 1942, 131–32.

15. Copeland 1938, 386.

16. Sapp 2009.

17. Lwoff 1957.

18. Stanier and van Niel 1962, 17.

19. Chatton 1925; Chatton 1938.

20. Stanier, Doudoroff, and Adelberg 1963, 85.

21. Ibid., 409.

22. Allsopp 1969; Stent 1971.

23. Whittaker 1969.

24. Ibid., 157.

25. Margulis 1970; Margulis 1971; Whittaker and Margulis 1978.

26. Margulis and Schwarz 1982, 1988, 1998; Margulis and Chapman 2009.

27. Sagan 1967.

28. Haeckel 1905.

29. Sapp, Carrapiço, and Zolotonosov 2002.

30. Merezhkowsky 1910; Merezhkowsky 1920.

31. Watasé 1893.

32. Kozo-Polyansky 2010.

33. Portier 1918.

34. Wallin 1927.

35. Wallin 1924.

36. D'Herelle 1926, 320.

37. Wallin 1927, 8.

38. Lederberg 1952; Sapp 1994.

39. Sapp 1994.

40. Wilson 1925, 739.

41. Nass and Nass 1963, 621.

42. Warren 1967.

43. Sapp 2006.

44. Margulis 1970; Margulis 1982.

45. Margulis 2005.

46. Ris and Plaut 1962, 390.

47. L. Sagan 1967.

48. Smith-Sonneborn and Plaut 1967.

49. Cleveland and Grimstone 1964.

50. Margulis 1967, 1970, 1981, 1993.

51. Uzzell and Spolsky 1974, 343.

52. Sapp 1998.

53. Margulis, 1981, 1993, 2006.

54. L. Sagan 1967.

55. Sapp 2009.

56. Woolhouse 1967.

57. Stanier 1970, 31.

58. Margulis 1975, 21.

59. Dubnau et al. 1965.

60. Sapp 2009.

61. Rivera et al. 1998.

62. Sapp 2009.

63. Crick 1958.

64. Woese 1987, 227.

65. Zablen et al. 1975; Bonen et al. 1977; Woese 1977; Gray and Doolittle 1982; Gray 1992.

66. Gray 1992.

67. Margulis and Fester (eds.) 1991.

68. Woese and Fox 1977; Woese 1982; Woese 1998b.

69. Carl Woese to Emile Zuckerkandl, March 29, 1977. Quoted in Sapp 2009, 173.

70. Lederberg 1952.

71. Woese and Fox 1977; Fox et al. 1980.

72. Woese, Kandler, and Wheelis 1990.

73. Ibid., 4576.

74. Ibid., 4577.

75. Sapp 2009.

76. Allsopp 1969; Stent 1971.

77. Balows et al 1992, vii.

78. Mayr 1991; Mayr 1998.

79. Sapp 2009.

80. Mayr 1998.

81. Mayr 1991.

82. Woese, Kandler, and Wheelis 1992, 2930.

83. Mayr 1998.

84. Woese 1998a, 11045.

85. Margulis 1981.

86. Margulis and Guerrero 1992, 48.

87. Mayr 1991.

88. Lynn Margulis to Carl Woese, January 15, 1978. Quoted in Sapp 2009, 198.

89. Margulis 1993, 92, 38.

90. Ibid., 22.

91. Cavalier-Smith, 1983, 1998.

92. Margulis and Chapman 2009.

93. Margulis and Fester (eds.) 1992; Margulis and Sagan 2002.

94. Hotopp et al. 2007.

95. Sapp 2003; Sapp 2010.

96. Sapp 1994.

97. Smith and Szathmáry 1995.

98. Gregory 1951.

99. Sapp 2003, 2007; Gilbert, Sapp, and Tauber 2012.

5. THE WORLD EGG AND THE OUROBOROS: TWO MODELS FOR THEORETICAL BIOLOGY
Susan Squier

1. Waddington 1969c; Waddington 1968a.

2. Waddington 1969a.

3. Bard 2008, 195.

4. Cited in Slack 2002, 894.

5. Kemp 1996, 29; "Conrad Hal Waddington" 2013; Hall 1992, 114.

6. Barad 2010.

7. Waddington 1972a.

8. Slack 2002.

9. Ibid., 889.

10. Waddington 1969b.

11. Waddington 1969c; Waddington 1968a.

12. The participants were the molecular biologist (Robin E. Monro), the theoretical biologist (Brian Goodwin), the three neuroscientists (Jack Cowan, Richard L. Gregory, and Karl Kornacker), the geneticist (John Maynard Smith), the seven physicists (theoretical physicist David Bohm, and physicists E.W. Bastin, W. M. Elsasser, Martin A. Garstens, Edward H. Kerner, Paul Lieber, and Howard H. Pattee), the theoretical chemist (Christopher Longuet-Higgins), the chemical engineer (L. E. "Skip" Scriven), the systems analyst (A. S. "Art" Iberall), the philosopher (Marjorie Grene), the automata theorist (Michael A. Arbib), and the secretary (Miss D. Manning). Waddington 1969, 339.

13. The concept of a theoretical biology had a precedent: formal scientific theory, whose accepted paradigm beginning in the 1920s or earlier was "the use of quantification and mathematical modeling to explain and predict phenomena" (Bard 2008, 194).

14. Waddington 1969b.

15. Ibid.

16. Waddington 1969d.

17. Bohm 1969a, 18

18. Ibid.

19. Bohm 1969b, 41.

20. All quotations in this paragraph are from Waddington 1969d, 72.

21. Ibid.

22. Ibid., 73.

23. Ibid.

24. Waddington 1968a, 525.

25. Margulis 1995, 132.

26. Ibid.

27. Waddington 1969d, 72–73.

28. Ibid., 74.

29. Gilbert, ed. 1991, 199.

30. Waddington 1969d, 74–75. As we will see, Waddington used a similar phrase to launch his argument in *Behind Appearance* to illustrate the scientific worldview surpassed by the "Third Science" as well as by Cubist and post-Cubist painting. There, however, the phrase omits the billiard balls: "if we believe that the entire universe is constructed of unequivocal impenetrable atoms, each simply located

at a precisely defined position in an unambiguous framework of space and time—existing, in Whitehead's phrase, in vacuous actuality" (ibid., x).

31. All quotations in this paragraph ibid., 75–76.

32. Ibid., 80.

33. Squier 2011, 37–41; Gilbert 1991a, 188–91.

34. Gilbert 1991a, 192.

35. Landecker 2007. For an excerpt of Gilbert 1991a in this connection, see also http://9e.devbio.com/article.php?ch=4&id=27.

36. Slack 2002, 893. See also Hall 1992, 119.

37. Waddington 1940.

38. Waddington 1969d, 81.

39. Ibid., 72–73.

40. Clarke and Hansen (eds.) 2009, 7.

41. Ibid., 6–7.

42. Clarke and Hansen (eds.) 2009; Waddington 1976, 249.

43. Gilbert 1991b, 151.

44. Waddington 1972b, 283.

45. Ibid, 284–87.

46. Ibid. 286–87.

47. Ibid.

48. Pigliucci 2011.

49. Clarke 2012, 197. See also Clarke's essay in this volume.

50. Waddington 1969a, x.

51. Ibid., ix–x.

52. Waddington 1969a, 1.

53. Waddington 1968b.

54. Waddington 1972b, 283.

55. Waddington 1969a, 1–8; Goodall 1965.

56. Ibid., 5.

57. Ibid., 99.

58. Ibid., 104.

59. Ibid., 13, 121–123, 113, 119, 6.

60. Ibid., 4.

61. Ibid., 70.

62. Waddington 1972b, 288. My thanks to Bruce Clarke for this observation.

63. Waddington 1969a, 6.

64. Ibid.

65. I discuss these gender dynamics at length in an earlier version of this essay published in *The Scholar and the Feminist Online* 11.3 (Summer 2013), http://sfonline.barnard.edu/life-un-ltd-feminism-bioscience-race.

66. Foreword to Waddington 1969a, n.p.

67. Schmitt 2008.

68. Schaechter 2012, 14.

69. Waddington 1969a, 238.

70. Ibid., 242.

71. Ibid., 6. Note how this anticipates developmental systems theory (DST). See Oyama in this volume.

72. Ibid., 241.

73. Ibid., 243.

74. Gilbert 1991.

75. Ibid., 151.

76. Jantsch and Waddington (eds.) 1976, 243.

77. Ibid., 248.

78. Ibid., 249.

79. Ibid., 243–44.

80. Waddington 1969a, 107, 119. Oyama's essay in this volume discusses just this problematic relation between "chance" and "determinism" in DST's concept of "contingency."

81. I am grateful to Nick Hopwood for pointing out the remarkably counterfactual nature of Waddington's caption for this image, and to Sarah Franklin, Nick Hopwood, and Martin Johnson for creating the occasion for that conversation, the IVF Histories and Cultures Workshop, Christ's College, Cambridge, England, June 22–24, 2014.

82. As this essay was going to press, I discovered philosopher Jan Baedke's sweeping survey of contemporary reinvigorations and revisions of the epigenetic landscape. While Baedke's essay also places the epigenetic landscape in the context of Waddington's work on *Behind Appearance*, he limits his analysis of the latter to confirming what Waddington himself termed his "life-long interest in paintings" (Baedke 2013, 757). It has been my goal in this essay to explore the epistemological impact of that interest on Waddington's scientific work.

83. Monk and Holding 2001; Monk 2002.

6. THE PLANETARY IMAGINARY: GAIAN ECOLOGIES FROM *DUNE* TO *NEUROMANCER*
Bruce Clarke

1. A historical sampling is collected at www.earthrise.org.uk.

2. Brand 1977, 186; Maher 2004; Kirk 2007, 40–42; Poole 2008.

3. For strong popular expositions of the planetary carbon cycle, see Volk 2003 and Harding 2006.

4. On planetary self-reference, see Clarke 2014, 126–30.

5. Bateson 1972, 454.

6. Herbert 1965, 13–14.

7. For a brilliant meditation on related aspects of Western globalization, see Sloterdijk 2013.

8. For instance, Morin 1977; Morin and Kern 1999; Serres 1995; Serres 2012; and Sloterdijk 2011; see also Clarke 2001.

9. Thompson 1974; Thompson (ed.) 1987; Thompson 1991. For background, see Clarke 2009.

10. Bateson 1972, 461.

11. For perhaps the earliest presentation of ecology through explicitly cybernetic concepts, see Hutchinson 1948, which also contains a discussion of the geobiological carbon cycle.

12. On the systems counterculture, see Clarke 2012b.

13. Hagen 1992, 192. See also Kingsland 2005, 179–85.

14. Margulis 1998, 7, 9.

15. Herbert 1965, 477.

16. Morton 2001.

17. Herbert 1965, 335.

18. Ibid., 336.

19. Ellis 1990.

20. Bateson 1972, xii.

21. Ibid., xv.

22. For "the phrase 'ecology of ideas,' I am indebted to Sir Geoffrey Vickers' essay 'The Ecology of Ideas' in *Value Systems and Social Process,* Basic Books, 1968" (Bateson 1972, 448). A figure of the British establishment, Vickers, too, like Herbert and Bateson, is a humanistic observer positioned outside the science of ecology.

23. Herbert 1965, 482.

24. Bateson 1972, 503–4. This article "represents afterthoughts" following a conference held in October 1970, with city planners in the office of John Lindsay, the mayor of New York (494).

25. Ibid., 504.

26. Herbert 1965, 477–78.

27. Bateson 1972, 460.

28. Ibid., 504.

29. Clarke 2012b.

30. Bateson 1972, 504.

31. Ibid., 505.

32. Hagen 1992, 192.

33. Clarke 2012a.

34. Sonea and Mathieu 2000.

35. For a detailed treatment, see Margulis and Sagan 2000.

36. Margulis 1998, 111.

37. Margulis and Sagan 2000, 189–90.

38. Margulis 1998, 119.

39. See Volk 2003, 124.

40. Lovelock 2009, 22; my italics.

41. James Lovelock to Lynn Margulis, October 22, 1986. Lynn Margulis papers.

42. Lotka 1925, 16; bracketed phrase in the original.

43. Sagan and Margulis 2007b, 77.

44. Ibid.

45. Sagan and Margulis 2007a, 183.

46. Gibson 1984.

47. Ibid., 52.

48. Ibid., 5, 52.

49. Ibid., 11, 46, 6.

50. Ibid., 31, 46.

51. Ibid., 85, 92, 96.

52. *CoEvolution Quarterly* debuted the Gaia hypothesis for a nonspecialist audience with Margulis and Lovelock 1975.

53. Gibson 1984, 101.

54. Brand, ed. 1977.

55. McCray 2012 provides a detailed study of O'Neill's projects. William Gibson is cited three times, the existence of *Neuromancer* once. However, McCray never opens the text of *Neuromancer* to discover its glorious literary repurposing of O'Neill's ideas and NASA's graphics.

56. O'Neill with Brand 1975, 20.

57. O'Neill 1977, 8.

58. Gibson 1984, 77.

59. Ibid., 107.

60. Gibson 1984, 101.

61. Ibid., 103, 109, 104.

62. Edwards 1996 develops a dichotomy between "closed worlds" and "green worlds" in the cyborg imagery of Cold War narratives: "the fate of *Neuromancer*'s Case and Molly," for instance, is to "remain within the closed world" whose operational boundaries are determined by the cybernetic control and communication systems that dominate the historical world of which *Neuromancer*'s storyworld is a reflection. The only escapes are "false exits into ersatz green worlds such as *Neuromancer*'s inverted worlds, the Zion and Freeside space stations" (309). The blind spot in Edwards's vision of cybernetics is its nonobservation of the second-order cybernetics of living systems, and in particular, of the neocybernetics of Gaia theory. Edwards's text remains stuck in the conceptual claustrophobia of Cold War control engineering, from which it cannot observe the ecological difference between Zion and Freeside.

63. Ibid., 225–26.

64. For other takes on this cultural episode, see Kirk 2007, 170–76, and Anker 2010, 113–25. Thanks to Christopher Witmore for the Anker reference.

65. Ehrlich 1977.

66. J. Todd 1977, 48.

67. This is what happened in 1992 with the most significant and large-scale experiment in closed ecological habitats to date, Biosphere 2. After seventeen months, the O_2 level of its closed atmosphere dipped precipitously, the habitat became unviable for its human inhabitants, and the closed environment had to be opened. See Anker 2010, chapter 8. For wider scientific contexts, see Morowitz, Allen, and Alling 2005; Sagan 1990.

7. BRINGING CELL ACTION INTO EVOLUTION
James A. Shapiro

1. Wier et al. 2010. While her early work on the mitochondrion and chloroplast as descended from endosymbiotic bacteria (Margulis 1970, Margulis 1971) has been amply substantiated by DNA sequence analysis (see http://shapiro.bsd.uchicago.edu/Origins_of_eukaryotic_cell.html), the work on eukaryotic organelles that lack their own genomes has not so far been widely accepted (Margulis, To, and Chase 1978; Margulis, Chase, and To 1979; Margulis 1980; Margulis, To, and Chase 1981; Chapman, Dolan, and Margulis 2000; Dolan et al. 2002; Margulis et al. 2006). Jan Sapp also touches on this point in this volume.

2. Shapiro 2009.

3. Huxley 1942.

4. Merezhkowsky 1920; Walin 1927; Kozo-Polyansky 2010.

5. Margulis 1981; Margulis 1996; Margulis, Dolan, and Guerrero 2000; Margulis et al. 2006. See http://shapiro.bsd.uchicago.edu/Origins_of_eukaryotic_cell.html.

6. Margulis and Bermudes 1985; Margulis 1993; Chapman and Margulis 1998; Margulis and Sagan 2002; Embley and Martin 2006; Margulis 2009. See http://shapiro.bsd.uchicago.edu/Secondary_and_tertiary_symbioses.html.

7. http://shapiro.bsd.uchicago.edu/Shapiro.1988.scientificamerican0688-82.pdf. See also Shapiro 1988; Shapiro and Dworkin 1997; Shapiro 1998; Shapiro 2007; http://shapiro.bsd.uchicago.edu/ExtraRefs.Symbiogenesis and the origin of eukaryotic cells.shtml; Margulis et al. 1986.

8. http://shapiro.bsd.uchicago.edu/Human_Microbiome.html.

9. http://shapiro.bsd.uchicago.edu/Gonadal_Symbiosis.html.

10. Avery, MacLeod, and McCarty 1944; Hershey and Chase 1952; Watson and Crick 1953; Watson 1953.

11. Crick 1958; Crick 1970.

12. Benzer 1956; Jacob and Monod 1961; Jacob and Wollman 1961; Benzer 1962; Tjian 1995.

13. Chambon 1981; Sharp 1994; Ast 2005.

14. http://shapiro.bsd.uchicago.edu/Regulatory_Networks.html.

15. http://shapiro.bsd.uchicago.edu/ExtraRefs.SystemsApproachGenerating-FunctionalNovelties.shtml.

16. Monod 1942. That was his day job. By night, he was also a leader of the Resistance.

17. Morange 2010; Ullmann 2010.

18. Lwoff 1954; Jacob and Monod 1961; Jacob and Wollman 1961.

19. Deutscher 2008; Gorke and Stulke 2008.

20. Kunkel and Bebenek 2000.

21. Radman and Wagner 1988; Rennie 1991.

22. Weigle 1953; Weigle and Bertani 1953.

23. Witkin 1975; Devoret 1979; Howard-Flanders 1981; Witkin 1991.

24. Sommer et al. 1993; Goodman 2002.

25. Huisman, D'Ari, and Gottesman 1984.

26. Hartwell 1989; Murray and Kirschner 1991; Weinberg 1996.

27. http://shapiro.bsd.uchicago.edu/ExtraRefs.CellCycleCheckpoints.shtml.

28. McIntosh and McDonald 1989; Musacchio 2011.

29. http://shapiro.bsd.uchicago.edu/Apoptosis.html.

30. Meredith, Fazeli, and Schwartz 1993; Ferrer and Planas 2003; Marini and Belka 2003.

31. Beadle 1948; Watson 1953; Crick 1958; Crick 1970; Shapiro 2009.

32. McClintock 1984.

33. Muller 1927.

34. McClintock 1932.

35. McClintock 1939; McClintock 1942.

36. McClintock 1984.

37. Bray 2009.

38. McClintock 1987.

39. http://shapiro.bsd.uchicago.edu/ExtraRefs.WholeGenomeDoublingCriticalStagesEvolution.shtml.

40. Shapiro 1983.

41. McClintock 1984.

42. Bukhari 1977; Shapiro 1983; Shapiro 2011.

43. McClintock 1953; McClintock 1961.

44. McClintock 1956.

45. http://shapiro.bsd.uchicago.edu/ExtraRefs.MolecularMechanismsNaturalGeneticEngineering.shtml.

46. Griffith 1928; Avery, MacLeod, and McCarty 1944.

47. http://shapiro.bsd.uchicago.edu/TableIII.1.shtml.

48. Lwoff 1953; Lwoff 1954; Lwoff 1966.

49. Forterre 2006; Koonin et al. 2006; Koonin 2010.

50. Lwoff 1957.

51. Groman 1953.

52. Zinder and Lederberg 1952; Zinder 1958.

53. Gilbert and Dupont 2011; http://shapiro.bsd.uchicago.edu/Viral_Composites.html.

54. Colson and Raoult 2010.

55. http://shapiro.bsd.uchicago.edu/Amoebal_Viruses.html.

56. Hallet and Sherratt 1997; Hallet and Sherratt 2010; http://www.huffingtonpost.com/james-a-shapiro/dna-as-poetry-multiple-me_b_1229190.html.

58. http://Shapiro.bsd.uchicago.edu/Specialized_Transduction_and_in_vivo_Cloning.html.

59. Shapiro et al. 1969; Shapiro 2009b.

60. Hayes 1968.

61. Gorini 1966.

62. Fauci 2001.

63. Watanabe 1967.

64. Clowes 1973; Novick 1980.

65. Davies 1979; Foster 1983; Levy 1998.

66. Sonea and Panisset 1983.

67. http://shapiro.bsd.uchicago.edu/TableIII.1.shtml.

68. Edwards and Rohwer 2005; http://shapiro.bsd.uchicago.edu/Viral_Composites.html.

69. Hacker and Carniel 2001; Juhas et al. 2009.

70. http://shapiro.bsd.uchicago.edu/Bacterial_Cell_Biology_In_Pathogenesis_and_Symbiosis.html.

71. http://shapiro.bsd.uchicago.edu/Competence_for_DNA_Uptake.html.

72. http://shapiro.bsd.uchicago.edu/Bacterial_Surface_Structures.html.

73. In terms of contemporary public controversies, the Intelligent Design (ID) advocates use the bacterial flagellum as their poster child for an "irreducibly complex" structure that could not have evolved by Darwinian evolutionary processes (Behe 1996; http://www.wesjones.com/darwin.htm). Both the ID and scientific evolution camps need to address how the flagellum and related biological inventions came to be diversified for so many different uses. Certainly, the ID argument is greatly undermined if it has to invoke supernatural intervention for the origin of each modified adaptive structure. At the same time, it is fair to recognize that evolutionary science faces the challenge to provide a plausible account for the origin and diversification of this intricate functional design for moving large molecules across the bacterial envelope.

74. http://shapiro.bsd.uchicago.edu/TableIII.1.shtml.

75. http://shapiro.bsd.uchicago.edu/Interkingdom_and_Eukaryotic_Horizon tal_Transfer.html.

76. http://shapiro.bsd.uchicago.edu/Intracellular_Horizontal_Transfer.html.

77. Embley and Martin 2006.

78. Bukhari 1977; Cohen and Shapiro 1980; Craig 2002; Shapiro 2009b.

79. http://shapiro.bsd.uchicago.edu/Bacterial_Transposons.html.

80. Polard and Chandler 1995.

81. Hedges and Jacob 1974.

82. Hall and Stokes 1993; Hall and Collis 1995.

83. http://shapiro.bsd.uchicago.edu/Integrons_Super-integrons.html.

84. http://shapiro.bsd.uchicago.edu/Legitimate_and_Illegitimate_Recombi nation.html.

85. Hayes 1968; Stahl 1987.

86. Aziz, Breitbart, and Edwards 2010.

87. Lander et al. 2001.

88. Ding, Haraguchi, and Hiraoka 2010; Szekvolgyi and Nicolas 2010.

89. Goldfarb and Lichten 2010.

90. Haber 1998; Klar 2007; Yamada-Inagawa, Klar, and Dalgaard 2007; Klar 2010.

91. Nasmyth 1993; Rusche and Rine 2010.

92. Parvanov, Kohli, and Ludin 2008.

93. http://shapiro.bsd.uchicago.edu/Antigenic_Variation.html; http://shapiro .bsd.uchicago.edu/TableII.5.shtml.

94. Temin 1972.

95. Temin and Mitzutani 1970; Varmus 1987.

96. Coffin, Hughes, and Varmus 1997.

97. Crick 1970.

98. http://shapiro.bsd.uchicago.edu/Retrovirus_Integration.html.

99. Varmus 1987.

100. Rous 1910.

101. Weinberg 1996.

102. Swain and Coffin 1992.

103. Nusse and Varmus 1982; Tsichlis 1987; Tsichlis et al. 1990.

104. Pilacinski et al. 1977; Errede et al. 1981.

105. http://shapiro.bsd.uchicago.edu/Endogenous_Retroviruses.html.

106. http://shapiro.bsd.uchicago.edu/Retroviral_involvement_in_placenta _evolution.html.

107. Coffin, Hughes, and Varmus 1997; McDonald et al. 1997.

108. Varmus 1987; Coffin, Hughes, and Varmus 1997.

109. http://shapiro.bsd.uchicago.edu/LTR_retrotransposons_and_genome _evolution.html.

110. http://shapiro.bsd.uchicago.edu/Genome_Size.html.

111. http://shapiro.bsd.uchicago.edu/Non-LTR_Retrotransposons.html.

112. Lander et al. 2001.

113. Sternberg and Shapiro 2005.

114. Kazazian 2000.

115. http://shapiro.bsd.uchicago.edu/RNA_Reverse_Transcription_and_Genome_Insertion.html.

116. http://shapiro.bsd.uchicago.edu/LINE_SINE_Retrotransduction.html.

117. http://shapiro.bsd.uchicago.edu/Exon_Shuffling.html; http://www.huffingtonpost.com/james-a-shapiro/genetic-engineering_b_1541180.html.

118. http://shapiro.bsd.uchicago.edu/NHEJ.html.

119. Since, as we have seen, genomes are replete with repetitive elements, it is also possible to generate many chromosome rearrangements by homologous recombination between repeats at different genomic locations. See http://shapiro.bsd.uchicago.edu/Chromosome_Rearrangements_by_Repeat_Recombination.html. Sequencing of the rearranged sites sometimes makes it possible to distinguish between the different mechanisms.

120. White 1945 (4th ed. 1973); White 1978; King 1995.

121. http://shapiro.bsd.uchicago.edu/Chimp_to_Human_Chromosome_Fusion.html.

122. http://www.huffingtonpost.com/james-a-shapiro/epigenetics-iii-epigeneti_b_1683713.html.

123. Shapiro 1992.

124. http://shapiro.bsd.uchicago.edu/Natural_genetic_engineering_and_multimolecular_networks.html.

125. http://shapiro.bsd.uchicago.edu/Evo21.Appendices.shtml; http://www.huffingtonpost.com/james-a-shapiro/genetic-engineering-immune-system-evolution_b_1255771.html; and http://www.huffingtonpost.com/james-a-shapiro/immune-cells-dna-engineering_b_1395040.html.

126. http://shapiro.bsd.uchicago.edu/VDJ_joining.html.

127. http://shapiro.bsd.uchicago.edu/Ig_Class_isotype_switching.html.

128. http://shapiro.bsd.uchicago.edu/Somatic_hypermutation.html.

129. http://shapiro.bsd.uchicago.edu/Fossil_Record.html.

130. Weismann 1893.

131. http://shapiro.bsd.uchicago.edu/TableII.7.shtml; http://shapiro.bsd.uchicago.edu/TableII.11.shtml.

132. Although documented cases have not yet occurred, I am confident we will eventually find examples of NGE targeted by RNA-RNA pairing as well. This is the mechanism behind RNA targeting of chromatin formatting.

133. http://shapiro.bsd.uchicago.edu/Hybrid_dysgenesis_interspecific_hybridization.html.

134. Stebbins 1951; Anderson 1954.

135. Woodruff and Thompson 2002.

136. http://shapiro.bsd.uchicago.edu/TableII.10.shtml; http://shapiro.bsd.uchi cago.edu/TableII.8.shtml.

137. Brown and O'Neill 2010.

138. http://shapiro.bsd.uchicago.edu/ExtraRefs.GenomeFormattingTransfer ToDaughterCells.shtml; http://shapiro.bsd.uchicago.edu/ExtraRefs.GenomeFor mattingAccessingStoredInformation.shtml.

139. http://shapiro.bsd.uchicago.edu/ExtraRefs.GenomeCompactionChroma tinFormattingEpigeneticRegultion.shtml.

140. Waddington 1942 (rpt. 1977); http://shapiro.bsd.uchicago.edu/ExtraRefs .GenomeCompactionChromatinFormattingEpigeneticRegultion.shtml.

141. http://shapiro.bsd.uchicago.edu/ExtraRefs.NaturalGeneticEngineering PartNormalLifeCycle.shtml.

142. http://shapiro.bsd.uchicago.edu/ExtraRefs.NaturalGeneticEngineering AndEvolutionaryGenomicInnovation.shtml.

143. http://shapiro.bsd.uchicago.edu/Transgenerational_Epigenetic_Effects .html.

8. SUSTAINABLE DEVELOPMENT: LIVING WITH SYSTEMS
Susan Oyama

Portions of this chapter were derived from talks at a panel at the July 14, 2011, meeting of the International Society for History, Philosophy, and Social Studies of Biology, in Salt Lake City, Utah, and at the Earth, Life & System Symposium on Environment and Evolution at Texas Tech University. I thank my panel mates from the 2011 conference, Lauren McCall, David McCandlish, and Kriti Sharma, as well as the participants in the 2012 TTU meeting. Others gave aid and comfort at vary-ing (sometimes recurrent) moments: Pat Bateson, Joan Hoffman, Tim Johnston, Bob Lickliter, Gregory Mengel, Armin Moczek, Barbara Herrnstein Smith, Peter Taylor, Denis Walsh, Rasmus Winther, and Lee Worden. My gratitude to you all.

1. Developmental systems include, that is, *but are not coextensive with* the en-tity whose development is being described: Oyama 1985; Oyama 2009. Several philosophers of biology have noted that this framing facilitates the bringing to-gether of a wide variety of approaches, making room in evolutionary accounts for often marginalized phenomena like symbioses, phenotypic plasticity, and epigenetic inheritance (indeed, I would add, like development itself). See God-frey-Smith 1999; Robert 2003; Robert 2006.

2. *Construct* as *constituting as* or *making relevant*, as well as *changing* or *making* in more usual ways.

3. Lewontin 1982.

4. By naming their book *Acquiring Genomes*, she and Dorion Sagan took a poke at a prime incarnation of the nature/nurture duality, that of innate (supposedly produced by genomes) vs. acquired characters: Margulis and Sagan 2002.

5. Both meanings are found in Margulis and Sagan's (1995, 117) declaration that life is "the ingenuity to make the most of contingency—to make animals, for example, out of a botched attempt at cannibalism." As suggested in the previous section, we should not just smile at the fortuity of failed predation (the first meaning) while missing the second: the numberless metabolic, morphological, and ecological customizations that make and maintain those animals. The eventual coordination of these dependencies over successive generations testifies to life's paradoxical juncture of mutability and repeatability, of surprise and reliable concatenation of causal contingencies.

6. Gilbert and Epel 2009, 31, whose large-spirited last section scans an array of thought-traditions for help in assimilating reciprocal inductions, context-sensitivity and intricate developmental co-construction. If the authors think their language does not do justice to their phenomena, I suspect it is a sense that we all recognize.

7. Oyama 2000, 126–127.

8. Sapp 1994, xvii; see note 6.

9. While we cannot say how much causal responsibility for a given *trait* is borne by one or the other, as many have noted, the sources of *differences* in a particular group can be studied *under defined conditions* (and generalized only under similar ones). Interpretations must thus be made with strict caution, as was evident in past debates over issues like race and intelligence. Quantifying contributions to population variation is thus distinct from partitioning organisms; attempts to use the former to draw conclusions about the latter have a long and sorry history: Lewontin 1974; Lickliter and Schneider 2006; Oyama 1985; Oyama 2000; Sober 1988; and Taylor 2005, 131, 243. For a sense of what may be finessed, ignored, assumed, and extrapolated by quantitative geneticists, as well as by their critics, see Taylor 2014, which offers ways of respecting (and inspecting) the gaps between traits and their underlying causal factors.

10. Taylor 2001, where he notes its articulations with DST; Taylor 2005.

11. Taylor 2005, chap. 5.

12. The *appear* does not imply that we are necessarily deceived if we detect regularity, but that fineness of focus and level of analysis affect that perception, as do other considerations. (This relativity becomes especially important when we discuss plasticity and redundancy.) See Ho 1998 and Ho and Ulanowicz 2005 for thermodynamic treatment. See also Margulis and Sagan 1995.

13. Slogans vary: one gene, one phenotype, trait, enzyme, or protein. In this volume, see critiques by Sapp (one germ plasm, one organism) and Shapiro (one

gene, one character). See also Lickliter and Schneider 2006; Newman and Müller 2000; Oyama 1985, chaps. 2, 6; Weiss and Fullerton, 2000; and West-Eberhard 2003, 20, 335. Susan Squier in this volume discusses mapping in Waddington's work, while Stern and Orgogozo 2009 present an intriguing alternative. Despite the metaphor's defects, I employ it here to make a point about the convention. In the end I think it withers before developmental variability and the incoherence of genetic representations in general.

14. Oyama 2000.

15. Griesemer and Wimsatt 1989.

16. To simplify a bit, organisms perish, so only genes are passed on. This fits the "Weismannian," "anti-Lamarckian" narratives of immortal genes and transient organisms that underwrite much of conventional biology.

17. Oyama 1985, chaps. 2, 5 (where redundancy was also termed *degeneracy*).

18. Somewhat different from those chosen by DST colleagues focusing on regularity (see Griffiths and Gray's 2001 negotiations with standard evolutionary theory and niche construction). See Oyama 1985; Oyama 2000, and brief discussion of the disagreement in Oyama 2009.

19. For a variety of transgenerational effects, some with delayed onsets, diminishing or cumulative effects, see Bateson and Gluckman 2011; S. Gilbert 2011.

20. Lewontin, Rose, and Kamin 1984; Oyama 1985.

21. Oyama 2002. Essences, some quite different from the sort adverted to here, do have advocates in philosophy: Devitt 2008; Griffiths 1999; Walsh 2006.

22. Burlein 2005; Fausto-Sterling 2000; Squier, this volume; Waldby and Squier 2003.

23. Mengel 2009; Moss 2003; Oyama 2010. Pradeu 2012 relies on neither preformation nor essence, presenting a definition of the organism that is immunological, developmental, symbiotic, and ecological.

24. Virtually a mini DS. Personal communication, July 15, 2011. (A close observer of plant-animal-soil interactions, Jackson is a pioneer in sustainable agriculture.) See also Krieger 2013; Neumann-Held 1999.

25. This interactive construction is not a social or cultural construction that contrasts with biology. It includes social influences, but is also completely biological, for it applies to the lives of organisms.

26. Evolutionary and/or ecological developmental biology, etc. See Gilbert and Epel 2009; Griffiths and Gray 2005; Hall 2000; Robert 2002; Robert 2006; Winther 2014.

27. Bateson and Gluckman 2011; Gilbert and Epel 2009; Ho 1998; Johnston and Gottlieb 1990; Waddington 1975; West-Eberhard 2003.

28. This last refers to the doctrine of the noninheritance of acquired characters, central to traditional evolutionary theory. For a sampling of multiple-channel theories seeking to accommodate nongenetic interactants, see Oyama, Griffiths,

and Gray (eds.) 2001; on Lamarck's legacy, see Gissis and Jablonka (eds.) 2011. However appealing some of these efforts may be, channels do not fit comfortably in the present framework.

29. Maynard Smith 2000, 189. For a response, see Griffiths 2001.

30. Maynard Smith 2000, 177.

31. The inclusion of the latter is crucial in these writings but is often absent from other developmentalists' invocations of canalization (in my experience, especially psychologists adopting a "biological" perspective), where buffering can be a virtual stand-in for developmental fixity: genetically protected pathways impervious to environmental fluctuations.

32. West-Eberhard 2003; West-Eberhard 2005. Moczek 2007 and Suzuki and Nijhout 2006 call this accumulation "developmental capacitance." The idea of cryptic genetic variants sounds exotic, but is present in the concept of recessive genes.

33. If full sets of interactants were used as mapping origins, there would be no ambiguity, just variant DSs. The question of ambiguous mapping arises because one DS component is singled out and its insufficiency to determine some result exposed, but the reasoning applies to any other interactant, including the organism itself. This is not to argue against plasticity studies but to note the entrenched practices that persist in them, and hence in my attempts to present them. For a meticulous treatment of some of these issues, see Nijhout 2001.

34. Newman and Müller 2000; Oyama 1985, chaps. 2, 6; Weiss and Fullerton 2000.

35. Any mention of constant or controlled environments must thus be read with caution, the present text included.

36. A landmark source is Maynard Smith et al 1985.

37. Maynard Smith 2000.

38. Like their stripped-down predecessors, my three expansions support each other. Indeed, it is hard to envision a world of ideational atoms. Conceptual connectivity is not always a bad thing, but we need to watch our commitments.

39. Laland, Odling-Smee, and Feldman 2001; Odling-Smee, Laland, and Feldman 2003.

40. How, for instance, do they handle interlevel slippages like Weiss and Fullerton's 2000 phenogenetic drift?

41. Definitions vary. My own usage here is informal, not attached to particular yields, tight homeostasis, or fixed system identities, though a rough criterion of individual or cross-generational viability informs my discussion of organisms. When sustainability is understood too simply and statically, the more open idea of resilience can serve as a corrective. Walker, Holling, Carpenter, and Kinzig 2004, and Zolli and Healy 2012 give distinctions, arguments, and case studies. Gregory Mengel (personal communication, June 30, 2011) notes that sustainability of

industrial or extraction practices can be so narrowly conceived as to ignore environmental degradation. I respect these observations, and although I employ *sustainable* here for its rhetorical suppleness and resonance, the spirit of my treatment is consistent with the resilience literature's proliferating consequences and sometimes surprising connections and possibilities.

42. Gilbert and Epel 2009 see development as central to evolution, but like others writing on such matters, appear to deem plasticity an evolutionary factor only when it leads to genetic change. See also West-Eberhard 2003, 17, 28. This literature could, I think, embrace a more systemic (and less genecentric) definition of evolution, as change in the constitution and distribution of developmental systems.

43. Agricultural capacitance? See note 32.

44. Levins and Lewontin 1985, 257.

45. Oyama 2000, chap. 8.

46. Lickliter 2000; Miller 2007. Often smudging the line between doing and being done to/for/with, such interactions are generally not parsable along dualistic lines. Among the wonderful examples: young chicks must see their toes if they are to peck at worms later on, Gottlieb 1992, 170. Notice that chicks do not *learn* to peck, as the instinct-learning opposition would have it, yet the behavior depends on particular (and common) experiences. In ominous counterpoint, Gilbert and Epel 2009 review a host of disturbed ontogenies and associated damaged (and damaging) environments. Both literatures reveal reliable (at least heretofore) and hence largely invisible interactants in innumerable DSs. See van der Weele 1999.

47. Both composed of biotic and abiotic features; on this feature of ecology in Gaian science, see James Strick and Bruce Clarke, both in this volume.

48. Taylor 2001.

49. Pradeu 2012. On ecosystems as organisms, see Ho and Ulanowicz 2005, and Margulis 1998, 106. On organisms as ecosystems, see Oyama 2001. Wade 2014 reports, "The sloth is not so much an animal as a walking ecosystem." For me, especially in the present broad context, recognition of systemic relations is more important than the direction of such analogizing.

50. International trade agreements, for instance, or a novel pathogen, can upend many complex and otherwise-stable arrangements.

51. For a hallucinatory hit on the confluence in the developed world of concerns with wellness, high architecture and design, environments (in several senses), and green virtue, see Finn 2013. Small wonder that social justice is on many environmentalists' agendas, and it is hard to imagine a social justice agenda that excludes developmental (including micro-, meso-, or macro-environmental) considerations. See, for instance, Crépin 2013, and the Pachamama Alliance (www .pachamama.org). Bateson and Gluckman 2011, 131 remark that the prevalence of

poverty in the world makes it imperative to join our understanding of organismic development with the "developmental economics of human societies."

52. Wilmers et al. 2012.

53. Laustsen 2008; Lawton 1994.

54. Bateson and Gluckman 2011; R. Gray 1992; Jablonka and Lamb 2005.

55. Work on niche construction or extended replicators, while couched in somewhat different theoretical terms, supply many examples of such evolutionary synergies: Odling-Smee, Laland, and Feldman 2003; Sterelny, Dickison, and Smith 1996; Stotz 2010.

56. See also Westbroek 2012b.

57. As moved by reasons, perhaps, rather than causes. Whatever Shapiro's and Westbroek's views on minds, bodies, choice and compulsion, agential language (or its opposite, that of passive matter) is tricky and easy to misinterpret. As demonstrated in these pages, my own efforts in this regard tend to center on contingency and systemic construction.

58. Westbroek 2012b.

59. Oyama 2006b.

60. See Johnston and Edwards 2002. For recent initiatives in anthropology, see Schultz 2013.

61. Latour 1986; Latour 2005, 11, 23; Oyama 2000, chap. 10; Oyama 2006a; Oyama 2006b. Karen Barad's 2012 *intra-action* seems at least a close relative to my constructivist interaction. For an audacious and astute analysis, see Sharma 2015.

62. Smith 1997, Smith 2006.

63. Just as constructivist accounts of personhood and action may be resisted in favor of internalist and individualist ones of agency. For an acute commentary on disciplinary and conceptual boundaries, see Ingold 2001.

64. While providing suggestive comparisons to developmental plasticity in evolution.

65. Gruen and Jamieson (eds.) 1994.

66. Leonard 2007.

67. Taylor 2005, 100–6.

68. Like a bad conscience, perhaps. In a prefatory note, the author credits a passage by William James. Le Guin 1976.

69. See Susan Squier's essay in this volume.

9. BOVINE URBANISM: THE ECOLOGICAL CORPULENCE OF *BOS URBANUS*
Christopher Witmore

I am grateful to Bruce Clarke for the ongoing conversation as well as the kind invitation to contribute to this volume, and to Michael San Francisco for his support.

I also thank Ewa Domanska, Matt Edgeworth, Alfredo González-Ruibal, David Gremillion, Corby Kelly, David Larmour, and Maria O'Connell for their advice in different iterations of this essay. Further comments provided by Tyler Volk and an anonymous peer reviewer went a long way toward improving this chapter. I am especially grateful to my Cattle City guide, who has chosen to remain unnamed.

1. Given their proportions, it is better to speak of CAFOs as cities. "Feedlots" or "industrial farms" are labels ill equipped to name an urban expanse accommodating 58,000 cattle.

2. Morton 2010.

3. In this essay, I maintain a fidelity to the units of measurement enrolled at Cattle City: US customary units.

4. See Gilbert, Sapp, and Tauber 2012.

5. Sloterdijk 2008, 49.

6. For more on the productive differences between archaeology and history, see Shanks and Witmore 2010.

7. For recent books in this vein, see Olivier 2011; Olsen 2010; Olsen et al 2012.

8. Enclosures have been around for ten thousand years, and with them have followed images of the good and bad herdsman. Every era has made its associations with slavery or concentration camps, with violence and the consumption of flesh. In order to truly grasp what is novel about this situation, in an effort to pose the problems differently, one needs to remember. It is archaeology, I suggest, that makes explicit this struggle against forgetting.

9. Steinfeld 2006.

10. Cary Wolfe provides a detailed discussion of this analogy in section IV of *Before the Law* (2013). A well-known example occurred in July 2003 when PETA sponsored an exhibition called "Holocaust on Your Plate," which juxtaposed images of the Holocaust with images of factory farming. The juxtaposition, it should be recalled, has etymological grounding—*holokauston* was the "whole burnt offering" in animal sacrifice. While authors routinely make this connection in the context of the "animal Holocaust" analogy, few here recall that the *holokauston* was not an honorific sacrifice, which would lead to a shared meal of Gods and mortals, but a piacular or expiatory one, where the victim was offered undivided and the flesh was not consumed—an important nuance in the historical use of the term. As with divining the omens, the human treatment of other animals as a portent for potential human futures is a long-standing topic of philosophical concern (see Patterson 2002; see also Wolfe 2013, 41).

11. See also Imhoff 2010.

12. See Global Perspectives Studies Unit 2006, 51. A 2012 FAO revision brought the consumption (not production) levels down to 458 million tons, a figure still 200 million tons more than consumption levels in 2005/2007 (Alexandratos and Bruinsma 2012, 77).

13. See Shiva 2008.

14. Cattle City sits upon flatlands where Borges's parable of cartographic perfection plays out; the ordered space of a mapmaker's table has been superimposed upon the land itself. A township is subdivided into thirty-six square miles. The lines of this mile-by-mile grid are inscribed as farm roads, some paved, most gravel. These square miles are subdivided into unitary and flat fields cultivated by tractors and watered through center-pivot irrigation. See Borges 1972.

15. Pollan 2006.

16. Witmore 2014.

17. Excavation at the former Roman fort of Binchester, England, suggests that the abandoned fort provided a readymade enclosure for cows, sheep, and pigs, which could be maintained until they were slaughtered in what may have been an abattoir. See Ferris 2011.

18. Whitaker 1975.

19. Cronon 1997, 223. See also Otter 2013.

20. On an excellent treatment of the role played by barbed wire in this transformation, see Netz 2004.

21. Haley 1967, 83.

22. This figure does not take into account the seven divisions of the XIT that, given the varying range conditions, were set aside for different types of cattle. Nonetheless, the wide spread provided for each XIT cow, steer, or heifer was not so much an issue of munificence as prudence. The arid grassland of the high plains is prone to fire and drought, blizzard and extreme heat. These are precarious combinations for any rancher.

23. The conditions have improved drastically from far worse situations. Nelson Morris, a major firm in the Chicago dressed beef trade, fed 16,000 cattle on distillery mash balanced with hay in 1880. Chained to feed bunks he would expand these Peoria, Illinois operations to 28,000 cattle in 1893. See Whitaker 1975, 53. It should be noted that Morris's operation qualified as a CAFO, according to the Environmental Protection Agency definition of concentrating more than a thousand cattle.

24. Here I am playing off the sights of Packingtown as seen by Jokubas Szedvilas and his friends in Sinclair 1971, chap. 3.

25. A consequence of the preference for only a few select breeds is the dwindling numbers of less popular domestic breeds. Many argue that this contributes to a loss of biodiversity. See, for example, Donald E. Bixby, "Old MacDonald Had Diversity: The Role of Traditional Breeds in a Dynamic Agricultural Future," in Imhoff 2010, 164–75.

26. Whittaker 1975.

27. Insofar as their relationship to Cattle City goes, "citizens" is an apposite label for members of its populace. I am not making any claims to entitlements

for Cattle City residents beyond their own urban enclosures. Still, I recognize how the term has a certain rhetorical resonance with respect to debates concerning animal rights, which exceed the purview of this article. See Wolfe 2013.

28. Cattle perceive the novelty of a bipedal human as a threat. In contrast, humans on horseback or humans in large trucks or humans in front loaders are given little more than a curious glance by residents and necessary space. See Grandin 1989, 1–11.

29. This transformation of herbivores into carnivores is not without its critics, but it is also not without its history; feeding cows animal fat was not unknown in the past. Norwegians, for example, are known to have collected fish scraps and boiled them into "nutritious mass" that they fed to their handful of cows in the winter. But here, locality is everything and humans live differently with animals in the high north. Unlike the rendered by-products of Norwegian fish, Cattle City tallow is a by-product of cattle "processing." For discussion of the Norwegian use of fish tallow, see Berglund 2010, 61. For criticism in the case of CAFOs, see Shiva 2000.

30. Water consumption varies seasonally. In summer, water consumption can peak at fifteen gallons per day, against seven gallons in winter.

31. New restrictions will soon limit Cattle City to fifteen acre-inches of water per acre annually.

32. While we may have a sense of a cow's care for self, while we may witness cow's concern for its own existence, a cow's sense of being a cow within these human-designed interiors is closed to those who run Cattle City, to me as a writer, and to you as a reader.

33. For a recent description of the transformation of cow into carcass into meat, see Conover 2013. It is worth comparing this process to the one described in *The Jungle*: "knockers" gained this name because they welded a sledge hammer, "and watch[ed] for a chance to deal a blow" (Sinclair 1971, 38). In the evolutionary sense of Leroi-Gourhan, the "knocker" process has moved from one involving the direct motive action of the hand, arm, and body with tool to one involving indirect mobility whereby the arms position the captive-bolt-gun and the hand intervenes by starting the motor process. These transformations modify the human relationship to killing animals both by increasing the number of death blows per minute and distancing the knocker from the act through a delusional veil of "indirect" force (Leroi-Gourhan 1993, 245).

34. The USDA quality grade is available at http://meat.tamu.edu/beefgrad ing. On the use of ultrasound, see www.sites.ext.vt.edu/newsletter-archive/live stock/aps-99_12/aps-0148.html.

35. Sloterdijk 2008, 45.

36. Ackerman 2012, 39.

37. Kamra 2005, cited in Gilbert, Sapp, and Tauber 2012.

38. Conover 2013.

39. Cattle and their former compatriots on the farm are not strangers to cities. In twelfth-century Constantinople, the Byzantine scholar John Tzetzes lived beneath a priest who kept pigs in his apartment and above a room in which a farmer kept hay. Giovanni Paolo Pannini's *View of the Roman Forum* (1735) depicts herdsmen with cattle across the middle ground in the vicinity of three standing columns, the remains of what had been the temple of Castor and Pollux. Free-roaming hogs were a ubiquitous feature of nineteenth-century life in New York City. Nonetheless, we may note that the final eviction notices were being carried out just when Louis Pasteur was working toward his discovery "nefarious disease bacteria" in the 1860s. Of course, all of these are very different "cities." In their erstwhile states, New York and Constantinople stood "at the center of a world planned in a circle around itself" (Sloterdijk 2008, 46). But this was very much a shared *oikos*—sharing warmth, waste, sustenance—with one's labor's companion (Harris 2007).

40. Sloterdijk 2012.

41. Hanson 1996, x–xi.

42. Sloterdijk 2012, 266.

43. Green 1990; Hanson 1995, 421.

44. Lonsdale 1979, 147.

45. I am borrowing this image of science, economy, technology, law as a four-fold of modernist graces, which act as beacons that guide modern decision making, from Bruno Latour and, before him, Michel Serres. See Latour 2007.

46. I did not grow up viewing these animals as cattle. For cattlemen, this distinction is an important one. For those who work in Cattle City, its residents are not cows. Cows are, as my Cattle City guide points out, "the class of cattle that have been bred and delivered a calf. It is jarring to a cattleman when steers and heifers are referred to as 'cows.' Most of the time cows are out on the pasture at a ranch, whereas cattle are raised to go to the feed yard." There may be more to this distinction than the issue of taxonomy, for it hints at a subtle disparity in the way bovine animals for food are framed and understood within a hierarchy of value vis-à-vis bovine animals for breeding (more than food) by workers within the beef industry.

47. Like a weighted Atlas, decisions made by our Cattle City guide, the operational manager, affect the beef industry for the whole of the United States. My argument here draws upon Serres 1995.

48. Appenzeller 2004; Shiva 2008.

49. Von Weizsäcker, Lovins, and Lovins 1997, 50, cited in Shiva, 97; see also Pollan 2009.

50. Taleb 2012.

51. See Bartholet 2011. The first official taste test of the $332,000 "moo-less burger" occurred on August 5, 2013 (see "A Lab-Grown Burger Gets a Taste Test" at www.nytimes.com/2013/08/06/science/a-lab-grown-burger-gets-a-taste-test.html).

52. Wolfe 2013 provides a similar assessment of synthetic meat, specifically in terms of exercising ever more control over life (95–99).

53. See McInerney 2010.

10. SYMBIOTISM: EARTH AND THE GREENING OF CIVILIZATION
Peter Westbroek

Bruce Clarke kindly invited me to the exciting meeting in honor of Lynn Margulis in September 2012 and carefully edited this paper. I thank Joop Goudsblom, Gijs Kuenen, Jaap Westbroek, Hans and Hanna Pieters, Mirjam Westbroek, and Eva Westbroek for their encouragement and criticisms. Joop Goudsblom and Frans Saris made me aware of Buckminster Fuller's idea of a worldwide web of energy. Cees van Nieuwburg made the drawings for Plates VII and VIII.

1. See http://img.timeinc.net/time/magazine/archive/covers/2012/1101121001_600.jpg.

2. Lovelock 1988.

3. Goudsblom and de Vries 2002.

4. Elias 1987; Elias 2000.

5. Personal communication.

6. Elias 2007.

7. Sternhell 2006.

8. Westbroek 2012a.

9. The introduction in 2000 of the term "Anthropocene" for the new geological epoch of civilization is completely in line with what I propose in this essay. We owe this initiative to Nobel laureate Paul Crutzen and Eugene F. Stoermer. This concept is now widely accepted in the scientific community.

10. Westbroek 2012a.

11. Westbroek 2012b.

12. Margulis 1998.

13. Oreskes and Conway 2010.

14. Fuller 1982.

REFERENCES

Abram, D. "A ferocious intelligence." In Sagan (ed.) 2012, 115–25.

Abramov, O., and S. J. Mojzsis. 2009. "Microbial habitability of the Hadean Earth during the Late Heavy Bombardment." *Nature* 459: 419–22.

Ackerman, J. 2012. "The ultimate social network." *Scientific American* 306 (6): 36–43.

Adams, M. B. 2000. "Last judgment: The visionary biology of J. B. S. Haldane." *Journal of the History of Biology* 33: 457–91.

Alberts, B., D. Bray, J. Lewis, M. Raff, K. Roberts, and J. D. Watson. 1994. *Molecular Biology of the Cell*. 3rd ed. New York: Garland Science.

Alexandratos, N., and J. Bruinsma. 2012. *World Agriculture: Towards 2030/2050*. The 2012 Revision. ESA Working Paper No. 12-03. www.fao.org/docrep/016/ap106e/ap106e.pdf.

Allsopp, A. 1969. "Phylogenetic relationships of the procaryota and the origin of the eucaryotic cell." *New Phytologist* 68: 59–612.

Anderson, E., and G. L. Stebbins Jr. 1954. "Hybridization as an evolutionary stimulus." *Evolution* 8: 378–88.

Anker, P. 2010. *From Bauhaus to Ecohouse: A History of Ecological Design*. Baton Rouge: Louisiana State University Press.

Appenzeller, T. 2004. "The end of cheap oil." *National Geographic* (June): 80–109.

Ast, G. 2005. "The alternative genome." *Scientific American* 292 (4): 58–65.

Atsatt, P. R. 1991. "Fungi and the origin of land plants." In Margulis and Fester (eds.) 1991, 301–18.

Avery, O. T., C. M. MacLeod, and M. McCarty. 1944. "Studies on the chemical nature of the substance inducing transformation of pneumococcal types." *Journal of Experimental Medicine* 79: 137–58.

Aziz, R. K., M. Breitbart, and R. A. Edwards. 2010. "Transposases are the most abundant, most ubiquitous genes in nature." *Nucleic Acids Research* 38 (13): 4207–17.

Balows, A. H., G. Trüper, M. Dworkin, W. Harder, and K. H. Schleifer (eds.). 1992. *The Prokaryotes*, vol. 1. 2nd ed. New York: Springer Verlag.

Barad, K. 2007. *Meeting the Universe Halfway: Quantum Physics and the Entanglement of Matter and Meaning.* Durham, NC: Duke University Press.

———. 2010. Personal communication. Society for Literature, Science, and the Arts Convention, Atlanta, GA.

———. 2012. "Intra-actions." Interview by Adam Kleinman. *Mousse* (34): 76–81.

Bard, J. B. L. 2008. "Waddington's legacy to developmental and theoretical biology." *Biological Theory* 3 (3): 188–97.

Bartholet, J. 2011. "Inside the meat lab." *Scientific American* 304(6): 64–69.

Bateson, G. 1972. *Steps to an Ecology of Mind.* New York: Ballantine.

Bateson, P., and P. Gluckman. 2011. *Plasticity, Robustness, Development and Evolution.* Cambridge: Cambridge University Press.

"Battle of Balliol." 2009. Homage to Darwin debate (May). http://www .voicesfromoxford.com/homagedarwin_part3.html.

Beadle, G. W. 1948. "The genes of men and molds." *Scientific American* 179 (3): 30–9.

Behe, M. J. 1996. *Darwin's Black Box: The Biochemical Challenge to Evolution.* New York: Free Press.

Benton, M. 2003. *When Life Nearly Died: The Greatest Mass Extinction of All Time.* London: Thames & Hudson.

Benzer, S. 1956. "Genetic fine structure and its relation to the DNA molecule." *Brookhaven Symposia in Biology* 8: 3–5.

———. 1962. "The fine structure of the gene." *Scientific American* 206: 70–84.

Berglund, J. 2010. "Did medieval Norse society in Greenland really fail?" In P. A. McAnany and N. Yoffee (eds.), *Questioning Collapse: Human Resilience, Ecological Vulnerability, and the Aftermath of Empire,* 45–70. Cambridge: Cambridge University Press.

Bernstein, M. P., S. A. Sandford, and L. J. Allamandola. 1999. "Life's far-flung raw material." *Scientific American* 263 (7): 42–49.

Billingham, J. (ed.). 1981. *Life in the Universe.* Cambridge, MA: MIT Press.

Bird, R. J. 2003. *Chaos and Life.* New York: Columbia University Press.

Bohm, D. 1969a. "Some remarks on the notion of order." In Waddington (ed.) 1969, 18–40.

———. 1969b. "Further remarks on order." In ibid., 41–60.

Bonen, L., R. S. Cunningham, M. W. Gray, and W. F. Doolittle. 1977. "Wheat mitochondrial 18s ribosomal RNA: Evidence of its prokaryotic nature." *Nucleic Acid Research* 4: 663–71.

Borges, J. L. 1972. "Of exactitude in science." In *A Universal History of Infamy.* New York: Dutton.

Brand, S. 1977. "The first whole Earth photograph." In M. Katz, W. P. Marsh, and G. G. Thompson (eds.), *Earth's Answer: Explorations of Planetary Culture at the Lindisfarne Conferences,* 184–88. New York: Harper & Row.

————. 1980. "Gaia." *The Next Whole Earth Catalog.* New York: Random House.

Brand, S. (ed.). 1977. *Space Colonies.* New York: Penguin.

Bray, D. 2009. *Wetware: A Computer in Every Living Cell.* New Haven: Yale University Press.

Breed, R. 1928. "The present status of systematic bacteriology." *Journal of Bacteriology* 15 (3): 143–63.

Brown, J. D., and R. J. O'Neill. 2010. "Chromosomes, conflict, and epigenetics: Chromosomal speciation revisited." *Annual Review of Genomics and Human Genetics* 11: 291–316.

Bruch, C. 1966. "Instrumentation for the detection of extraterrestrial life." In Pittendrigh et al. (eds.) 1966, 487–502.

Bukhari, A. I., J. A. Shapiro, and S. L. Adhya (eds.). 1977. *DNA Insertion Elements, Plasmids and Episomes.* Cold Spring Harbor, NY: Cold Spring Harbor Press.

Burlein, A. 2005. "The productive power of ambiguity: Rethinking homosexuality through the virtual and developmental systems theory." *Hypatia* 20 (1): 21–53.

Bybee, J. 2012. "No subject too sacred." In D. Sagan (ed.) 2012, 156–62.

Cairns-Smith, A. G. 1985. "The first organisms." *Scientific American* 252 (6): 90–100.

Callahan, M. P., K. E. Smith, H. J. Cleaves, J. Ruzicka, J. C. Stern, D. P. Glavin, C. H. House, and J. P. Dworkin. 2011. "Carbonaceous meteorites contain a wide range of extraterrestrial nucleobases." *Proceedings of the National Academy of Sciences* 108: 13995–98.

Callicott, J. B. 2010. "From the land ethic to the Earth ethic: Aldo Leopold and the Gaia hypothesis." In Crist and Rinker (eds.) 2010, 177–93.

Cameron, R. E. 1966. "Properties of desert soils." In Pittendrigh et al. (eds.) 1966, 164–86.

"Can exploration of a Chilean desert assist in the search for life on Mars?" 1966. Jet Propulsion Laboratory (May 1).

Cavalier-Smith, T. 1983. "A 6-kingdom classification and a unified phylogeny." In H. E. A. Schenk and W. Schwemmler (eds.), *Endocytobiology II*, 1027–34. Berlin: de Gruyter.

————. 1998. "A revised 6-kingdom system of life." *Biological Reviews* 73: 203–66.

Cech, T. R. 1986. "RNA as an enzyme." *Scientific American* 255 (5): 64–75.

————. 2000. "The ribosome is a ribozyme." *Science* 289: 878–79.

Chambon, P. 1981. "Split genes." *Scientific American* 244 (5): 60–71.

Chapman, M. J., M. F. Dolan, and L. Margulis. 2000. "Centrioles and kinetosomes: Form, function, and evolution." *Quarterly Review of Biology* 75 (4): 409–29.

Chapman, M. J., and L. Margulis. 1998. "Morphogenesis by symbiogenesis." *International Microbiology* 1 (4): 319–26.

Charlson, R. J., J. Lovelock, M. Andreae, and S. Warren. 1987. "Oceanic phytoplankton, atmospheric sulphur, cloud albedo and climate." *Nature* 326 (16 April): 655–61.

Chatterjee, S., N. Guven, A. Yoshinobu, and R. Donofrio. 2006. "Shiva structure: A possible KT boundary impact crater on the western shelf of India." *Special Publications of the Museum of Texas Tech University* 50: 1–39.

Chatton, E. 1925. "*Pansporella perplexa*. Réflexions sur la biologie et la phylogénie des protozoaires." *Annales des Sciences Naturelles—Zoologie et Biologie Animale*, Series 10, 7: 1–84.

———. 1938. *Titre et travaux scientifique de Edouard Chatton 1906–1937*. Sottano: Sette.

Chyba, C. F., and C. Sagan. 1992. "Endogenous production, exogenous delivery and impact-shock synthesis of organic molecules: An inventory for the origin of life." *Nature* 355: 125–32.

Chyba, C. F., P. J. Thomas, L. Brookshaw, and C. Sagan. 1990. "Cometary delivery of organic molecules to early Earth." *Nature* 249: 366–73.

Clarke, B. 2001. "Science, theory, and systems." *Interdisciplinary Studies in Literature and Environment* 8 (1): 149–65.

———. 2009. "Neocybernetics of Gaia: The emergence of second-order Gaia theory." In Crist and Rinker (eds.) 2009, 293–314.

———. 2012a. "'Gaia is not an organism': The early scientific collaboration of Lynn Margulis and James Lovelock." In D. Sagan (ed.) 2012, 32–43.

———. 2012b. "From information to cognition: The systems counterculture, Heinz von Foerster's pedagogy, and second-order cybernetics." *Constructivist Foundations* 7 (3): 196–207.

———. 2014. "Mediations of Gaia." In S. Neef, H. Sussman, and D. Boschung (eds.), *Astroculture: Figurations of Cosmology in Media and Arts*, 119–41. Munich: Wilhelm Fink.

Clarke, B., and M. B. N. Hansen. 2009. "Neocybernetic emergence: Retuning the posthuman." *Cybernetics and Human Knowing* 16(1–2): 83–99.

Clarke, B., and M. B. N. Hansen (eds.). 2009. *Emergence and Embodiment: New Essays in Second-Order Systems Theory*. Durham, NC: Duke University Press.

Cleveland, L. R., and A. V. Grimstone. 1964. "The fine structure of the flagellate *Mixotricha paradoxa* and its associated micro-organisms." *Proceedings of the Royal Society B* 159: 668–86.

Clowes, R. C. 1973. "The molecule of infectious drug resistance." *Scientific American* 228 (4): 19–27.

Cockell, C. S. 2006. "The origin and emergence of life under impact bombardment." *Philosophical Transactions of the Royal Society* B 361: 1845–56.

Codell, C. K. "The development of Pasteur's concept of disease causation and the emergence of specific causes in nineteenth-century medicine." *Bulletin of the History of Medicine* 65 (1991): 528–48.

Coffin, J. M., S. H. Hughes, and H. E. Varmus. 1997. *Retroviruses*. Cold Spring Harbor, NY: Cold Spring Harbor Laboratory Press.

Cohen, B. A., T. D. Swindle, and D. A. Kring. 2000. "Support for the lunar cataclysmic hypothesis from lunar meteorite impact melt ages." *Science* 290: 1754–56.

Cohen, S. N., and J. A. Shapiro. 1980. "Transposable genetic elements." *Scientific American* 242 (2): 40–9.

Colson, P., and D. Raoult. 2010. "Gene repertoire of amoeba-associated giant viruses." *Intervirology* 53 (5): 330–43.

Conover, T. 2013. "The way of all flesh." *Harper's Magazine* (May): 31–49.

"Conrad Hal Waddington." 2013. http://en.wikipedia.org/wiki/Conrad_Hal _Waddington.

Conrad, P., and K. Nealson. 2001. "A non-Earthcentric approach to life detection." *Astrobiology* 1: 15–24.

Cooks, R. G., D. Zhang, and K. J. Koch. 2001. "Chiroselective self-directed octamerization of serine: Implications for homochirogenesis." *Analytical Chemistry* 73: 3645–55.

Cooper, H. S. F. 1980. *The Search for Life on Mars*. New York: Holt, Rinehart and Winston.

Copeland, H. F. 1938. "The kingdoms of organisms." *Quarterly Review of Biology* 13: 383–420.

———. 1956. *The Classification of Lower Organisms*. Palo Alto, CA: Pacific Books.

Craig, N. L., R. Craigie, M. Gellert, and A. M. Lambowitz. 2002. *Mobile DNA II*. Washington, DC: American Society for Microbiology Press.

Crépin, A.-S. 2013. "Global and cross-level dynamics." Stockholm Resilience Center. www.stockholmresilience.org/21/research/research-themes/ global-and-cross-level-dynamics.html.

Crick, F. H. 1958. "On protein synthesis." *Symposia of the Society for Experimental Biology* 12: 138–63.

———. 1968. "The origin of the genetic code." *Journal of Molecular Biology* 38: 367–79.

———. 1970. "Central dogma of molecular biology." *Nature* 227: 561–63.

———. 1981. *Life Itself: Its Origin and Nature*. New York: Simon & Schuster.

Crist, E., and H. B. Rinker (eds.). 2009. *Gaia in Turmoil: Climate Change, Biodepletion, and Earth Ethics in an Age of Crisis*. Cambridge, MA: MIT Press.

Cronon, W. 1997. *Nature's Metropolis: Chicago and the Great West*. New York: Norton.

Darwin, C. 1859. *On the Origin of Species*. Facsimile edition. Cambridge, MA: Harvard University Press, 1964.

Darwin Correspondence Project. 2013. University of Cambridge. www .darwinproject.ac.uk.

Davies, J. 1979. "General mechanisms of antimicrobial resistance." *Reviews of Infectious Diseases* 1 (1): 23–29.

Davis, B. K. 2002. "Molecular evolution before the origin of species." *Progress in Biophysics & Molecular Biology* 79: 77–133.

Deamer, D. W. 2011. *First Life*. Berkeley: University of California Press.

Deamer, D. W., J. P. Dworkin, S. A. Sandford, M. P. Bernstein, and L. J. Allamandola. 2002. "The first cell membranes." *Astrobiology* 2: 371–82.

Deamer, D. W., and G. R. Fleischaker (eds.). 1994. *Origins of Life*. Boston: Jones & Bartlett.

De Duve, C. 1995. "The beginnings of life on Earth." *American Scientist* 83: 428–437.

Delsemme, A. H. 2001. "An argument for the cometary origin of the biosphere." *Scientific American* 89 (9): 431–42.

de Marcellus, P., C. Meinert, M. Nuevo, J.-J. Filippi, G. Danger, D. Deboffle, L. Nahon, L. Le Sergeant d'Hendecourt, and U. J. Meierhenrich. 2011. "Non-racemic amino acid production by ultraviolet radiation of achiral interstellar ice analogs with circularly polarized light." *Astrophysical Journal* 727: L27.

Deutscher, J. 2008. "The mechanisms of carbon catabolite repression in bacteria." *Current Opinion in Microbiology* 11 (2): 87–93.

Devitt, M. 2008. "Resurrecting biological essentialism." *Philosophy of Science* 75: 344–82.

Devoret, R. 1979. "Bacterial tests for potential carcinogens." *Scientific American* 241 (2): 40–9.

De Wit, M. J., and L. D. Ashwal. 1997. *Greenstone Belts*. Oxford: Clarendon Press.

d'Herelle, Félix. 1926. *The Bacteriophage and Its Behavior*. Trans. George H. Smith. Baltimore: Williams and Wilkins.

Dick, S. J., and J. E. Strick. 2004. *The Living Universe: NASA and the Development of Astrobiology*. New Brunswick, NJ: Rutgers University Press.

Ding, D. Q., T. Haraguchi, and Y. Hiraoka. 2010. "From meiosis to postmeiotic events: Alignment and recognition of homologous chromosomes in meiosis." *Federation of the Societies of Biochemistry and Molecular Biology Journal* 277 (3): 565–70.

Dolan, M. F. 2012. "Lynn Margulis and Stephen Jay Gould." In D. Sagan (ed.) 2012, 50–53.

Dolan, M. F., H. Melnitsky, L. Margulis, and R. Kolnicki. 2002. "Motility proteins and the origin of the nucleus." *Anatomical Record* 268 (3): 290–301.

Doolittle, W. F. 2000. "Uprooting the tree of life." *Scientific American* 282 (2): 90–95.

Dotto, L., and H. Schiff. 1978. *The Ozone War*. New York: Doubleday.

Doyle, R. "Gaiadelic: Lynn Sagan and LSD." In D. Sagan (ed.) 2012, 143–49.

Dubnau, D., I. Smith, P. Morell, and J. Marmur. 1965. "Gene conservation in *Bacillus* species. I. Conserved genetic and nucleic acid base sequence homologies." *Proceedings of the National Academy of Sciences* 54: 491–98.

Dyson, F. 2004. *Origins of Life*. Cambridge: Cambridge University Press.

Edwards, P. N. 1996. *The Closed World: Computers and the Politics of Discourse in Cold War America*. Cambridge, MA: MIT Press.

Edwards, R. A., and F. Rohwer. 2005. "Viral metagenomics." *Nature Reviews Microbiology* 3 (6): 504–10.

Ehrlich, P., and A. Ehrlich. 1977. Correspondence. In Brand (ed.) 1977, 43.

Elias, N. 1987. *Involvement and Detachment*. Dublin: University College Dublin Press.

———. 2000. *The Civilizing Process*. London: Blackwell.

Ellis, R. J. 1990. "Frank Herbert's *Dune* and the discourse of apocalyptic ecologism in the United States." In R. Garnett and R. J. Ellis (eds.), *Science Fiction Roots and Branches: Contemporary Critical Approaches*, 104–24. Basingstoke, UK: Palgrave Macmillan.

Embley, T. M., and W. Martin. 2006. "Eukaryotic evolution, changes and challenges." *Nature* 440 (7084): 623–30.

Errede, B., T. S. Cardillo, G. Wever, F. Sherman, J. I. Stiles, L. R. Friedman, and F. Sherman. 1981. "Studies on transposable elements in yeast." *Cold Spring Harbor Symposia on Quantitative Biology* 45 (Part 2): 593–607.

Ezell, E. C., and L. N. Ezell. 1984. *On Mars: Exploration of the Red Planet, 1958–1978*. Washington, DC: NASA.

Fauci, A. S. 2001. "Infectious diseases: Considerations for the 21st century." *Clinical Infectious Diseases* 32 (5): 675–85.

Fausto-Sterling, A. 2000. *Sexing the Body: Gender Politics and the Construction of Sexuality*. New York: Basic Books.

Fedo, C. M., and M. J. Whitehouse. 2002. "Metasomatic origin of quartz-pyroxene rock, Akilia, Greenland, and implication for Earth's life." *Science* 296: 1448–52.

Fenster, B. 2000. *Duh! The Stupid History of the Human Race*. Kansas City, MO: Andrews McMeel.

Ferrer, I., and A. M. Planas. 2003. "Signaling of cell death and cell survival following focal cerebral ischemia: Life and death struggle in the penumbra." *Journal of Neuropathology and Experimental Neurology* 62 (4): 329–39.

Ferris, I. M. 2011. *Vinovia: The Buried Roman City of Binchester in Northern England*. Stroud, UK: Amberley.

Ferris, J. P., A. R. Hill, R. Liu, and L. E. Orgel. 1996. "Synthesis of long prebiotic oligomers on mineral surfaces." *Nature* 381: 59–61.

Finn, R. 2013. "For health nuts who've squirreled away a lot." *New York Times* (June 30).

Fleck, L. 1979. *The Genesis and Development of a Scientific Fact*. Chicago: University of Chicago Press.

Forterre, P. 2005. "The two ages of the RNA world, and the transition to the DNA world: a story of viruses and cells." *Biochimie* 87: 793–803.

———. 2006. "The origin of viruses and their possible roles in major evolutionary transitions." *Virus Research* 117 (1): 5–16.

Forterre, P., J. Filee, and H. Myllykallio. 2004. "Origin and evolution of DNA and DNA replication machineries." In Ll. R. de Pouplana (ed.), *The Genetic Code and the Origin of Life*, 145–68. Amsterdam: Springer.

Foster, T. J. 1983. "Plasmid-determined resistance to antimicrobial drugs and toxic metal ions in bacteria." *Microbiological Reviews* 47 (3): 361–409.

Foucault, M. 1977. "What is an author?" Trans. D. F. Bouchard and S. Simon. In *Language, Counter-Memory, Practice: Selected Essays and Interviews*, 113–38. Ithaca, NY: Cornell University Press.

Fox, G. E. et al. 1980. "The phylogeny of prokaryotes." *Science* 209: 457–63.

Fox, S. 1998. *The Emergence of Life*. New York: Basic Books.

Francis, B. R. 2011. "An alternative to the RNA world hypothesis." *Trends in Evolutionary Biology* 3: e2.

Fuller, R. B. 1982. *Critical Path*. New York: St. Martin's Press.

Gauchy, E. A., J. M. Thomson, M. F. Burgan, and S. A. Benner. 2003. "Inferring the paleoenvironment of ancient bacteria on the basis of resurrected proteins." *Nature* 425: 285–88.

Gibson, W. 1984. *Neuromancer*. New York: Ace.

Gilbert, J. A., and C. L. Dupont. 2011. "Microbial metagenomics: Beyond the genome." *Annual Review of Marine Science* 3: 347–71.

Gilbert, S. F. 1991. "Epigenetic landscaping: Waddington's use of cell fate bifurcation diagrams." *Biology and Philosophy* 6: 135–54.

———. 2011. "Expanding the temporal dimensions of developmental biology: The role of environmental agents in establishing adult-onset phenotypes." *Biological Theory* 6: 65–72.

Gilbert, S. F. (ed.). 1991a. *A Conceptual History of Modern Embryology*. Baltimore: Johns Hopkins University Press.

Gilbert, S. F., and D. Epel. 2009. *Ecological Developmental Biology: Integrating Epigenetics, Medicine, and Evolution*. Sunderland, MA: Sinauer Associates.

Gilbert, S. F., J. Sapp, and A. I. Tauber. 2012. "A symbiotic view of life: We have never been individuals." *Quarterly Review of Biology* 87 (4): 325–41.

Gilbert, W. 1986. "The RNA World." *Nature* 319: 618.

Gissis, S. B., and E. Jablonka (eds.). 2011. *Transformations of Lamarckism: From Subtle Fluids to Molecular Biology*. Cambridge, MA: MIT Press.

Glikson, A. Y. 2008. "Field evidence of Eros-scale asteroids and impact forcing of Precambrian geodynamic episodes, Kaapvaal (South Africa) and Pilbara (Western Australia) Cratons." *Earth and Planetary Science Letters* 267: 558–70.

———. 2010. "Archean asteroid impacts, banded iron formations and MIF-S anomalies." *Icarus* 207: 39–44.

Global Perspectives Studies Unit. 2006. *World Agriculture: Towards 2030/2050*. Interim Report. Rome: Food and Agriculture Organization of the United Nations. www.fao.org/fileadmin/user_upload/esag/docs/Interim_report _AT2050web.pdf.

Godfrey-Smith, P. 1999. "Genes and codes: Lessons from the philosophy of mind?" In V. Hardcastle (ed.), *Biology Meets Psychology: Constraints, Connections, Conjectures*, 305–31. Cambridge, MA: MIT Press.

Goldfarb, T., and M. Lichten. 2010. "Frequent and efficient use of the sister chromatid for DNA double-strand break repair during budding yeast meiosis." *PLoS Biology* 8 (10): e1000520.

Goodall, M. C. 1965. *Science and the Politician*. Cambridge, MA: Schenkman.

Goodman, M. F. 2002. "Error-prone repair DNA polymerases in prokaryotes and eukaryotes." *Annual Review of Biochemistry* 71: 17–50.

Gorini, L. 1966. "Antibiotics and the genetic code." *Scientific American* 214: 102–9.

Gorke, B., and J. Stulke. 2008. "Carbon catabolite repression in bacteria: Many ways to make the most out of nutrients." *Nature Reviews Microbiology* 6 (8): 613–24.

Gottlieb, G. 1992. *Individual Development and Evolution*. Oxford: Oxford University Press.

Goudsblom, J. 1995. *Fire and Civilization*. London: Allen Lane.

Goudsblom, J., and B. de Vries. 2002. *Mappae Mundi: Humans and Their Habitats in a Long-Term Socio-Ecological Perspective*. Amsterdam: Amsterdam University Press.

Gould, S. J. 1997. "Darwinian fundamentalism." *New York Review of Books* (June 12): 1–2.

Grandin, T. 1989. "Behavioral principles of livestock handling." *Professional Animal Scientist* (December): 1–11.

Gray, M. W. 1992. "The endosymbiont hypothesis revisited." *International Review of Cytology* 141: 233–57.

Gray, M. W., and W. F. Doolittle. 1982. "Has the endosymbiont hypothesis been proven?" *Microbiological Reviews* 46: 1–42.

Gray, R. D. 1992. "Death of the gene: Developmental systems strike back." In P. E. Griffiths (ed.), *Trees of Life: Essays in Philosophy of Biology*, 165–209. Dordrecht: Kluwer Academic.

Green, D. H. 1972. "Archean greenstone belts may include terrestrial equivalents of lunar maria." *Earth and Planetary Science Letters* 15: 263–70.

Green, P. 1990. *Alexander to Actium: The Historical Evolution of the Hellenistic Age*. Berkeley: University of California Press.

Griesemer, J., and W. Wimsatt. 1989. "Picturing Weismannism: A case study of conceptual evolution." In M. Ruse (ed.), *What the Philosophy of Biology Is: Essays for David Hull*, 75–137. Dordrecht: Kluwer Academic.

Griffith, F. 1928. "The significance of pneumococcal types." *Journal of Hygiene (London)* 27 (2): 113–59.

Griffiths, P. E. 1999. "Squaring the circle: Natural kinds with historical essences." In R. A. Wilson (ed.), *Species: New Interdisciplinary Essays*, 208–28. Cambridge, MA: MIT Press.

———. 2001. "Genetic information: A metaphor in search of a theory." *Philosophy of Science* 68: 394–412.

Griffiths, P. E., and R. D. Gray. 2001. "Darwinism and developmental systems." In Oyama, Griffiths, and Gray (eds.) 2001, 195–218.

———. 2005. "Discussion: Three ways to misunderstand developmental systems theory." *Biology and Philosophy* 20: 417–25.

Groman, N. B. 1953. "The relation of bacteriophage to the change of *Corynebacterium diphtheriae* from avirulence to virulence." *Science* 117 (3038): 297–99.

Gruen, L., and D. Jamieson (eds.). 1994. *Reflecting on Nature: Readings in Environmental Philosophy.* New York: Oxford University Press.

Gupta, R. S., and D. W. Mathews. 2010. "Signature proteins for the major clades of cyanobacteria." *BMC Evolutionary Biology* 10 (January 25): 24.

Haber, J. E. 1998. "Mating-type gene switching in Saccharomyces cerevisiae." *Annual Review of Genetics* 32: 561–99.

Hacker, J., and E. Carniel. 2001. "Ecological fitness, genomic islands and bacterial pathogenicity: A Darwinian view of the evolution of microbes." *EMBO Reports* 2 (5): 376–81.

Haeckel, E. 1866. *Generelle Morphologie der Organismen.* Berlin: Reimer.

———. 1905. *The Evolution of Man: A Popular Scientific Study.* Trans. and ed. Joseph McCabe. New York: G. P. Putnam.

Hagen, J. B. 1992. *An Entangled Bank: The Origins of Ecosystem Ecology.* New Brunswick, NJ: Rutgers University Press.

Haley, J. E. 1967. *The XIT Ranch of Texas and the Early Days of the Llano Estacado.* Norman: University of Oklahoma Press.

Hall, B. K. 1992. "Waddington's legacy in development and evolution." *American Zoologist* 32 (1): 113–22.

———. 2000. "Guest editorial: Evo-devo or devo-evo—Does it matter?" *Evolution and Development* 2: 177–78.

Hall, J. L. 2011. "Spirochete contributions to the eukaryotic genome." *Symbiosis* 54: 119–29.

Hall, R. M., and C. M. Collis. 1995. "Mobile gene cassettes and integrons: Capture and spread of genes by site-specific recombination." *Molecular Microbiology* 15 (4): 593–600.

Hall, R. M., and H. W. Stokes. 1993. "Integrons: Novel DNA elements which capture genes by site-specific recombination." *Genetica* 90 (2–3): 115–32.

Hallet, B., and D. J. Sherratt. 1997. "Transposition and site-specific recombination: Adapting DNA cut-and-paste mechanisms to a variety of genetic rearrangements." *Federation of European Microbiological Societies Microbiology Reviews* 21 (2): 157–78.

Hansen, V. L. 2007. "Subduction on early Earth: A hypothesis." *Geology* 35: 1059–62.

Hanson, V. D. 1995. *The Other Greeks: The Family Farm and the Agrarian Roots of Western Civilization.* New York: Free Press.

Harding, S. 2006. *Animate Earth: Science, Intuition, and Gaia.* White River Junction, VT: Chelsea Green.

Harish, A., and G. Caetano-Anollés. 2012. "Ribosomal history reveals origin of modern protein synthesis." *PLoS ONE* 7 (3): e32776.

Harris, J. 2007. *Constantinople: Capital of Byzantium*. London: Hambledon Continuum.

Hartwell, L. H., and T. A. Weinert. 1989. "Checkpoints: Controls that ensure the order of cell cycle events." *Science* 246:629–634.

Hayes, W. 1968. *The Genetics of Bacteria and Their Viruses*. 2nd ed. London: Blackwell.

Hazen, R. M. 2001. "Life's rocky start." *Scientific American* 284 (4): 76–85.

———. 2005. *Genesis: The Scientific Quest for Life*. Washington, DC: Joseph Henry Press.

Hedges, R. W., and A. E. Jacob. 1974. "Transposition of ampicillin resistance from RP4 to other replicons." *Molecular and General Genetics* 132 (1): 31–40.

Herbert, F. 1965. *Dune*. New York: Ace, 2005.

Hershey, A. D., and M. Chase. 1952. "Independent functions of viral protein and nucleic acid in growth of bacteriophage." *Journal of General Physiology* 36 (1): 39–56.

Hitchcock, D. R., and J. E. Lovelock. 1967. "Life detection by atmospheric analysis." *Icarus* 7: 149–59.

Ho, M.-W. 1998. "Evolution." In G. Greenberg and M. M. Haraway (eds.), *Comparative Psychology: A Handbook*, 107–19. New York: Garland.

Ho, M.-W., and R. Ulanowicz. 2005. "Sustainable systems as organisms?" *BioSystems* 82: 39–51.

Hogg, J. 1860. "On the distinction of a plant and an animal, and on a fourth kingdom of nature." *Edinburgh New Philosophical Journal*, new series. 12: 216–25.

Holm, N. G. 1992. "Why are hydrothermal vent systems proposed as plausible environments for the origin of life?" *Origins of Life and Evolution of Biospheres* 22: 5–14.

Horowitz, N., R. E. Cameron, and J. S. Hubbard. 1972. "Microbiology of the dry valleys of Antarctica." *Science* 176 (21 April): 242–45.

Horowitz, N., R. P. Sharp, and R. W. Davies. 1967. "Planetary contamination I: The problem and the agreements." *Science* 155 (24 March): 1501–5.

Howard-Flanders, P. 1981. "Inducible repair of DNA." *Scientific American* 245 (5): 72–80.

Huisman, O., R. D'Ari, and S. Gottesman. 1984. "Cell-division control in Escherichia coli: Specific induction of the SOS function SfiA protein is sufficient to block septation." *Proceedings of the National Academy of Sciences* 81 (14): 4490–4.

Hutchinson, G. E. 1948. "Circular causal systems in ecology." *Annals of the New York Academy of Sciences* 50 (October): 221–46.

Huxley, J. S. 1942. *Evolution: The Modern Synthesis*. London: Allen and Unwin.

Imhoff, D. 2010. *The CAFO Reader: The Tragedy of Industrial Animal Factories*. Healdsburg, CA: Watershed Media.

Ingold, T. 2001. "From complementarity to obviation: On dissolving the boundaries between social and biological anthropology, archaeology and psychology." In Oyama, Griffiths, and Gray (eds.) 2001, 255–79.

Jablonka, E., and M. J. Lamb. 2005. *Evolution in Four Dimensions: Genetic, Epigenetic, Behavioral, and Symbolic Variation in the History of Life.* Cambridge, MA: MIT Press.

Jacob, F., and J. Monod, 1961. "Genetic regulatory mechanisms in the synthesis of proteins." *Journal of Molecular Biology* 3: 318–56.

Jacob, F., and E. L. Wollman. 1961. "Viruses and genes." *Scientific American* 204 (6): 92–110.

Jantsch, E., and C. H. Waddington (eds.). 1976. *Evolution and Consciousness: Human Systems in Transition.* Reading, MA: Addison-Wesley.

Johnston, T. D., and L. Edwards. 2002. "Genes, interactions, and the development of behavior." *Psychological Review* 109 (1): 26–34.

Johnston, T. D., and G. Gottlieb. 1990. "Neophenogenesis: A developmental theory of phenotypic evolution." *Journal of Theoretical Biology* 147: 471–95.

Joyce, G. F. 2002. "The antiquity of RNA-based evolution." *Nature* 418: 214–21.

Juhas, M., J. R. van der Meer, M. Gaillard, R. M. Harding, D. W. Hood, and D. W. Crook. 2009. "Genomic islands: Tools of bacterial horizontal gene transfer and evolution." *Federation of European Microbiological Societies Microbiology Reviews* 33 (2): 376–93.

Kamra, D. N. 2005. "Rumen microbial ecosystem." *Current Science* 89: 124–35.

Kazazian, H. H. 2000. "L1 retrotransposons shape the mammalian genome." *Science* 289: 1152–53.

Kemp, M. 1996. "Doing what comes naturally: Morphogenesis and the limits of the genetic code." *Art Journal* 55 (1): 27–32.

Kerr, R. 1988. "No longer willful." *Science* 240 (22 April): 393–95.

King, M. 1995. *Species Evolution: The Role of Chromosome Change.* Cambridge: Cambridge University Press.

Kingsland, S. E. 2005. *The Evolution of American Ecology, 1890–2000.* Baltimore: John Hopkins University Press.

Kirk, A. G. 2007. *Counterculture Green: The Whole Earth Catalog and American Environmentalism.* Lawrence: University Press of Kansas.

Kite, G. 1986. "Evolution by endosymbiosis: The inside story." *New Scientist* (3 July): 50–52.

Klar, A. J. 2007. "Lessons learned from studies of fission yeast mating-type switching and silencing." *Annual Review of Genetics* 41: 213–36.

———. 2010. "The yeast mating-type switching mechanism: A memoir." *Genetics* 186 (2): 443–49.

Koonin, E. V. 2010. "Taming of the shrewd: Novel eukaryotic genes from RNA viruses." *BioMed Central Biology* 8 (2).

Koonin, E. V., T. G. Senkevich, and V. V. Dolja. 2006. "The ancient virus world and the evolution of cells." *Biology Direct* 1: 29.

Kozo-Polyansky, B. M. 2010. *Symbiogenesis: A New Principle of Evolution.* Ed. V. Fet and L. Margulis. Cambridge, MA: Harvard University Press.

Krieger, N. 2013. "History, biology, and health inequities: Emergent embodied phenotypes and the illustrative case of the breast cancer estrogen receptor." *American Journal of Public Health* 103 (1): 22–27.

Kring, D. A. 2000. "Impact events and their effect on the origin, evolution, and distribution of life." *GSA Today* 10 (8): 1–7.

Kuhn, T. 1970. *The Structure of Scientific Revolutions.* 2nd ed. Chicago: University of Chicago Press.

Kunkel, T. A., and K. Bebenek. 2000. "DNA replication fidelity." *Annual Review of Biochemistry* 69: 497–529.

Laland, K. L., F. J. Odling-Smee, and M. W. Feldman. 2001. "Niche construction, ecological inheritance, and cycles of contingency in evolution." In Oyama, Griffiths, and Gray (eds.) 2001, 117–26.

Lamarck, J. B. 1809. *Zoological Philosophy.* Trans. Hugh Elliot. Chicago: University of Chicago Press, 1984.

Landecker, H. 2007. *Culturing Life: How Cells Became Technologies.* Cambridge, MA: Harvard University Press.

Lander, E. S., et al. 2001. "Initial sequencing and analysis of the human genome." *Nature* 409 (6822): 860–921.

Latour, B. 1986. "The powers of association." In J. Law (ed.), *Power, Action and Belief: A New Sociology of Knowledge?* 264–79. London: Routledge and Kegan Paul.

———. 2005. *Reassembling the Social: An Introduction to Actor-Network-Theory.* Oxford: Oxford University Press.

———. 2007. "The recall of modernity: Anthropological approaches." *Cultural Studies Review* 13 (1): 11–30.

Laustsen, P. 2008. "Alaska's sea-otter decline affects health of kelp forests and diet of eagles." November. http://soundwaves.usgs.gov/2008/11/research.html.

Lawton, J. H. 1994. "What do species do in ecosystems?" *Oikos* 71 (3): 367–74.

———. 2001. "Earth system science." *Science* 292 (15 June): 1965–66.

Le Guin, U. K. 1976. "The ones who walk away from Omelas." In *The Wind's Twelve Quarters,* 251–59. New York: Bantam Books.

Lederberg, J. 1952. "Cell genetics and hereditary symbiosis." *Physiological Reviews* 32: 403–30.

Lem, S. 2013. *Summa Technologia.* Minneapolis: University of Minnesota Press.

Lenton, T. 1998. "Gaia and natural selection." *Nature* 394: 439–47.

Leonard, A. 2007. *The Story of Stuff.* Web video. www.storyofstuff.com.

Leroi-Gourhan, A. 1993. *Gesture and Speech*. Trans. A. B. Berger. Cambridge, MA: MIT Press.

Leu, K., B. Obermayer, S. Rajamani, U. Gerland, and I. A. Chen. 2011. "The prebiotic evolutionary advantage of transferring genetic information RNA to DNA." *Nucleic Acids Research* 39 (18): 8135–47.

Levins, R., and R. C. Lewontin. 1985. *The Dialectical Biologist*. Cambridge, MA: Harvard University Press.

Levy, M., and S. L. Miller. 1998. "The stability of RNA bases: Implications for the origin of life." *Proceedings of the National Academy of Sciences* 95: 7933–38.

Levy, S. B. 1998. "The challenge of antibiotic resistance." *Scientific American* 278 (3): 46–53.

Lewontin, R. C. 1974. "The analysis of variance and the analysis of causes." *American Journal of Human Genetics* 26: 400–11.

———. 1982. "Organism and environment." In H. C. Plotkin (ed.), *Learning, Development, and Culture*, 151–70. New York: Wiley.

Lewontin, R. C., S. Rose, and L. J. Kamin. 1984. *Not in Our Genes*. New York: Pantheon.

Lickliter, R. 2000. "An ecological approach to behavioral development: Insights from comparative psychology." *Ecological Psychology* 12: 319–34.

Lickliter, R., and S. M. Schneider. 2006. "Role of development in evolutionary change: A view from comparative psychology." *International Journal of Comparative Psychology* 19: 151–69.

Lonsdale, S. H. 1979. "Attitudes towards animals in ancient Greece." *Greece & Rome* 26 (2): 146–59.

Lotka, A. J. 1925. *Elements of Physical Biology*. Baltimore: Williams and Wilkins.

Lovelock, J. 1965. "A physical basis for life-detection experiments." *Nature* 207 (August 7): 568–70.

———. 1972. "Gaia as seen through the atmosphere." *Atmospheric Environment* 6: 579–80.

———. 1975. "Thermodynamics and the recognition of alien biospheres." *Proceedings of the Royal Society (London) B* 189: 167–81.

———. 1979a. *Gaia: A New Look at Life on Earth*. Oxford: Oxford University Press.

———. 1979b. "The independent practice of science." *New Scientist* 83 (September 6): 714–17.

———. 1981. "James Lovelock responds." *CoEvolution Quarterly* 29 (Spring): 62–63.

———. 1986. "Geophysiology: A new look at earth science." *Bulletin of the American Meteorological Society* 67 (April): 392–97.

———. 1988. *The Ages of Gaia: A Biography of Our Living Earth*. New York: Norton.

———. 1995. *The Ages of Gaia: A Biography of Our Living Earth*. 2nd ed. New York: Norton.

———. 2000. *Homage to Gaia: The Life of an Independent Scientist.* London: Oxford University Press.

———. 2001. "A way of life for agnostics?" *Skeptical Inquirer* 25: 40–42.

———. 2009. "Our sustainable retreat." In Crist and Rinker (eds.) 2009, 21–24.

Lovelock, J., and C. E. Giffin. 1968. "Planetary atmospheres: Compositional and other changes associated with the presence of life." In O. L. Tiffany and E. Zaitzeff (eds.), *Advanced Space Experiments* 25, 179–93. Washington, DC: American Astronautical Society.

Lovelock, J., and L. Margulis. 1973. "Atmospheric homeostasis by and for the biosphere." *Tellus* 26: 2–10.

Lovelock, J., and A. J. Watson. 1982. "The regulation of carbon dioxide and climate: Gaia or geochemistry." *Planetary and Space Science* 30: 795–802.

Lowe, D. R., Byerly, G. R., Asaro, F., and Kyte, F. J. 1989. "Geological and geochemical record of 3400-million-year-old terrestrial meteorite impacts." *Science* 245: 959–62.

Lupi, O., P. Dadalti, E. Cruz, and P. R. Sanberg. 2006. "Are prions related to emergence of early life?" *Medical Hypothesis* 67: 1027–33.

Lwoff, A. 1953. "Lysogeny." *Bacteriological Reviews* 4: 269–337.

———. 1954. "The life cycle of a virus." *Scientific American* 190 (3): 34–37.

———. 1957. "The concept of virus." *Journal of General Microbiology* 17 (2): 239–53.

———. 1966. "Interaction among virus, cell, and organism." *Science* 152 (726): 1216–20.

Maher, N. 2004. "Shooting the moon." *Environmental History* 9 (3): 526–31.

Mann, C. 1991. "Lynn Margulis: Science's unruly earth mother." *Science* 252 (April 19): 378–81.

Manwarring, W. H. 1934. "Environmental transformation of bacteria." *Science* 79: 466–70.

Margulis, L. 1968. "Evolutionary criteria in thallophytes: A radical alternative." *Science* 161: 1020–22.

———. 1970. *Origin of Eukaryotic Cells.* New Haven: Yale University Press.

———. 1971a. "The origin of plant and animal cells." *Scientific American* 59 (2): 230–35.

———. 1971b. "Whittaker's five kingdoms of organisms: Minor revisions suggested by considerations of mitosis." *Evolution* 25: 242–45.

———. 1975. "Symbiotic theory of the origin of eukaryotic organelles: Criteria for proof." *Symposia of the Society for Experimental Biology* 29: 21–38.

———. 1980. "Undulipodia, flagella and cilia." *Biosystems* 12 (1–2): 105–8.

———. 1981. *Symbiosis in Cell Evolution: Life and Its Environment on the Early Earth.* San Francisco: W. H. Freeman.

———. 1993a. "Origins of species: Acquired genomes and individuality." *Biosystems* 31 (2–3): 121–25.

———. 1993b. *Symbiosis in Cell Evolution: Microbial Communities in the Archean and Proterozoic Eons.* 2nd ed. New York: W. H. Freeman.

———. 1995. "Gaia is a tough bitch." In J. Brockman (ed.), *The Third Culture: Beyond the Scientific Revolution,* 129–46. New York: Simon & Schuster.

———. 1996. "Archaeal-eubacterial mergers in the origin of Eukarya: Phylogenetic classification of life." *Proceedings of the National Academy of Sciences* 93 (3): 1071–76.

———. 1997. "Big trouble in biology." In Margulis and Sagan 1997b, 265–82.

———. 1998. *Symbiotic Planet: A New Look at Evolution.* New York: Basic Books.

———. 2005. "Hans Ris (1914–2004): Genophore, chromosomes and the bacterial origin of chloroplasts." *International Microbiology* 8: 145–48.

———. 2007. "Speculation on speculation." In Margulis and Sagan 2007, 48–56.

———. 2009. "Genome acquisition in horizontal gene transfer: Symbiogenesis and macromolecular sequence analysis." *Methods in Molecular Biology* 532: 181–91.

Margulis, L. (ed.). 1970. *Origins of Life II.* New York: Gordon and Breach.

Margulis, L., C. A. Asikainen, and W. E. Krumbein (eds.). 2011. *Chimeras and Consciousness: Evolution of the Sensory Self.* Cambridge, MA: MIT Press.

Margulis, L., and D. Bermudes. 1985. "Symbiosis as a mechanism of evolution: Status of cell symbiosis theory." *Symbiosis* 1:101–24.

Margulis, L., and E. Case. 2006. "The germs of life." *Orion.*

Margulis, L., and M. J. Chapman. 2009. *Kingdoms and Domains: An Illustrated Guide to the Phyla of Life on Earth.* New York: W. H. Freeman.

Margulis, L., M. Chapman, R. Guerrero, and J. Hall. 2006. "The last eukaryotic common ancestor (LECA): Acquisition of cytoskeletal motility from aerotolerant spirochetes in the Proterozoic Eon." *Proceedings of the National Academy of Sciences* 103 (35): 13080–85.

Margulis, L., D. Chase, and L. P. To. 1979. "Possible evolutionary significance of spirochaetes." *Proceedings of the Royal Society B: Biological Sciences* 204 (1155): 189–98.

Margulis, L., and J. E. Cohen. 1994. "Combinatorial generation of taxonomic diversity: Implication of symbiogenesis for the proterozoic fossil record." In S. Bengtson, J. Bergström, G. Vidal, and A. Knoll (eds.), *Early Life on Earth,* 327–33. New York: Columbia University Press.

Margulis, L., and M. Dolan. 2007. "Advances in biology reveal truth about prokaryotes." *Nature* 445: 24.

Margulis, L., M. F. Dolan, and R. Guerrero. 2000. "The chimeric eukaryote: Origin of the nucleus from the karyomastigont in amitochondriate protists." *Proceedings of the National Academy of Sciences* 97 (13): 6954–59.

———. 1981. "Microtubules, undulipodia and pillotina spirochetes." *Annals of the New York Academy of Sciences* 361: 356–68.

Margulis, L., and R. Fester (eds.). 1991. *Symbiosis as a Source of Evolutionary Innovation: Speciation and Morphogenesis.* Cambridge, MA: MIT Press.

Margulis, L., and R. Guerrero. 1991. "Kingdoms in turmoil." *New Scientist* 23: 46–50.

Margulis, L., L. Lopez Baluja, S. M. Awramik, and D. Sagan. 1986. "Community living long before man: fossil and living microbial mats and early life." *Science of the Total Environment* 56: 379–97.

Margulis, L., and J. Lovelock. 1974. "Biological modulation of the atmosphere." *Icarus* 21: 471–89.

———. 1975. "The atmosphere as circulatory system of the biosphere—The Gaia hypothesis." *CoEvolution Quarterly* 5 (Summer): 31–40.

Margulis, L., and D. Sagan. 1995. *What Is Life?* New York: Simon & Schuster.

———. 1997a. *Microcosmos: Four Billion Years of Microbial Evolution.* Berkeley: University of California Press,

———. 1997b. *Slanted Truths: Essays on Gaia, Symbiosis and Evolution.* New York: Springer Verlag.

———. 2000. *What Is Life?* Berkeley: University of California Press.

———. 2002. *Acquiring Genomes: A Theory of the Origins of Species.* New York: Basic Books.

———. 2007. *Dazzle Gradually: Reflections on the Nature of Nature.* White River Junction, VT: Chelsea Green.

Margulis, L., and K. V. Schwartz. 1988. *Five Kingdoms: An Illustrated Guide to the Phyla of Life on Earth.* New York: W. H. Freeman.

Margulis, L., L. P. To, and D. Chase. 1978. "Microtubules in prokaryotes." *Science* 200 (4346): 1118–24.

Marini, P., and C. Belka. 2003. "Death receptor ligands: New strategies for combined treatment with ionizing radiation." *Current Medicinal Chemistry— Anti-Cancer Agents* 3 (5): 334–42.

Martin, W., J. Baross, D. Kelley, and M. J. Russells. 2008. "Hydrothermal vents and the origin of life." *Nature Review Microbiology* 6: 805–14.

Maynard Smith, J. 2000. "The concept of information in biology." *Philosophy of Science* 67: 177–94.

Maynard Smith, J., R. M. Burian, S. Kauffman, P. Alberch, J. H. Campbell, B. Goodwin, R. Lande, D. Raup, and L. Wolpert. 1985. "Developmental constraints and evolution." *Quarterly Review of Biology* 60: 265–87.

Mayr, E. 1991. "More natural classification." *Nature* 353: 122.

———. 1998. "Two empires or three?" *Proceedings of the National Academy of Sciences* 95: 9720–23.

McClintock, B. 1932. "A correlation of ring-shaped chromosomes with variegation in *Zea mays.*" *Proceedings of the National Academy of Sciences* 18 (12): 677–81.

———. 1939. "The behavior in successive nuclear divisions of a chromosome broken at meiosis." *Proceedings of the National Academy of Sciences* 25 (8): 405–16.

———. 1942. "The fusion of broken ends of chromosomes following nuclear fusion." *Proceedings of the National Academy of Sciences* 28 (11): 458–63.

———. 1953. "Induction of instability at selected loci in maize." *Genetics* 38 (6): 579–99.

———. 1956. "Intranuclear systems controlling gene action and mutation." *Brookhaven Symposia in Biology* (8): 58–74.

———. 1961. "Some parallels between gene control systems in maize and in bacteria." *American Naturalist* 95:265–77.

———. 1984. "The significance of responses of the genome to challenge." *Science* 226 (4676): 792–801.

———. 1987. *Discovery and Characterization of Transposable Elements: The Collected Papers of Barbara McClintock*. New York: Garland.

McCollom, T. M., and E. L. Shock. 1997. "Geochemical constraints on chemo-lithoautotrophic metabolism by microorganisms in seafloor hydrothermal systems." *Geochemica et Cosmochemica Acta* 61: 4375–91.

McCray, W. P. 2012. *The Visioneers: How a Group of Elite Scientists Pursued Space Colonies, Nanotechnology, and a Limitless Future*. Princeton: Princeton University Press.

McDermott, J. 1980. "Lynn Margulis: Unlike most microbiologists." *CoEvolution Quarterly* 25 (Spring): 31–38.

McDonald, J. F., L. V. Matyunina, S. Wilson, I. K. Jordan, N. J. Bowen, and W. J. Miller. 1997. "LTR retrotransposons and the evolution of eukaryotic enhancers." *Genetica* 100 (1–3): 3–13.

McFall-Ngai, M., et al. 2013. "Animals in a bacterial world: A new imperative for the life sciences." *Proceedings of the National Academy of Sciences* 110 (9): 3229–36.

McInerney, J. 2010. *The Cattle of the Sun*. Princeton: Princeton University Press.

McIntosh, J. R., and K. L. McDonald. 1989. "The mitotic spindle." *Scientific American* 261 (4): 48–56.

Mengel, G. A. 2009. Reconstructing a legacy: On overcoming biological preformationism, dualism, and the inheritance paradigm. Ph.D. thesis. California Institute of Integral Studies. San Francisco.

Merchant, C. 1980. *The Death of Nature: Women, Ecology, and the Scientific Revolution*. New York: Harper & Row.

Meredith, J. E., Jr., B. Fazeli, and M. A. Schwartz. 1993. "The extracellular matrix as a cell survival factor." *Molecular Biology of the Cell* 4 (9): 953–61.

Merezhkowsky, C. 1910. "Theorie der zwei Plasmaarten als Grundlage der Symbiogenesis, einer neuen Lehre von der Entstehung der Organismen." *Biologisches Zentralblatt* 30: 277–303, 321–347, 353–367.

———. 1920. "La plante considérée comme une complexe symbiotique." *Bulletin de la Société des Science Naturelles de l'Ouest de la France* 6: 17–98.

Midgley, M. 2001. *Science and Poetry*. London: Routledge.

Miller, D. B. 2007. "From nature to nurture, and back again." *Developmental Psychobiology* 49: 770–79.

Milne, D., D. Raup, J. Billingham, K. Niklas, and K. Padian (eds.). 1985. *The Evolution of Complex and Higher Organisms*. NASA SP-478. Moffet Field, CA: NASA Ames.

Mira, A., et al. 2010. "The bacterial pan-genome: A new paradigm in microbiology." *International Microbiology* 13 (2): 45–57.

Mitteldorf, J. 2012. "Neo-Darwinism and the group selection controversy." In D. Sagan (ed.) 2012, 86–94.

Mitteldorf, J., and J. Pepper. 2009. "Senescence as an adaptation to limit the spread of disease." *Journal of Theoretical Biology* 260:186–95.

Moczek, A. P. 2007. "Developmental capacitance, genetic accommodation, and adaptive evolution." *Evolution and Development* 9: 299–305.

Mojzsis, S. J., G. Arrhenius, K. D. McKeegan, T. M. Harrison, A. P. Nutma, and C. R. L. Friend. 1996. "Evidence for life on Earth before 3,800 million years ago." *Nature* 384: 55–59.

Mojzsis, S. J., and T. M. Harrison. 2000. "Vestiges of a beginning: Clues to the emergent biosphere recorded in the oldest known sedimentary rocks." *GSA Today* 10: 1–5.

Mojzsis, S. J., T. M. Harrison, and T. T. Pidgeon. 2001. "Oxygen-isotope evidence from ancient zircons for liquid water at the Earth's surface 4,300 Myr ago." *Nature* 409: 178–81.

Monod, J. 1942. *Recherches sur la croissance des cultures bactériennes*. Paris: Hermann.

Morange, M. 2010. "The scientific legacy of Jacques Monod." *Research in Microbiology* 161 (2): 77–81.

Morin, E. 1977. *La méthode, 1: La nature de la nature*. Paris: Éditions du Seuil.

Morin, E., and A. B. Kern. 1999. *Homeland Earth: A Manifesto for the New Millennium*. New York: Hampton Press.

Morowitz, H. 1992. *Beginnings of Cellular Life*. New Haven: Yale University Press.

Morowitz, H., J. P. Allen, M. Nelson, and A. Alling. 2005. "Closure as a scientific concept and its application to ecosystem ecology and the science of the biosphere." *Advances in Space Research* 36: 1305–11.

Morton, T. 2001. "Imperial measures: *Dune*, ecology and romantic consumerism." *Romanticism on the Net* 21 (February). http://id.erudit.org/iderudit/005966ar.

———. 2010. *The Ecological Thought*. Cambridge, MA: Harvard University Press.

Moss, L. 2003. *What Genes Can't Do*. Cambridge, MA: MIT Press/Bradford.

Muller, H. J. 1927. "Artificial transmutation of the gene." *Science* 66 (1699): 84–87.

Murray, A., and M. Kirschner. 1991. "What controls the cell cycle." *Scientific American* 264 (3): 56–63.

Musacchio, A. 2011. "Spindle assembly checkpoint: The third decade." *Philosophical Transactions of the Royal Society of London, Series B: Biological Sciences* 366 (1584): 3595–604.

NASA. 1962. Sixth Semiannual Report to Congress, 1 July–31 Dec. 1961. Washington, DC: NASA.

Nasmyth, K. 1993. "Regulating the HO endonuclease in yeast." *Current Opinion in Genetics & Development* 3 (2): 286–94.

Nass, M., and S. Nass. 1963. "Intramitochondrial fibers with DNA characteristics." *Journal of Cell Biology* 19: 613–28.

Nealson, K. 1999. "Post-Viking microbiology: New approaches, new data, new insights." *Origins of Life and Evolution of Biospheres* 29: 73–93.

Needleman, J. 2012. *An Unknown World: Notes on the Meaning of the Earth.* New York: Tarcher Penguin.

Netz, R. 2004. *Barbed Wire: An Ecology of Modernity.* Middletown, CT: Wesleyan University Press.

Neumann-Held, E. M. 1999. "The gene is dead—long live the gene! Conceptualizing genes the constructionist way." In P. Koslowski (ed.), *Sociobiology and Bioeconomics: The Theory of Evolution in Biological and Economic Theory,* 105–37. Berlin: Springer.

Newman, S. A., and G. B. Müller. 2000. "Epigenetic mechanisms of character origination." *Journal of Experimental Zoology, Part B: Molecular and Developmental Evolution* 288: 304–17.

Nijhout, H. F. 2001. "The ontogeny of phenotypes." In Oyama, Griffiths, and Gray (eds.) 2001, 129–40.

Nisbet, E. G., and N. H. Sleep. 2001. "The habitat and nature of early life." *Nature* 409: 1083–91.

Novick, R. P. 1980. "Plasmids." *Scientific American* 243 (6).

Nusse, R., and H. E. Varmus. 1982. "Many tumors induced by the mouse mammary tumor virus contain a provirus integrated in the same region of the host genome." *Cell* 31 (1): 99–109.

O'Neill, G. K. 1977. "The high frontier." In Brand (ed.) 1977, 8–21.

O'Neill, G. K., with S. Brand. 1975. "Is the surface of a planet really the right place for an expanding technological civilization?" *CoEvolution Quarterly* 6 (Fall): 20–28; reprinted in Brand (ed.) 1977, 22–30.

Odling-Smee, F. J., K. L. Laland, and M. W. Feldman. 2003. *Niche Construction: The Neglected Process in Evolution.* Princeton: Princeton University Press.

Olivier, L. 2011. *The Dark Abyss of Time: Archaeology and Memory.* Lanham, MD: AltaMira Press.

Olsen, B. 2010. *In Defense of Things: Archaeology and the Ontology of Objects.* Lanham, MD: AltaMira Press.

Olsen, B., M. Shanks, T. Webmoor, and C. Witmore. 2012. *Archaeology: The Discipline of Things.* Berkeley: University of California Press.

Oparin, A. I. 1957. *The Origin of Life on Earth*. Edinburgh: Oliver and Boyd.

Oreskes, N., and E. M. Conway. 2010. *Merchants of Doubt*. Bloomsbury: New York.

Orgel, L. E. 1994. "The origin of life on Earth." *Scientific American* 271 (4): 77–83.

Osinski, G. R. 2011. "The role of meteorite impacts in the origin and evolution of life." *GSA Abstracts with Programs* 43 (5): 73.

Otter, C. 2013. "Planet of meat: A biological history." In T. Bennett (ed.), *Challenging (the) Humanities*, 33–49. Canberra: Australian Academy of the Humanities.

Overbye, Dennis. 2002. "NASA presses its search for extraterrestrial life." *New York Times* (June 4).

Owen, R. 1860. *Paleontology or a systematic summary of extinct animals and their geological relations*. Edinburgh: Adam and Charles Black.

Oyama, S. 1985. *The Ontogeny of Information: Developmental Systems and Evolution*. Cambridge: Cambridge University Press.

———. 2000. *Evolution's Eye: A Systems View of the Biology-Culture Divide*. Durham, NC: Duke University Press.

———. 2001. "Terms in tension: What do you do when all the good words are taken?" In Oyama, Griffiths, and Gray (eds.) 2001, 177–93.

———. 2002. "The nurturing of natures." In A. Grunwald, M. Gutmann, and E. M. Neumann-Held (eds.), *On Human Nature: Anthropological, Biological, and Philosophical Foundations*, 163–70. New York: Springer Verlag.

———. 2006a. "Boundaries and (Constructive) Interaction." In E. M. Neumann-Held and C. Rehmann-Sutter (eds.), *Genes in Development: Re-Reading the Molecular Paradigm*, 272–89. Durham, NC: Duke University Press.

———. 2006b. "Speaking of Nature." In C. Dyke and Y. Haila (eds.), *How Does Nature Speak? Dynamics of the Human Ecological Condition*, 49–65. Durham, NC: Duke University Press.

———. 2009. "Friends, neighbors, and boundaries." *Ecological Psychology* 21: 147–54.

———. 2010. "Biologists behaving badly: Vitalism and the language of language." *History and Philosophy of the Life Sciences* 32: 401–24.

Oyama, S., P. E. Griffiths, and R. D. Gray (eds.). 2001. *Cycles of Contingency: Developmental Systems and Evolution*. Cambridge, MA: MIT Press.

Pace, N. 2006. "Time for a change." *Nature* 441: 289.

Palme, H. 2004. "The giant impact formation of the Moon." *Science* 304: 977–79.

Parvanov, E., J. Kohli, and K. Ludin. 2008. "The mating-type-related bias of gene conversion in *Schizosaccharomyces pombe*." *Genetics* 180 (4): 1859–68.

Patterson, C. 2002. *Eternal Treblinka: Our Treatment of Animals and the Holocaust*. New York: Lantern Books.

Pigliucci, M. 2011. "The meaning of 'theory' in biology." *Rationally Speaking*. http://www.rationallyspeaking.blogspot.com/2011/07/meaning-of-theory-in-biology.html.

Pilacinski, W., E. Mosharrafa, R. Edmundson, J. Zissler, M. Fiandt, and W. Szybalski. 1977. "Insertion sequence IS2 associated with int-constitutive mutants of bacteriophage lambda." *Gene* 2 (2): 61–74.

Pirozynski, K. A. 1991. "Galls, flowers, fruits, and fungi." In Margulis and Fester (eds.) 1991, 364–79.

Pittendrigh, C. S. et al. (eds.). 1966. *Biology and the Exploration of Mars.* Washington, DC: NASA.

Pizzarello, S., and E. Shock. 2012. "The organic composition of carbonaceous meteorites: The evolutionary story ahead of biochemistry." *Proceedings of the National Academy of Sciences* 109 (48): E3288.

Polard, P., and M. Chandler. 1995. "Bacterial transposases and retroviral integrases." *Molecular Microbiology* 15 (1): 13–23.

Pollan, M. 2006. *The Omnivore's Dilemma: A Natural History of Four Meals.* New York: Penguin.

———. 2008. "Farmer in Chief." *New York Times Magazine* (October 12).

Poole, R. 2008. *Earthrise: How Man First Saw the Earth.* New Haven: Yale University Press.

Portier, P. 1918. *Les symbiotes.* Paris: Masson.

Pradeu, T. 2012. *The Limits of the Self: Immunology and Biological Identity.* New York: Oxford University Press.

Prusiner, S. B. 1998. "Prions." *Proceedings of the National Academy of Sciences* 95: 13363–83.

Pyne, S. 1986. *The Ice: A Journey to Antarctica.* Iowa City: University of Iowa Press.

Radman, M., and R. Wagner. 1988. "The high fidelity of DNA duplication." *Scientific American* 259 (2): 40–46.

Razeto-Barry, P. 2012. "Autopoiesis 40 years later: A review and a reformulation." *Origins of Life and Evolution of Biospheres* 42 (6): 543–67.

Rennie, J. 1991. "Proofreading genes." *Scientific American* 264 (5): 28–32.

Reysenbach, A. L., and S. L. Cady. 2001. "Microbiology of ancient and modern hydrothermal systems." *Trends in Microbiology* 9: 79–86.

Ribo, J. M., J. Crusats, F. Sagues, J. Claret, and R. Rubires. 2001. "Chiral sign induction by vortices during the formation of mesophases in stirred solutions." *Science* 292: 2063–66.

Ris, H, and W. Plaut. 1962. "Ultrastructure of DNA-containing areas in the chloroplasts of *Chlamydomonas.*" *Journal of Cell Biology* 13: 383–91.

Rivera, M.C., R. Jain, J.E. Moore, and J.A. Lake. 1998. "Genomic evidence for two functionally distinct gene classes." *Proceedings of the National Academy of Sciences* 95: 6239–44.

Robert, J. S. 2002. "How developmental is evolutionary developmental biology?" *Biology & Philosophy* 17: 591–611.

————. 2003. "Developmental systems and animal behavior." *Biology & Philosophy* 18: 477–89.

————. 2006. "Review of *Environment, Development, and Evolution: Toward a Synthesis.*" *American Journal of Human Biology* 18 (2): 230–31.

Rosing, M. T. 1999. "^{13}C-depleted carbon microparticles in >3700-Ma sea-floor sedimentary rocks from West Greenland." *Science* 283: 674–76.

Rous, P. 1910. "A transmissable avian neoplasm." *Journal of Experimental Medicine* 12: 696–705.

Rusche, L. N., and J. Rine. 2010. "Switching the mechanism of mating type switching: a domesticated transposase supplants a domesticated homing endonuclease." *Genes & Development* 24 (1): 10–14.

Ryan, F. P. 2011. *The Mystery of Metamorphosis: A Scientific Detective Story.* White River Junction, VT: Chelsea Green.

Sagan, C., E. Levinthal, and J. Lederberg. 1968. "Contamination of Mars." *Science* 159 (March 15): 1191–96.

Sagan, D. 1990. *Biospheres: Metamorphosis of Planet Earth.* New York: McGraw-Hill.

————. 2011. "The human is more than human: Interspecies communities and the new 'facts of life.'" Culture@Large, Keynote, Society for Cultural Anthropology, American Anthropology Association (November 18). http://culanth.org/?q=node/509. Also in Sagan 2013a, 17–32.

————. 2012. "Lynn Margulis and the pursuit of knowledge: Rebel with a cause." *Empirical* (September): 10–14.

————. 2013a. *Cosmic Apprentice: Dispatches from the Edge of Science.* Minneapolis: University of Minnesota Press

————. 2013b. "On Doyle on drugs." In D. Sagan 2013a, 199–223.

————. 2013c. "On star stuff, 'science's unruly earth mother,' and the scientific art of empirical rebellion." http://www.uminnpressblog.com/2013/05/on-star-stuff-sciences-unruly-earth.html.

————. 2013d. "Priests of the modern age." In D. Sagan 2013a, 133–63.

————. 2013e. "The problem of intelligence: Bad philosophy, overpopulation, and the questionability of human intelligence." *Empirical* (May): 24–38.

————. 2014. "Nabokov's fictional realism: Antinomy, temporality, and the unforeclosed." In S. Blackwell and K. Johnson (eds.), *Fine Lines: Nabokov's Science and Art.* New Haven: Yale University Press.

Sagan, D. (ed.). 2012. *Lynn Margulis: The Life and Legacy of a Scientific Rebel.* White River Junction, VT: Chelsea Green.

Sagan, D., and L. Margulis. 2007a. "Gaia and philosophy." In Margulis and Sagan 2007, 172–84.

————. 2007b. "Welcome to the machine." In Margulis and Sagan 2007, 76–88.

————. 2011. "Foreword." In Ryan 2011, xiii–xvii.

Sagan [Margulis], L. 1967. "On the origin of mitosing cells." *Journal of Theoretical Biology* 14: 225–74.

Sapp, J. 1987. *Beyond the Gene: Cytoplasmic Inheritance and the Struggle for Authority in Genetics.* New York: Oxford University Press.

———. 1991. "Living together: Symbiosis and cytoplasmic inheritance." In Margulis and Fester (eds.) 1991, 15–25.

———. 1994. *Evolution by Association: A History of* Symbiosis. New York: Oxford University Press.

———. 1998. "Freewheeling centrioles." *History and Philosophy of the Life Sciences* 20: 255–90.

———. 2003. *Genesis: The Evolution of Biology.* New York: Oxford University Press.

———. 2005. "The prokaryote-eukaryote dichotomy: Meanings and mythology." *Microbiology and Molecular Biology Reviews* 69: 292–305.

———. 2006. "Mitochondria and their host: Morphology to molecular phylogeny." In W. Martin and M. Müller (eds.), *Mitochondria and Hydrogenosomes,* 57–84. Heidelberg: Springer Verlag.

———. 2007. "The structure of microbial evolutionary theory." *Studies in History and Philosophy of Biological and Biomedical Sciences* 38: 780–95.

———. 2009. *The New Foundations of Evolution: On the Tree of Life.* New York: Oxford University Press.

———. 2010. "Saltational symbiosis." *Theory in Bioscience* 129: 125–33.

———. 2012. "Too fantastic for polite society: A brief history of symbiosis theory." In D. Sagan (ed.) 2012, 54–67.

Sapp, J., F. Carrapiço, and M. Zolotonosov, 2002. "The hidden face of Constantin Merezhkowsky." *History and Philosophy of the Life Sciences* 24: 421–49.

Schaechter, M. 2012. "Erudition." In D. Sagan (ed.) 2012, 14–16.

Schmitt, Gail K. 2008. "Sager, Ruth." *Complete Dictionary of Scientific Biography. Encyclopedia.com.* http://www.encyclopedia.com.

———. 2001. "A goddess of Earth or the imagination of a man?" *Science* 291 (9 March): 1906–7.

Schneider, S., and P. Boston (eds.). 1991. *Scientists on Gaia.* Cambridge, MA: MIT Press.

Schneider, S., and R. Londer. 1984. *The CoEvolution of Climate and Life.* San Francisco: Sierra Club.

Schneider, S., J. R. Miller, E. Crist, and P. Boston (eds.). 2004. *Scientists Debate Gaia: The Next Century.* Cambridge, MA: MIT Press.

Schopf, J. W. 1999. *Cradle of Life: The Discovery of the Earth's Earliest Fossils.* Princeton: Princeton University Press.

Schultz, E. A. 2013. "New perspectives on organism-environment interactions in anthropology." In G. Barker, E. Desjardins, and T. Pearce (eds.), *Entangled*

Life: Organism and Environment in the Biological and Social Sciences, 79–102.
Dordrecht: Springer.

Serres, M. 1995. *The Natural Contract*. Trans. E. MacArthur and W. Paulson. Ann
Arbor: University of Michigan Press.

———. 2012. *Biogea*. Trans. R. Burks. Minneapolis: Univocal.

Shanks, M., and C. Witmore. 2010. "Memory practices and the archaeological
imagination in risk society: design and long term community." In S.
Koerner and I. Russell (eds.), *Unquiet Pasts: Risk Society, Lived Cultural
Heritage, Re-designing Reflexivity*, 269–90. Burlington, VT: Ashgate.

Shapiro, J. A. 1983. *Mobile Genetic Elements*. New York: Academic Press.

———. 1988. "Bacteria as multicellular organisms." *Scientific American* 256 (6):
82–89.

———. 1992. "Natural genetic engineering in evolution." *Genetica* 86 (1–3): 99–111.

———. 1998. "Thinking about bacterial populations as multicellular organisms." *Annual Review of Microbiology* 52: 81–104.

———. 2007. "Bacteria are small but not stupid: Cognition, natural genetic
engineering and socio-bacteriology." *Studies in History and Philosophy of
Biological and Biomedical Sciences* 38 (4): 807–19.

———. 2009a. "Letting *Escherichia coli* teach me about genome engineering."
Genetics 183 (4): 1205–14.

———. 2009b. "Revisiting the central dogma in the 21st century." *Annals of the
New York Academy of Sciences* 1178: 6–28.

———. 2011. *Evolution: A View from the 21st Century*. Upper Saddle River, NJ: FT
Press Science.

Shapiro, J. A., and M. Dworkin. 1997. *Bacteria as Multicellular Organisms*. New
York: Oxford University Press.

Shapiro, J. A., L. Machattie, L. Eron, G. Ihler, K. Ippen, and J. Beckwith. 1969.
"Isolation of pure lac operon DNA." *Nature* 224 (5221): 768–74.

Shapiro, R. 2007. "A simpler origin for life." *Scientific American* 296: 24–31.

Sharma, K. 2015. *Interdependence: Biology and Beyond*. New York: Fordham
University Press.

Sharp, P A. 1994. "Split genes and RNA splicing." *Cell* 77:805–15.

Shiva, V. 2000. *Stolen Harvest: The Hijacking of the Global Food Supply*. Cambridge,
MA: South End Press.

———. 2008. *Soil Not Oil: Environmental Justice in an Age of Climate Crisis*.
Cambridge, MA: South End Press.

Sinclair, U. 1971. *The Jungle*. Cambridge, MA: R. Bentley.

Slack, J. M. W. 2002. "Conrad Hal Waddington: The last renaissance biologist?"
Nature Reviews Genetics 3: 889–95.

Sloterdijk, P. 2008. "Excerpts from *Spheres III*: Foams." *Harvard Design Magazine*
29: 38–52.

———. 2011. *Bubbles—Spheres Volume 1: Microspherology*. Los Angeles: Semiotext(e).

———. 2012. "Voices for animals: A fantasy on animal representation." In G. R. Smulewicz-Zucker (ed.), *Strangers to Nature: Animal Lives and Humans Ethics*, 263–69. Lanham, MD: Lexington Books.

———. 2013. *In the World Interior of Capital: For a Philosophical Theory of Globalization*. Trans. W. Hoban. Malden, MA: Polity Press.

Smith, B. H. 1997. *Belief and Resistance: Dynamics of Contemporary Intellectual Controversy*. Cambridge, MA: Harvard University Press.

———. 2006. *Scandalous Knowledge: Science, Truth, and the Human*. Durham, NC: Duke University Press.

Smith-Sonneborn, J., and W. Plaut. 1967. "Evidence for the presence of DNA in the pellicle of *Paramecium*." *Journal of Cell Science* 2: 225–34.

Sober, E. 1988. "Apportioning causal responsibility." *Journal of Philosophy* 85: 303–18.

Sommer, S., J. Knezevic, A. Bailone, and R. Devoret. 1993. "Induction of only one SOS operon, umuDC, is required for SOS mutagenesis in *Escherichia coli*." *Molecular and General Genetics* 239 (1–2): 137–44.

Sonea, S., and L. G. Mathieu. 2000. *Prokaryotology: A Coherent View*. Montreal: University of Montreal Press.

Sonea, S., and M. Panisset. 1983. *A New Bacteriology*. Boston: Jones and Bartlett.

Squier, S. 2011. *Poultry Science, Chicken Culture: A Partial Alphabet*. New Brunswick, NJ: Rutgers University Press.

Stahl, F. W. 1987. "Genetic recombination." *Scientific American* 256 (2): 90–101.

Stanier, R. 1970. "Some aspects of the biology of cells and their possible evolutionary significance." In H. P. Charles and B. C. Knight (eds.), *Organization and Control in Prokaryotic Cells. Twentieth Symposium of the Society for General Microbiology*, 1–38. Cambridge: Cambridge University Press.

Stanier, R., M. Douderoff, and E. Adelberg. 1963. *The Microbial World*. 2nd ed. Englewood Cliffs, NJ: Prentice-Hall.

Stebbins, G. L., Jr. 1951. "Cataclysmic evolution." *Scientific American* 184 (4): 54–59.

Steig, W. 1987. *The Zabajaba Jungle*. New York: Farrar, Straus and Giroux.

Steinfeld, H. 2006. *Livestock's Long Shadow: Environmental Issues and Options*. Rome: Food and Agriculture Organization of the United Nations.

Steitz, T. A., and P. B. Moore. 2003. "RNA, the first macromolecular catalyst: The ribosome is a ribozyme." *Trends in Biochemical Sciences* 28: 411–18.

Stent, G. 1971. *Molecular Genetics: An Introductory Narrative*. San Francisco: W. H. Freeman.

Sterelny, K., M. Dickison, and K. C. Smith. 1996. "The extended replicator." *Biology and Philosophy* 11: 377–403.

Stern, D. L. and V. Orgogozo. 2009. "Is genetic evolution predictable?" *Science* 323 (5915): 746–51.

Sternberg, R. V., and J. A. Shapiro. 2005. "How repeated retroelements format genome function." *Cytogenetic and Genome Research* 110: 108–16.

Sternhell, Z. 2006. *Les anti-lumières: Une tradition du XVIIIe siècle à la guerre froide.* Paris: Fayard.

Stotz, K. 2010. "Human nature and cognitive-developmental niche construction." *Phenomenology and the Cognitive Sciences* 9: 483–501.

Strick, J. 2004. "Creating a cosmic discipline: The crystallization and consolidation of exobiology, 1957–1973." *Journal of History of Biology* 37: 131–80.

Suzuki, Y., and H. F. Nijhout. 2006. "Evolution of a polyphenism by genetic accommodation." *Science* 311: 650–52.

Swain, A., and J. M. Coffin. 1992. "Mechanism of transduction by retroviruses." *Science* 255 (5046): 841–45.

Szekvolgyi, L., and A. Nicolas. 2010. "From meiosis to postmeiotic events: Homologous recombination is obligatory but flexible." *Federation of the Societies of Biochemistry and Molecular Biology Journal* 277 (3): 571–89.

Szostak, J. W., D. P. Bartel, and P. Luigi Luisi. 2001. "Synthesizing life." *Nature* 409: 387–90.

Taleb, N. N. 2012. *Antifragile: Things That Gain from Disorder.* New York: Random House.

Taylor, P. J. 2001. "Distributed agency within intersecting ecological, social, and scientific processes." In Oyama, Griffiths, and Gray (eds.) 2001, 313–32.

———. 2005. *Unruly Complexity: Ecology, Interpretation, Engagement.* Chicago: University of Chicago Press.

———. 2014. *Nature-Nurture? No: Moving the Sciences of Variation and Heredity Beyond the Gaps.* Arlington, MA: The Pumping Station.

Temin, H. M. 1972. "RNA-directed DNA synthesis." *Scientific American* 226 (1): 24–33.

Temin, H. M., and S. Mizutani. 1970. "RNA-dependent DNA polymerase in virions of Rous sarcoma virus." *Nature* 226: 1211–1213.

Thompson, W. I. 1974. *Passages about Earth: An Exploration of the New Planetary Culture.* New York: Harper & Row.

———. 1991. "The imaginary of a new science and the emergence of a planetary culture." In Thompson (ed.) 1991, 11–29.

Thompson, W. I. (ed.) 1987. *Gaia—A Way of Knowing: Political Implications of the New Biology.* Great Barrington, MA: Lindisfarne Press.

———. 1991. *Gaia 2: Emergence, The New Science of Becoming.* Hudson, NY: Lindisfarne Press.

Tjian, R. 1995. "Molecular machines that control genes." *Scientific American* 272 (2): 54–61.

Todd, J. 1977. Correspondence. In Brand (ed.) 1977, 48–49.

Todd, R. M., W. A. Cunningham, A. K. Anderson, and E. Thompson. 2012. "Affect-biased attention as emotion regulation." *Trends in Cognitive Sciences* 16 (7): 365–72.

Trefil, J., H. J. Morowitz, and E. Smith. 2009. "The origin of life." *American Scientist* 97: 206–13.

Tsichlis, P. N. 1987. "Oncogenesis by Moloney murine leukemia virus." *Anticancer Research* 7 (2): 171–80.

Tsichlis, P. N., J. S. Lee, S. E. Bear, P. A. Lazo, C. Patriotis, E. Gustafson, S. Shinton, N. A. Jenkins, N. G. Copeland, K. Huebner, et al. 1990. "Activation of multiple genes by provirus integration in the Mlvi-4 locus in T-cell lymphomas induced by Moloney murine leukemia virus." *Journal of Virology* 64 (5): 2236–44.

"Twenty 20 most influential scientists living today." 2011. *SuperScholar.* www.superscholar.org/features/20-most-influential-scientists-alive-today.

Ullmann, A. 2010. "Jacques Monod, 1910–1976: His life, his work and his commitments." *Research in Microbiology* 161 (2): 68–73.

Uzell, T., and C. Spolsky. 1974. "Mitochondria and plastids as endosymbionts: A revival of special creation?" *American Scientist* 62: 334–43; 343.

van der Weele, C. 1999. *Images of Development: Environmental Causes in Ontogeny.* Albany: State University of New York Press.

Varela, F. J., H. M. Maturana, and R. Uribe. 1974. "Autopoiesis: The organization of living systems, its characterization and a model." *BioSystems* 5: 187–96.

Varmus, H. 1987. "Reverse transcription." *Scientific American* 257 (3): 56–64.

Vernadsky, V. I. 1998. *The Biosphere: Complete Annotated Edition.* Trans. D. B. Langmuir. Ed. M. McMenamin. New York: Copernicus.

Villarreal, L. P. 2005. *Viruses and the Evolution of Life.* Washington, DC: American Society for Microbiology Press.

Volk, T. 2003. *Gaia's Body: Toward a Physiology of Earth.* Cambridge, MA: MIT Press.

von Weizsäcker, E. U., A. Lovins, and H. Lovins. 1997. *Factor Four: Doubling Wealth, Halving Resource Use.* London: Earthscan.

Wacey, D., M. R. Kilburn, J. Cliff, and M. D. Brasier. 2011. "Microfossils of sulfur-metabolizing cells in 3.4-billion-year-old rocks of Western Australia." *Nature Geoscience* 4: 698–702.

Wächtershäuser, G. 1993. "The cradle of chemistry of life: On the origin of natural products in a pyrite-pulled chemoautotrophic origin of life." *Pure and Applied Chemistry* 65: 1343–48.

Waddington, C. H. 1940. *Organisers and Genes.* Cambridge: Cambridge University Press.

———. 1942. "The epigenotype." *Endeavour* 1: 18–20.

————. 1957. *The Strategy of the Genes: A Discussion of Some Aspects of Theoretical Biology*. London: George Allen & Unwin Ltd.

————. 1968a. "Towards a theoretical biology." *Nature* 218: 525–27.

————. 1968b. "Preface." In Waddington (ed.) 1968, n.p.

————. 1969a. *Behind Appearance: A Study of the Relations between Painting and the Natural Sciences in This Century*. Edinburgh: Edinburgh University Press.

————. 1969b. "Preface." In Waddington (ed.) 1969, n.p.

————. 1969c. "Sketch of the second Serbelloni symposium." In Waddington (ed.) 1969, 1–9.

————. 1969d. "The practical consequences of metaphysical beliefs on a biologist's work: An autobiographical note." In Waddington (ed.) 1969, 72–81.

————. 1972a. "Preface." In Waddington (ed.) 1972, n.p.

————. 1972b. "Epilogue." In Waddington (ed.) 1972, 283–89.

————. 1975. *The Evolution of an Evolutionist*. Ithaca, NY: Cornell University Press.

————. 1976. "Concluding remarks." In Jantsch and Waddington (eds.) 1976, 243–50.

Waddington, C. H. (ed.). 1968. *Towards a Theoretical Biology 1: Prolegomena*. Edinburgh: Edinburgh University Press.

————. 1969. *Towards a Theoretical Biology 2: Sketches*. Edinburgh: Edinburgh University Press.

————. 1972. *Towards a Theoretical Biology 4. Essays*. Edinburgh: Edinburgh University Press.

Wade, N. 2014. "The sloth's busy inner life." *The New York Times* (January 28).

Waldby, C., and S. Squier. 2003. "Ontogeny, ontology and phylogeny: Embryonic life and stem cell technologies." *Configurations* 11: 27–46.

Walker, B., C. S. Holling, S. R. Carpenter, and A. Kinzig. 2004. "Resilience, adaptability and transformability in social-ecological systems." *Ecology and Society* 9 (2): 5.

Wallin, I. E. 1924. "On the nature of mitochondria, VII. The independent growth of mitochondria in culture media." *American Journal of Anatomy* 33: 147–73.

————. 1927. *Symbionticism and the Origin of Species*. Baltimore: Williams and Wilkins.

Walsh, D. M. 2006. "Evolutionary essentialism." *British Journal for the Philosophy of Science* 57: 425–48.

Walsh, M. 1992. "Microfossils and possible microfossils from the early Archean Onverwacht Group, Barberton Mountain Land, South Africa." *Precambrian Research* 54: 271–93.

Warren, K. (ed.). 1967. *Formation and Fate of Cell Organelles*. New York: Academic Press.

Watanabe, T. 1967. "Infectious drug resistance." *Scientific American* 217 (6): 19–28.

Watasé, S. 1893. "On the nature of cell organization." *Woods Hole Biological Lectures.* 83–103.

Watson, A. J., and J. E. Lovelock. 1983. "Biological homeostasis of the global environment: The parable of Daisyworld." *Tellus* 35B: 284–89.

Watson, J. D., and F. H. Crick. 1953a. "Molecular structure of nucleic acids: A structure for deoxyribose nucleic acid." *Nature* 171 (4356): 737–38.

———. 1953b. "Genetical implications of the structure of deoxyribonucleic acid." *Nature* 171: 964–967.

Weigle, J. J. 1953. "Induction of mutations in a bacterial virus." *Proceedings of the National Academy of Sciences* 39 (7): 628–36.

Weigle, J. J., and G. Bertani. 1953. "Variations des bacteriophages conditionnées par les bacteries hôtes." *Annales de l'Institut Pasteur* 84 (1): 175–79.

Weinberg, R. A. 1996. "How cancer arises." *Scientific American* 275 (3): 62–70.

Weismann, A. 1893. *The Germ-Plasm: A Theory of Heredity.* New York: Charles Scribner's Sons.

Weiss, K. M., and S. M. Fullerton. 2000. "Phenogenetic drift and the evolution of genotype-phenotype relationships." *Theoretical Population Biology* 57: 187–95.

Welsh, J. 2011. "Eating plants may change our cells." *LiveScience* (September 20). www.livescience.com/16137-plants-animals-micorna-genes.html.

Westbroek, P. 1991. *Life as a Geological Force.* New York: Norton.

———. 2009. *Terre! Des menaces globales à l'espoir planétaire.* Paris: Le Seuil.

———. 2012a. "Civilizing Earth." *Human Figurations.* http://hdl.handle.net/2027 /spo.11217607.0001.108.

———. 2012b. "Fishermen in the maelstrom: Big history, symbiosis, and Lynn Margulis as a modern-day Copernicus." In D. Sagan (ed.) 2012, 126–39.

West-Eberhard, M. J. 2003. *Developmental Plasticity and Evolution.* Oxford: Oxford University Press.

———. 2005. "Developmental plasticity and the origin of species differences." *Proceedings of the National Academy of Sciences* 102: 6543–49.

Wheelis, M. L., O. Kandler, and C. R. Woese. 1992. "On the nature of global classification." *Proceedings of the National Academy of Sciences* 89: 2930–34.

Whitaker, J. W. 1975. *Feedlot Empire: Beef Cattle Feeding in Illinois and Iowa, 1840–1900.* Ames: Iowa State University Press.

White, M. J. D. 1945. *Animal Cytology and Evolution.* Cambridge: Cambridge University Press.

———. 1978. *Modes of Speciation.* San Francisco: Freeman.

Whitehead, A. N. 1948. *Essays in Science and Philosophy.* New York: Philosophical Library.

Whittaker, R. H. 1969. "New concepts of kingdoms of organisms." *Science* 163:150–60.

Whittaker, R. H., and L. Margulis. 1978. "Protist classification and the kingdoms of organisms." *BioSystems* 10: 3–18.

Wiegel, J., and M. W. W. Adams (eds.). 1998. *Thermophiles: The Keys to the Molecular Evolution and the Origin of Life?* London: Taylor and Francis.

Wier, A. M., L. Sacchi, M. F. Dolan, J. Bandi, J. Macallister, and L. Margulis. 2010. "Spirochete attachment ultrastructure: Implications for the origin and evolution of cilia." *Biological Bulletin* 218 (1): 25–35.

Wills, C., and J. Bada. 2001. *The Spark of Life: Darwin and the Primeval Soup.* New York: Basic Books.

Wilmers, C. C., J. A. Estes, M. Edwards, K. L. Laidre, and B. Konar. 2012. "Do trophic cascades affect the storage and flux of atmospheric carbon? An analysis of sea otters and kelp forests." *Frontiers in Ecology and the Environment* 10: 409–15.

Wilson, E. B. 1925. *The Cell in Development and Heredity.* New York: Macmillan.

Wilson, E. O. 2013. *The Social Conquest of Earth.* New York: Liveright.

Winther, R. G. 2014. "Evo-Devo as a trading zone." http://philpapers.org/archive/WINEAA-4.4.pdf.

Witkin, E. M. 1975. "Elevated mutability of polA derivatives of Escherichia coli B/r at sublethal doses of ultraviolet light: Evidence for an inducible error-prone repair system ("SOS repair") and its anomalous expression in these strains." *Genetics* 79 (Suppl.): 199–213.

———. 1991. "RecA protein in the SOS response: Milestones and mysteries." *Biochimie* 73 (2–3): 133–41.

Witmore, C. 2014. "Archaeology and the new materialisms." *The Journal of Contemporary Archaeology,* forthcoming.

Woese, C. R. 1977. "Endosymbionts and mitochondrial origins." *Journal of Molecular Evolution* 10: 93–96.

———. 1982. "Archaebacteria and cellular origins: An overview." In O. Kandler (ed.), "Archaebacteria," *Proceedings of the 1st International Workshop on Archaebacteria,* 1–17. Stuttgart: Gustav Fischer.

———. 1987. "Bacterial evolution." *Microbiological Reviews* 51: 221–71.

———. 1998a. "Default taxonomy: Ernst Mayr's view of the microbial world." *Proceedings of the National Academy of Sciences* 95: 11043–46.

———. 1998b. "The universal ancestor." *Proceedings of the National Academy of Sciences* 95: 6854–59.

———. 2000. "Interpreting the universal phylogenetic tree." *Proceedings of the National Academy of Sciences* 97: 8392–96.

Woese, C. R., and G. E. Fox. 1977a. "Phylogenetic structure of the prokaryote domain: The primary kingdoms." *Proceedings of the National Academy of Sciences* 75: 5088–90.

———. 1977b. "The concept of cellular evolution." *Journal of Molecular Evolution* 10: 1–6.

Woese, C. R., O. Kandler, and M. Wheelis. 1990. "Towards a natural system of organisms: Proposal for the domains Archaea, Bacteria, and Eucarya." *Proceedings of the National Academy of Sciences* 87: 4576–79.

Wolfe, C. 2013. *Before the Law: Humans and Other Animals in a Biopolitical Frame.* Chicago: University of Chicago Press.

Woodhead, G. S. 1891. *Bacteria and Their Products.* London: Walter Scott.

Woodruff, R. C., and J. N. Thompson, Jr. 2002. "Mutation and premating isolation." *Genetica* 116 (2–3): 371–82.

Woolhouse, W. H. 1967. "A review of *The Plastids* by J. T. O. Kirk and R. A. E. Tilney-Bassett." *New Phytologist* 66: 832–33.

Wrangham, R. 2010. *Catching Fire: How Cooking Made Us Human.* New York: Basic Books.

Yamada-Inagawa, T., A. J. Klar, and J. Z. Dalgaard. 2007. "Schizosaccharomyces pombe switches mating type by the synthesis-dependent strand-annealing mechanism." *Genetics* 177 (1): 255–65.

Young, R. M. 1985. "Darwin's metaphor: Does nature select?" In *Darwin's Metaphor,* 79–125. Cambridge: Cambridge University Press.

Zablen, L. B., M. S. Kissle, C. R. Woese, and D. E. Butow. 1975. "The phylogenetic origin of the chloroplast and the prokaryotic nature of its ribosomal RNA." *Proceedings of the National Academy of Sciences* 72: 2418–22.

Zhang, C.-Y. et al. 2011. "Exogenous Plant MIR168a specifically targets mammalian LDLRAP1: Evidence of cross-kingdom regulation by microRNA." *Cell Research* 22: 107–26.

Zinder, N. D. 1958. "Transduction in bacteria." *Scientific American* 199 (5): 38–43.

Zinder, N. D., and J. Lederberg. 1952. "Genetic exchange in *Salmonella*." *Journal of Bacteriology* 64 (5): 679–99.

Zolli, A., and A. M. Healy. 2012. *Resilience: Why Things Bounce Back.* New York: Free Press.

CONTRIBUTORS

SANKAR CHATTERJEE is Paul Whitfield Horn Professor of Geosciences and Curator of Paleontology at the Museum of Texas Tech University. He has led expeditions to India, China, Antarctica, and the American Southwest in search of dinosaurs and early birds and has discovered, named, and described several new taxa. His current research focuses on Mesozoic vertebrates, flight of pterosaurs and birds, origin of flight, mass extinction, macroevolution, plate tectonics, and paleobiogeography. His work on macroevolution encompasses the study of large-scale patterns of evolution above the species level, and mass extinctions and their long-term consequences in biodiversity. He has published *The Rise of Birds* (1997), as well as more than one hundred scientific papers, several monographs, and two coedited collections, *New Concepts in Global Tectonics* (1992) and *Posture, Locomotion, and Paleoecology of Pterosaurs* (2004). His research work has been featured in several documentary films. Recently he has received the prestigious Fulbright-Nehru Academic and Professional Excellence Award.

BRUCE CLARKE is Paul Whitfield Horn Professor of Literature and Science and chair of the Department of English at Texas Tech University. His research focuses on nineteenth- and twentieth-century literature and science, with special interests in systems theory, narrative theory, and ecology. In 2010–11 he was senior fellow at the International Research Institute for Cultural Technologies and Media Philosophy, Bauhaus-University Weimar. His books are *Allegories of Writing* (1995), *Dora Marsden and Early Modernism* (1996), *Energy Forms* (2001), *Posthuman Metamorphosis* (2008), and *Neocybernetics and Narrative* (2014). He has coedited *From Energy to Information* (2002), *Emergence and Embodiment* (2009), and the *Routledge Companion to Literature and Science* (2010). He is now writing a cultural history of the American locations, transnational authors, and key concepts of the systems discourses gathered in the *Whole Earth Catalog* and *CoEvolution Quarterly*.

SUSAN OYAMA trained at Harvard University and is now professor emerita at the John Jay College of Criminal Justice and the Graduate School and University Center (both of the City University of New York). She has written widely on the nature/nurture opposition and on the concepts of development, evolution, and genetic information. She is best known for her work on developmental systems theory, to which many readers were introduced by *The Ontogeny of Information: Developmental Systems and Evolution* (1985, 2000). She is also the author of *Evolution's Eye: A Systems View of the Biology-Culture Divide* (2000). With Paul Griffiths and Russell Gray, she edited *Cycles of Contingency* (2003), a book of essays on developmental systems by scholars from many fields. She is especially interested in the role of theory in the relations among disciplines and between academia and society at large.

DORION SAGAN worked closely with Lynn Margulis for more than thirty years, co-authoring works such as *Microcosmos* (1987), *What Is Life?* (2000), and *Dazzle Gradually* (2007). He is the author or coauthor of twenty-four books translated into thirteen languages, including *Into the Cool* (2005), on the thermodynamics of life, and *Death and Sex* (2009), winner of the Bookbinder's Guild of New York award for best nonfiction hardcover. Sagan has lectured widely, including at the American Museum of Natural History, the Smithsonian Institution, the Artist's Institute in Manhattan, and Xian, China. His writings have appeared in the *New York Times*, the *New York Times Book Review*, *Wired*, *Cabinet*, and *Empirical*. His most recent book publications include the essay collection *Lynn Margulis: The Life and Legacy of a Scientific Rebel* (2012), and *Cosmic Apprentice: Dispatches from the Edges of Science* (2013).

JAN SAPP is professor in the Department of Biology at York University in Toronto. He was Andrew Mellon Fellow at the Rockefeller University, 1991–92, and held the Canada Research Chair in the History of the Biological Sciences at the University of Québec at Montréal from 2001 to 2003. His writings focus especially on evolutionary biology and ecology, emphasizing the fundamental importance of symbiosis and horizontal gene transfer in heredity and evolution. He is the author of numerous papers and several books, including *What Is Natural? Coral Reef Crisis* (1999), *Genesis: The Evolution of Biology* (2003), and *The New Foundations of Evolution: On the Tree of Life* (2009).

JAMES A. SHAPIRO is professor of microbiology at the University of Chicago. He has a BA in English literature from Harvard (1964) and a PhD in genetics from Cambridge (1968). In 1979, Shapiro formulated the first precise molecular model for transposition and replication of phage Mu and other transposons. Since 1992, he has been writing about the importance of biologically regulated natural genetic engineering as a fundamental new concept in evolution science, work he summarized in *Evolution: A View from the Twenty-First Century* (2011). He is editor of *Mobile Genetic Elements* (1983) and coeditor of *DNA Insertion Elements, Episomes and Plasmids* (1977), and *Bacteria as Multicellular Organisms* (1997). For twenty years he has been a leading scientific critic of orthodox evolutionary theory. He is the recipient of an honorary OBE from Queen Elizabeth II for services to higher education in the United Kingdom and United States. In April 2014, he received the Ide and Luella Trotter Prize from Texas A&M University for his contributions to the study of evolution.

SUSAN MERRILL SQUIER is Brill Professor of Women's Studies and English at Penn State University. She is the author or editor of eight books, including *Babies in Bottles: Twentieth-Century Visions of Reproductive Technology* (1984), *Liminal Lives: Imagining the Human at the Frontiers of Biomedicine* (2004), and *Poultry Science, Chicken Culture: A Partial Alphabet* (2011). She was scholar in residence at the Bellagio Study and Conference Center (2001); Visiting Distinguished Fellow, LaTrobe University, Melbourne, Australia; and Fulbright Senior Research Scholar, Melbourne, Australia. She codirected an NEH Institute on Medicine, Literature, and Culture at the Penn State College of Medicine, Hershey Medical Center. She coedits the *Graphic Medicine* book series at Penn State University Press and serves on the editorial boards of *Literature and Medicine, Configurations*, and *Journal of the Medical Humanities*.

JAMES E. STRICK is associate professor in the Program in Science, Technology and Society and chair of the Department of Earth and Environment at Franklin and Marshall College. Originally trained in microbiology and later in history of science, he has published extensively on the history of ideas and experiments about the origin of life, including *Sparks of Life: Darwinism and the Victorian Debates over Spontaneous Generation* (2000) and, with Steven Dick, *The Living Universe: NASA and the Development of Astrobiology* (2004). He is also the editor of two six-volume collections of primary sources: *Evolution and the Spontaneous Generation Debate* (2001) and *The Origin of Life Debate: Molecules, Cells, and Generation* (2004). His *Wilhelm Reich, Biologist*, on Reich's bion experiments on the origin of life, is forthcoming from Harvard University Press.

PETER WESTBROEK took his PhD in paleobiology at Leiden University in the Netherlands. As professor of geophysiology at Leiden University, he published on biomineralization in living coccolithophores and the serology of fossil macromolecules and gave regular lecture courses on Earth system science. He is a member of the Royal Netherlands Academy of Sciences and Arts, and was the first Netherlander after Erasmus to be nominated as a professor at the Collège de France (Chaire Européenne 1996–97). He honors include the first Vladimir Ivanovich Vernadsky Medal of the European Geophysical Society in 2003. His books are *Life As* a *Geological Force* (1991), *Terre!* (Seuil, 2009, in French), and the bestselling *De Ont-dekking van de Aarde; het Grote Verhaal van een Kleine Planeet* (Balans, 2012, in Dutch), to be published in English as *Discovering the Earth: Big History of a Small Planet.*

CHRISTOPHER WITMORE is associate professor of archaeology and classics at Texas Tech University. His research interests include archaeology and the long term, things and the new materialisms, land and chorography, the history of archaeology, and science, technology, and media studies. In recent years he has become a leading figure in archaeological theory, engaging "old" things (*ta archaia*) in their relevance to present life and ecology. He has written more than fifty articles and is coauthor of *Archaeology: The Discipline of Things* (2012), coeditor of *Archaeology in the Making* (2013), and coeditor of the Routledge book series Archaeological Orientations. His current book project draws insights from the material memories of human and nonhuman inhabitation over ten millennia in the Eastern Peloponnese, Greece, and situates them in light of radical upheavals in regional space and current ecological crises.

INDEX

meteorites, 40; Murchison meteorite, 54, 56

microbes, 4, 29–33, 48–49, 85, 105, 175, 192; cell membranes of, 58; cognition in, 23; ecological role of, 96, 226; evolutionary role of, 106, 113–14, 120; gene transfer in, 97, 126; humans as colonies of, 13, 36; Margulis as spokeswoman for, 17; as mere germs, 27; phylogeny of, 107, 109, 239–40, 244; as planetary phenomenon, 161–62; as source of Earth's atmosphere, 19; symbiosis and, 116–17; viruses as, 77

microbial mats, 29, 32, 35; as archetypal biosphere, 260

microbiome, 16, 33, 240

Microcosmos (L. Margulis and D. Sagan), 19

Midgley, Mary, 90, 101–2

mitochondria, 14–15, 176, 189–90; DNA in, 115–16; as symbionts, 113, 119–20

mitosis, 15

Mittledorf, Josh, 20, 33

Mixotricha paradoxa, 16, 29

mobile genetic element, 183–84, 190, 192–94, 200; activation of, 183, 199, 200; targeting of, 199, 202

modern synthesis, the, 109–10, 199. *See also* neo-Darwinism

modernism, 254–57, 259, 261–63, 265, 267; Earth's view of, 261–66

Monod, Jacques, 178, 185

monomers, 58–59

morality: agrarian, 242; nomadic, 242

Morowitz, Harold, 90

Muller, Hermann Joseph, 181–82

Museum of Comparative Zoology (Harvard), 17

mutagenesis (x-ray), 181–82

mutator DNA polymerase, 180

mycoplasma, 14, 94. See also *Wolbachia*

mythology, 141, 143; and Gaia, 143; and Mother of all Being, 146; and Diana Polymastogos, 145

Nabokov, Vladimir, 17–18

NASA, 166; exobiology program, 80–104; images, 168–69

Nass, Sylan and Margit, 115

natural genetic engineering (NGE), 176, 197–202

natural selection, 212–16, 218–19

naturalists, 16

nature: as Dionysian, 20–25, 27; distinction from nurture, 205–13, 217; thermodynamic, 28–33

Needleman, Jacob, 103–4

nematodes, 15

neo-Darwinism, 6, 15, 35, 126; Margulis's perspective on, 30–33. *See also* selfish gene

networks, 177, 181, 184, 194, 197, 202

Neuromancer (W. Gibson), 163–74

A New Bacteriology (S. Sonea and M. Panisset), 36

New York City, 299n39

niche construction, 215, 219, 295n55

non-homologous end-joining (NHEJ), 196

Nowak, Martin, 33

nucleoprotein complex, 201

O'Neill, Gerard K., 166–72

Odum, Eugene and Howard, 28

ontogeny, 203, 209, 216–17, 222; disturbed, 294n46

Onychophora, 38

operon, 137

Ophrydium, 29

Woese, Carl, 36, 90, 94–96; on origin of genetic code, 119–20; on the progenote, 120; on prokaryote-eukaryote dichotomy, 121–22; on ribosomal RNA phylogenetics, 119–20; on symbiosis, 119–21; on three domains, 105, 121–24
Wolbachia, 14, 126

"Wolf Trap," 81, 83–85
Wolfe, Cary, 296n10, 300n52

XIT Ranch, 231–32, 297n22

Zhang, Chen-Yu, 36
zoocentrism, 30, 36
Zuckerkandl, Emile, 120

MEANING SYSTEMS

Heinz von Foerster, *The Beginning of Heaven and Earth Has No Name: Seven Days with Second-Order Cybernetics*. Edited by Albert Müller and Karl H. Müller. Translated by Elinor Rooks and Michael Kasenbacher.

Bernhard Siegert, *Cultural Techniques: Grids, Filters, Doors, and Other Articulations of the Real*. Translated by Geoffrey Winthrop-Young.

Bruce Clarke, ed., *Earth, Life, and System: Evolution and Ecology on a Gaian Planet*.

Kriti Sharma, *Interdependence: Biology and Beyond*.